Marrow of Tragedy

Marrow of Tragedy

The Health Crisis of the American Civil War

Margaret Humphreys

THE JOHNS HOPKINS UNIVERSITY PRESS   *Baltimore*

© 2013 The Johns Hopkins University Press
All rights reserved. Published 2013
Printed in the United States of America on acid-free paper

Johns Hopkins Paperback edition, 2017
2 4 6 8 9 7 5 3 1

The Johns Hopkins University Press
2715 North Charles Street
Baltimore, Maryland 21218-4363
www.press.jhu.edu

The Library of Congress has cataloged the hardcover edition of this book as follows:

Humphreys, Margaret, 1955–
Marrow of tragedy : the health crisis of the American Civil War /
Margaret Humphreys.
pages cm
Includes bibliographical references and index.
ISBN 978-1-4214-0999-3 (hardcover : alk. paper) — ISBN 978-1-4214-
1000-5 (electronic) — ISBN 1-4214-0999-2 (hardcover : alk. paper) —
ISBN 1-4214-1000-1 (electronic)
1. United States—History—Civil War, 1861–1865—Medical
care.   2. United States—History—Civil War, 1861–1865—
Casualties.   3. Medicine, Military—United States—History—19th
century.   4. Public health—United States—History—19th
century.   I. Title.
E621.H86 2013
973.7'75—dc23   2012044604

A catalog record for this book is available from the British Library.

ISBN-13: 978-1-4214-2277-0
ISBN-10: 1-4214-2277-8

Special discounts are available for bulk purchases of this book.
For more information, please contact Special Sales at 410-516-6936 or
specialsales@press.jhu.edu.

The Johns Hopkins University Press uses environmentally friendly book
materials, including recycled text paper that is composed of at least 30
percent post-consumer waste, whenever possible.

For
Will and Ted

And in memory of
William D. Humphreys (1836–186?), Fifth Tennessee Infantry
Hugh Donnelly (1828–1868), Eighth Minnesota Infantry

CONTENTS

ACKNOWLEDGMENTS

To properly thank all of those who have contributed to this book, I'd have to begin with the Brownie Scout leaders who took me to Fort Donelson in the second grade and continue from there. Raised in the South by a mother from Minnesota, I have long realized that the war was *the* key event in American history and that sooner or later I would have to engage it professionally. This effort formally began in 1997, when I little thought that it would consume me for the next fifteen years of my professional life.

In the words used long ago by television advertisers, this work is brought to you by many sponsors. First and foremost is the support of Mary Duke Biddle Trent Semans (1920–2012), the Trent Family, the Trent Foundation, and their donations to Duke University, which allowed me the luxury of an academic position as a base for all of my scholarly endeavors. The Trent Foundation funded my chair in the history of medicine, and research funds from that chair have in turn directly supported my research (and enabled me to hire research help). I must also thank Ross McKinney Jr., director of the Trent Center for Bioethics, Humanities, and History of Medicine for his patience in channeling funds to help "get that book done."

Duke is a wonderful location from which to explore Civil War topics. Not only is the Duke Library itself rich in resources, but within a day's drive are the Southern Historical Collection (at the University of North Carolina–Chapel Hill), the South Carolina State Library, the Museum of the Confederacy (in Richmond), and the National Archives in Washington, D.C. Librarians at all these places have offered cheerful and very welcome support. Librarians Suzanne Porter, Rachel Ingold, and Elizabeth Dunn at Duke have been especially helpful. Duke has allowed sabbatical leave, supported the work with small grants, and supplied me with valuable colleagues who have enriched my perspectives. I'm particularly grateful to Lil Fenn, Laura Edwards, Ed Balleisen, Karin Shapiro, Anna Krylova, John Martin, Thavolia Glymph, Peter Wood, Adriane Lentz-Smith, Nancy McLean, Sally Deutsch, Sy Mauskopf, Tom Robisheaux, Peter English, Jeff Baker, and Anne Firor Scott for conversations and encouragement. And even though Duke freshpersons are routinely given shirts saying "Go to Hell Carolina" at orientation, I want to thank my colleagues at the University of North Carolina–Chapel Hill, including Michael McVaugh, Wayne Lee, Joseph Glatthaar, and Dick Kohn, for their enlightenment and support over the years. Susanna Lee at North Carolina State rounds out the roster of my academic village. Thanks to you all.

I had the luxury of two years at the National Humanities Center, years that gave me not only the gift of time but the stimulation of colleagues across the humanities, whose conceptions redirected my work in fundamental ways. My first year was supported by an American Council of Learned Societies Burkhardt Fellowship and my second by Duke sabbatical leave. I also received funding from a National Library of Medicine Publication Grant (G13-LM008350-02). This is the second book to emerge from the time supported by these research funds, and all the people thanked in the preface to the first one (*Intensely Human: The Health of the Black Soldier in the American Civil War* [Baltimore: Johns Hopkins University Press, 2008]) deserve equal credit here.

Historians are always thinking of their audience, and for works on Civil War medicine that audience is particularly diverse. I appreciate the support and feedback from the Society of Civil War Surgeons (Pete D'Onofrio, president) and the meetings sponsored by the National Museum of Civil War Medicine (George Wunderlich, executive director), where I both met interesting colleagues engaged in public history and often learned about valuable local sources previously unknown to me. The Society of Civil War Historians brought me into the circle of professional historians interested in the Civil War. I'm particularly grateful to Judy Giesberg for welcoming me into the group and introducing me to other gender historians of the war. Finally, my friends in the American Association for the History of Medicine continue to "give me the strength to get up and do what needs to be done" (with apologies to Powdermilk Biscuits). So thanks, as always, to Jacalyn Duffin, Nancy Tomes, Norman Gevitz, Caroline Hannaway, Gerry Grob, John Eyler, Richard Meckel, Judith Leavitt, and other friends for community over many years.

I was fortunate to have research assistants for this project. Kelly Kennington, now assistant professor of history at Auburn University, worked for me during her graduate training at Duke. She managed the many mundane tasks of editing the *Journal of the History of Medicine,* acted as teaching assistant for my course on the history of public health in the United States (and, on one memorable snow day, as emergency babysitter), and otherwise helped lighten my workload. Marie Hicks, also a Duke history PhD, served as managing editor for the journal, assisted me in planning a new course, and ably edited this book manuscript. During this past year, Shauna Devine, a recent PhD from the University of Western Ontario, herself a scholar of Civil War medicine, assisted me in revisions while likewise doing journal and course support work. And she planned a one-day conference on the Civil War that was enlightening and fun (especially once it was all done). All of this help has been much appreciated.

I have been lucky to have Jackie Wehmueller as my editor at the Johns Hopkins University Press. My colleagues have marveled at her hand-edited pages, both for the detail and for the retro style (it's much easier for me to follow penciled suggestions than to see the equivalent changes in a marked-up Word document). Jackie has been endlessly patient with me as the deadline for this book's submission has been pushed back and then pushed back again. Every year at American Association for the History of Medicine meetings for quite a while now we've had a glass of wine and heartening conversation that helped keep me going. Thanks, Jackie.

In 2005 my knees began to creak and I acquired an orthopedic surgeon, a physical therapist, and finally a personal trainer who whacked me into some kind of shape (thanks, Mary Ann!). I even have a Pilates instructor (thanks, Nicola!) and know what a downward facing dog is (thanks, David!). This book might have been done sooner without their intervention, but they have taught me much about the balance of health and work.

Finally, I come to family. When this project began, I lived a quiet life with two kitties in a small bungalow near Duke's East Campus. During the years that followed, I married my college sweetheart and bore a son who now at twelve keeps me from being too ignorant about all the important Internet memes (and what Tumblr is). Will, you are my joy and continuing challenge. Through it all, my husband, Ted, has been a constant source of love, patience, tech support, and enhanced graphics. I think we make a pretty good team. Thanks, guys.

Many Civil War scholars report that their work began with a genealogical project that revealed a Civil War ancestor. Mine did not begin there, but in the process I found soldier ancestors on both sides of the family. My father told me of family lore about "cousin Willie who was kidnapped into the Confederate army and never seen again." The census records for 1860 list a William D. Humphreys, 26, who lived with his widowed mother and three brothers in Henry County, Tennessee, home to my father's people. No doubt, the conscription agents grabbed him, since this section of Tennessee, on the Kentucky border, tended to be pro-Union and he made a reluctant rebel. William is on the rolls of the Fifth Tennessee Infantry, which attempted to keep up its numbers by repeatedly taking in new men, amounting to 2,235 in all. Of those, only 173 remained by the end of 1863, and by 1865 William's regiment had merged with the Fourth Tennessee and had been reduced to a company, a unit usually comprising 100 men. That remnant was part of Joseph Johnston's army, which surrendered to William T. Sherman on April 26, 1865, at a farmhouse called Bennett Place, a few miles from where I am writing these words. When or where cousin Willie died is unknown.

Hugh Donnelly left Carlingford, Ireland, and immigrated with his brother Peter in the mid-1850s. They arrived at Anoka, Minnesota, and began farming. Hugh enrolled in the Eighth Minnesota Infantry and was present at Bennett Place when Johnston surrendered. He made his way home, weakened by disease and the hardship of war, and died in 1868, most likely from tuberculosis. His brother Peter was my grandfather's grandfather, and Peter's son Hugh (b. 1869) was the "papa" fondly remembered by my great Aunt Ellie. Even now, 150 years later, the war is still alive in family memory.*

*For the histories of these two regiments, see Fifth Tennessee, www.cwnorthandsouth.com/5thTNHst.htm and Eighth Minnesota, www.ac-hs.org/eighthregiment.htm (accessed July 21, 2012).

Introduction

Call and Response

The American Civil War was the greatest health disaster that this country has ever experienced, killing more than a million Americans and leaving many others invalided or grieving.[1] It does not seem to have occurred to the leadership on either side that the war would produce casualties and that the gathering of men in great camps would provide a field day for infectious diseases. But it did not require the rising casualty rates of summer and fall 1861 for citizens on both sides to realize that their governments needed help in delivering health care to their soldiers; women were meeting in New York City seventeen days after the firing of Fort Sumter to plan for nursing and re-lief of the wounded. Their men had heard President Abraham Lincoln's call for volunteers, and these women had responded with foresight of the suffer-ing to come. Southern men were similarly eager to take arms, and southern women likewise were concerned about their soldiers' subsequent well-being.

"Call and Response," a custom traditionally associated with African Ameri-can religious gatherings, involves active participation of the congregation (or the political crowd) upon hearing questions put by the preacher; they answer "Amen!" or "Tell it brother!" or "You know that's right!" in affirma-tion of the message delivered. I borrow the phrase here to emphasize that the condition of the war's wounded and ill soldiers sent forth a massive call for help that demanded an answer from their anguished compatriots. The answer that came was not just from the military hierarchy but was society-wide. As the provision of health care in the antebellum period had been so much the province of women, so in the early years of the war it was the women of both sides who most clearly saw what was needed and set about getting it done. Doctors were important, and their role is explicated in what follows, but far more central for the mid-nineteenth century patient was the healing environment epitomized in the comfortable sickroom overseen by competent women. They may have lacked antibiotics, but they had other essential ingredients for cure: nutritious food, adequate hydration, cleanliness, mental stimulation, and moral encouragement. These aspects of health care had been essential elements of the feminine sphere before the war; the great challenge of the war years was how to administer them in providing proper care for the soldier who was separated from his family by war.

This book takes a *gendered* approach to understanding how the story of health care unfolded in the armies of the Civil War. Using language employed by historians LeeAnn Whites and Nina Silber, I see medical care in the Civil War as a "crisis in gender relations" or, as Silber put it, an arena that stretched "the broader ideological constructions made by northerners and southerners when they thought about masculinity and femininity."[2] I do not focus primarily on women's work during war, although their work was crucially important. That story has already been told ably by other historians. Rather, I draw on works that examine how concepts of masculinity played out in the new social sphere of war, to explore how perceptions of what behaviors and attitudes were masculine or feminine affected the structure of medical care. My account begins with the perception that, far from glorious, the "real war" (in Walt Whitman's memorable phrase) was fought as much in hospital beds and barns turned camp hospitals as on Little Round Top with flags waving. Whitman referred to "an unending, universal mourning-wail of women, parents, orphans—the marrow of the tragedy concentrated in those Army Hospitals—(it seem'd sometimes as if the whole interest of the land, North and South, was one vast central hospital, and all the rest of the affair but flanges)."[3] The war, for those who fought it, was less about heroism and more about the daily grind of disease, hunger, death, and disability. And for their women? They handed over nurturance to a clueless military authority and sent their men off to die.

Margaret Mitchell, a journalist who spent her childhood listening to Confederate veterans describe their experiences, likewise had a deep understanding of that reality. In her novel *Gone with the Wind,* a Mexican war veteran announces to the young bucks all whooping with war fever at the initial barbecue at Twelve Oaks, "You all don't know what war is." He cautioned anyone who thought it was all about coming home as a hero: "Well it ain't. No, sir! It's going hungry, and getting the measles and pneumonia from sleeping in the wet. And if it ain't measles and pneumonia, it's your bowels. Yes sir, what war does to a man's bowels—dysentery and things like that."[4] In the book his relatives shush the old man, but as in Rhett Butler's speech about the South's inadequate ironworks, Mitchell's knowledge of the war's grim truths is conveyed through these unheeded jeremiahs. Her novel does not glorify the lost cause, even if it does present an overly rosy view of slavery; instead, seen through Scarlett's ruthlessly pragmatic eyes, her saga depicts the death and destruction wrought by the war's terrible passions. Americans of both North and South were unprepared for this disaster. The Mexican War had occurred two decades before, and on the distant borders of the United States and in Mexico; in the Civil War, the two sides lost more men in one day at Antietam Creek than had been killed in the entire previous conflict. They had no idea what was coming.

Harvard psychologist Steven Pinker has described a "humanitarian revolution" that began with Enlightenment philosophers who emphasized the rights of individuals and promoted a humanism that swept aside "fetishized virtues, such as manliness, dignity, heroism, glory, and honor," even if it "did not, at first, carry the day."[5] Certainly, the Civil War challenged all Americans to face the reality of violence and the question of whether they were civilized at all. The revolution had made greater strides in the northern states, where women's rights and abolitionism had attracted followings, than it had in the South. A culture of honor may sound grand at first, but it leads to the barbarities of dueling over received insults (or killing over a pair of sneakers). Already in the United States of the antebellum era, the idea was emerging that "gentlemen" should settle differences less violently, that a "gentleman" should display self-control and employ his reasoning faculties, not fisticuffs, to overcome obstacles. These were elite ideals, of course, not pervasive attitudes, as can be seen in Lorien Foote's depiction of interactions between officers and enlisted men in *The Gentleman and the Roughs*.[6] But as a gentlemanly culture grew, the ideal male was becoming more and more, well, feminine. Neuroscience, as well as common knowledge, recognizes aggression and violence as traits more commonly associated with males, especially adolescent and young adult males, who require significant societal quelling for peace to prevail. Pinker's persuasive argument is that humankind has become less violent, less aggressive, less cruel to one another (and to animals) in the past five hundred years. In the American Civil War, the contending forces of this transition are in abundant evidence. Honor, benevolence, Christianity, bloodlust, and the moral high ground all jostled for position as foremost in each nation's character.

A major influence on this wartime mentality was the widespread Christianity that infused American culture. Historians have shown how ministers on both sides preached sermons justifying the war, with only the Quakers and their like remaining as outright pacifists. But whether or not they were motivated by the human-rights ideas of the humanitarian revolution, those seeking to contain the horrors generated by the war increasingly tried to find ways to limit suffering and offer solace to the afflicted.[7] The treatment of prisoners of war crystallized these sentiments. This was the first American war in which the articles of war (rules for military behavior) were expanded by a code of ethics that explicitly detailed the humane treatment of prisoners. As explained in chapter 9, brutality toward prisoners, such as the massacre at Fort Pillow in 1864, was no longer a cause for pride, but instead a cause for accusation. The mistreatment of prisoners of war in northern and southern camps called forth rhetoric and public discussion that revealed common assumptions about the civilized behavior of governments even amid a war that belied such basic principles. Americans of both North and South had trouble facing their own acts of cruelty.

Many of these themes can be followed in the story of the U.S. Sanitary Commission (USSC), a northern organization. It grew from a meeting of women in April 1861 to become a national organization that prided itself on channeling the healing power of women to the wounded and sick soldiers who so desperately needed it. Although ultimately headed by men, the group remained a force for teaching the military medical leadership about the essentially feminine aspects of healing. Its inspectors visited camps to drive home the necessity of sanitation to slovenly troops. The organization responded to the emergency evacuation crisis after the battles of spring 1862 with rented steamers, beds, nurses, and supplies. It initially offered to organize a female nursing corps, but that responsibility was ultimately taken over by the government. Above all it supplied the material needs of the hospital sickroom and necessary foods for the maintenance of health and the promotion of healing. As its work was gradually assumed by the increasingly competent Union medical management, the USSC turned to the plight of prisoners of war and then, postwar, to the promotion of the Geneva Convention and the amelioration of battlefield conditions in future conflicts. Many USSC workers were women, laboring behind the lines to gather or manufacture items needed by the troops, although the group's agents who were closest to the fighting or military camps were mostly male. Nevertheless, it is not too much to argue that the USSC represented the feminine side of the northern war effort and that its duties encompassed responding to the disastrous mess created by men.

The United States General Hospital in Georgetown, D.C., formerly the Union Hotel. Note the many women on the wards. Louisa M. Alcott, Hannah Ropes, and Walt Whitman all wrote about their experiences in this hospital. Paul F. Mottelay and T. Campbell-Copeland, eds., *Frank Leslie's "The Soldier in Our Civil War": A Pictorial History of the Conflict, 1861–1865, Illustrating the Valor of the Soldier as Displayed on the Battle-field,* Columbian Memorial edition, 2 vols. (New York: S. Bradley, 1893), 1:108.

It is at times difficult to reconcile seeing the war as a disaster with the fascination that Civil War history inspires.[8] It draws us in spite of the slaughter, or perhaps because of it. Like viewers of a horror movie who want to shout to the actors, "Don't go into the basement!" historians reliving the war, especially through archival sources, can enter into this greatest event in American history and wish they could share their hindsight with the wartime participants. William Faulkner famously captured this sense of immediacy in his novel *Intruder in the Dust,* when he wrote of the power that the Battle of Gettysburg had on the southern imagination. For every southern boy, he claimed, "not once but whenever he wants it," there are times "when it's still not yet two o'clock on that July afternoon in 1863, the brigades are in position behind the rail fence, the guns are laid and ready in the woods. . . . It's all in the balance, it hasn't happened yet, it hasn't even begun yet."[9] Faulkner pictured Major General George Pickett there, waiting for his orders to advance, before the moment when this man's name began to label the most famous and disastrous charge in the Civil War, in that battle at Gettysburg that some have called the "high-water mark" of the Confederacy.[10] Faulkner knew, "It doesn't need even a fourteen-year-old boy to think *This time. Maybe this time* with all this much to lose and all this much to gain." He can imagine the war still in process, with "Pennsylvania, Maryland, the world, the golden dome of Washington itself to crown . . . the desperate gamble, the cast made two years ago."[11] While southern girls fantasized about the "party at Twelve Oaks that might have gone on forever," as Pat Conroy remembers from his southern childhood, few wax rhapsodic about the hospital scenes during the battle of Atlanta.[12] The medical side rarely figures in fond memories of the war.

Most historians of medicine that have approached the American Civil War have done so with a focus on the surgeons, the men (with a few exceptions) who performed the operations and oversaw medical care for wounds and disease within the vast Civil War hospitals. This account hopes to open that view up to the broader cast of characters who answered the call to heal the sick and wounded. Two decades ago historian Maris Vinovskis urged scholars to write the "social history" of the war, and numerous scholars have taken on the challenge to explore the many lived experiences of that conflict by those who did not take part directly in battles, including women, blacks, and those who stayed home.[13] Jane Schultz has admirably described the stories of women who served as nurses in those hospitals. Others have explored the ways in which the war infiltrated fundamental cultural rituals such as religious and death practices.[14] In some ways, to approach the history of medicine during the war is to illuminate one more aspect hitherto little heeded by those who focus so precisely on battles and tactics. But it is more than that. Armies were obviously central to the war story, and the quality of medical care, broadly conceived, in turn determined how many men were in those armies and how effectively they could serve in any given battle.[15]

No one at the start of the war knew how to organize a medical system to care for the millions of men who suddenly found themselves under government guardianship, much less how to care for the 4 million slaves who traveled erratically toward freedom as the war progressed. Some have said the American Civil War occurred at the end of the medical "middle ages"; recently one account of the emergency medical response to the Haitian earthquake called that response practicing "Civil War medicine" to denote how primitive the conditions were.[16] Like Haiti today, the United States in 1861 had minimal health infrastructure, at least in terms of hospital beds and trained medical professionals who knew how to organize health care on a large scale. Almost all health care occurred in the home; the hospitals that

existed served as institutions for the poor or those on the road. The importance of this fact for Civil War medicine cannot be underestimated.

Imagine the situation of the 3 million or so men who would serve in
the war as the decade of the 1860s dawned. Most were pursuing some occupation that created value and earned their living. They or their families
purchased their food, clothing, and shelter. If they got sick, health care was
provided in the home by family members, many of them female (wives,
mothers, sisters), or perhaps by a charitable lady who visited the impoverished sick.[17] A doctor might well be called, but his contribution to healing
was relatively small compared to the nursing work done by family members.
In retrospect, we can recognize that most mid-nineteenth century medicaments were ineffective or even harmful, although there are some exceptions
(quinine, opium, anesthetics for surgery). What made a difference in the
outcome of illness was the provision of nursing care.

Now consider the vast differences brought by war. With the force of a
flood or an earthquake, the men were washed away from their homes and
deposited in crowded, dirty camps; suddenly their food, clothing, shelter,
and medical care became the responsibility of their military leaders or themselves. As the green troops inevitably sickened from the epidemic diseases
that spread through the camps, the military surgeons, almost all of them
also wrenched from civilian life, had to invent an army medical system to
provide care with little prior experience and few concrete models to draw
upon. They had certainly never performed the sort of caring work that had
been almost entirely the province of women just a few months earlier. When
the first battles created casualties by the thousands, war physicians were
similarly unprepared to organize transport or care en route, or even to feed
the wounded. Army medical leaders drew desperately on the experience of
European armies in the Crimea, a war fought just five years earlier, but they
were often making procedures up as they went along, with predictable lack
of success.

As a result, the last half of 1861 and all of 1862 were a time of sore trial
for the wounded and sick soldiers of both sides. An amalgamated system
evolved, with women pushing in to provide the sort of care that they knew so
well how to perform, while the medical services of both sides tried to grow
the infrastructure to deal with the miserable tide. In the North, women organized sanitary associations by state and joined together to create the U.S.
Sanitary Commission. The loose system of nursing by convalescent soldiers
gave way to an organized female nursing corps. In the South, women like-

wise staffed hospitals and provided relief, albeit in a less nationally organized fashion. Together this medley of doctors and skilled women helped create a new sort of health system to care for the troops, one that recreated as much as possible the home sickroom while maintaining the military discipline at the core of army medical goals.

But those five years, 1861 to 1865, were still a disaster on a major scale, in spite of the modest successes of Civil War medical leaders. The case was worse in the South, where one estimate notes that one in four men of military age died in the conflict. Only (only!) 10 percent of Union men suffered this outcome. No one can write a history of Civil War medicine as a triumphalist narrative. It is worth noting, however, that where the military was able to recreate the healing environment of the home sickroom most effectively, more men survived. Sufficient food, water, and clean clothing and bedding, all provided quickly in the aftermath of disease or wounds, were the factors that made the most difference for the mortality rate of hospitalized Civil War troops. And the ability to provide these needs was not equal between the two sides. The Union not only had more cannons, rifles, factories, and men; it had the affluence and access to build an effective military health care system.

Years after the Gettysburg battle, the governor general of Canada asked General George Pickett "to what he attributed the failure of the Confederates at Gettysburg. With a twinkle in his eyes, he replied: 'I think principally to the Yankees.'"[18] Other historians have rendered this famous answer as "I always thought the Yankees had something to do with it."[19] Similarly, Gary Gallagher addressed an audience of Civil War historians in summer 2010 on the topic of slave liberation during the war. While acknowledging the bravery and sacrifice of those African Americans who freed themselves by escaping from their owners, Gallagher's argument can be summarized as reminding his audience that "the Yankees had something to do with it."[20] He pointed out that slaves escaped mainly in areas occupied by the Union Army, and then he drew the larger conclusion that the role of armies in the vast complex of events that made up the American Civil War should not be shunted aside by cultural historians.

Although many factors contributed to the Confederacy's defeat, aside from the Yankees, one undeniable cause was that the Confederate states ran out of men to put in the ranks. The disproportionate losses of southern men of military age could not be answered. There were few new immigrants to the region to replace them, and many adult men were tied to the maintenance of the slave system and so were unavailable for battle. The adult African American males remained mostly in bondage, although the Confederacy impressed some to war work, while the Union put nearly two hundred thousand black men into the ranks. Also important, however, was the inability of the Confederacy to create the sort of healing hospital system that ultimately prevailed in the northern cities. The South also lacked the reformist force of the U.S. Sanitary Commission that so powerfully met immediate medical needs, pushed for necessary reforms in the Union Army medical system, and funneled the beneficence of northern women to the care of Union troops. The Union both had broader resources at its command to keep its men healthy and created a more effective system to put those men back in the ranks after they became sick or wounded. Armies were central to the war, and health care was central to the armies' numbers and effectiveness.[21] There can be no doubt that the Union marshaled and employed the tools available in mid-nineteenth century medicine to greater advantage than did the Confederacy.

This book documents the ways in which government and philanthropic institutions on both sides of the conflict met the unexpectedly heavy demands of the war wounds and diseases. The historian can choose to emphasize the gruesomeness and deficiencies of the war's medical practice or to illustrate how much progress had been made over prior wars. Chroniclers at the time liked to quote the statistics from the European war in the Crimea a few years earlier to indicate how much more humane and effective health care had become in the American case. The difference was not in an alteration in medical theory or access to different medicines. The difference was in the systematic application of the principles of hospital design advanced by Florence Nightingale and in the systematic application of sanitary principles likewise promoted by English sanitarians. Regulating the disposal of wastes and providing patients with clean, airy hospital wards staffed by compassionate nurses who supplied appropriate food and water made a difference in how likely a man was to survive either a wound or microbial invasion. In one story from Haiti's earthquake of 2010, doctors described amputations done with a hacksaw with only vodka for an anesthetic; surgeons in the Civil War had ether and chloroform, and they had surgical instruments that were kept honed, if not as clean as modern medicine would command. The surgeons had opium and quinine as well. Conditions had improved markedly since earlier wars, and the men were better off because of it.

It may seem counterintuitive to emphasize the progress in medical care made during a war that killed so many people. Indeed, some historians have denigrated the very activity of seeking innovation in the midst of the horror of war as a means of adding glory to this most evil of human activities.[22] But it is undeniable that army surgeons did a better job of caring for their sick and wounded men as the war went on. Their focus on hospital design (spearheaded by Surgeon General William Hammond in the North and his counterpart Samuel P. Moore in the South) probably saved lives wherever the hospital plans could be implemented. The improvements in transport from battlefield to hospital were lauded at the time and remain significant developments. The dissemination of information about health and disease promoted by the medical departments of both armies as well as the Sanitary Commission taught many surgeons about the latest information on disease prevention. The practice of medicine and surgery in large hospitals allowed the most able practitioners to influence the tyros, and some hospitals formalized the education process. Women entered the hospitals as respected nurses, signaling the beginning of professional nursing in the United States. The list could go on, and various outcomes are explored in this book. The war did change American medicine, even if that fact was an unexpected side effect of a struggle over states' rights and the fate of slavery.

A further aspect of Civil War medicine deserves mention here. Medical care in 1861 was not the same as in 1865. There was no single entity that could be tagged "Civil War medicine." The medical officers of both sides began the war ill equipped to handle the onslaught of casualties shortly to come, but they did not remain inert as the war went on. They built new hospitals, increased the medical officer's training, evolved systems of moving the sick and wounded more efficiently, and built laboratories to manufacture and develop medications. In addition, there was no single place where Civil War medicine was administered. The makeshift field hospital in a barn near a battlefront had much different ambience than the distant general field hospital run by quietly competent nuns. Food supplies varied by place and time of year, directly affecting the life of the hospitalized patient and his chances for recovery. All of these factors meant that scenes of wretchedness are easy to find all through the war, but increasingly you hear of Union men, at least, writing letters home about how happy they are to be in the hospital compared to life back in camp. For Confederates, the cumulative effect of southern shortages meant that over the course of the war, hospitals became increasingly places to starve, although at least no one was shooting at them while they were there.

One aspect of wartime had a particular influence on the growth of medical thought and practice. In both countries, surgeons who served as military officers had access to the federal treasuries. When they responded to the call for military service, they were able not only to serve their country but also to help promote causes they held dear. They could marshal this federal money to fund therapeutic and anatomical research, publish and distribute literature, build pharmaceutical laboratories, and construct model hospitals. Those who had sought in vain to acquire money from civilian sources for medical reform now had their hands in the federal pocket. Leaders of the U.S. Sanitary Commission had a bank account built from successful philanthropic appeals to likewise promote their various agendas. Never before had so much money been at hand to be used in medical reform efforts, and physicians made good use of it where they could. This was especially true for the Union side of the conflict, but even in the cash-strapped South, the surgeon general pursued research aims and the "elevation of the profession." Leaders of both sides sought to disseminate the latest ideas and best practices to their less-enlightened brethren, using government and philanthropic money to promote these ends. While government funding of public health reform gradually increased in the late nineteenth and twentieth centuries, it was not until the founding of the National Institutes of Health in 1948 that the U.S. Government invested significant funds in medical research outside of the military setting.[23]

It bears repeating that the Civil War occurred in the middle of the nineteenth century, not in some isolated cocoon of time that began in 1860, say, and stopped in 1865. The physicians who served in the war were shaped by the decades that came before, and their wartime experiences in turn shaped the decades that followed. The 1840s and 1850s were a period of widespread medical debate and tumult around such issues as proper medical training, the importance of quarantine and sanitation in the prevention of epidemic diseases, the role of government licensing in the determination of appropriate medical practice, the necessity of gathering and analyzing vital statistics, states'-rights medicine, and the admission of women and blacks to medical schools. The leaders of northern and southern medicine brought these various agendas with them into the war, and not a few sought to extend the reach of their preconceived notions once in a position of government power (and funding).

The American Medical Association, for example, had been founded in 1847 to meet the sectarian challenges of homeopathy and botanical practitioners, to elevate the financial stability of the profession, and to promote education through the publication of its house journal. The word *sectarian* is used here to refer to groups of medical practitioners who today might be styled "alternative practitioners." Homeopaths followed the writings of Samuel Hahnemann, a late-eighteenth-century physician who argued that minuscule dilutions of drugs (for example, ipecac) that normally caused symptoms (such as vomiting) would instead be curative. Medical practitioners who rejected chemical drugs such as mercury and antimony and instead favored botanical remedies were known by various labels, such as Thomsonians, botanics, or eclectics. These medical sects in turn labeled "regular" physicians allopaths and fought to overturn licensing laws that limited medical practice to those with orthodox training. In a period when much of medicine was "heroic" and relied on purging, puking, or bloodletting, these alternative practitioners found adherents. In response some orthodox physicians called for a dialing back of heroic therapies and a greater dependence on the healing power of nature. It was time of great ferment in medical therapeutics, but one in which the general trend among the most educated physicians was toward milder therapies. The war offered these elites the chance to both solidify the professional legitimacy of regular practitioners and spread the gospel of gentler therapeutics.[24]

Others with reform agendas likewise found opportunities during the war to promote their goals. The southern surgeon general used his office to further the "elevation" of the southern medical profession through education and research. Women's-rights activists expanded the clarion call of the Seneca Falls Convention of 1848 to push for women to have a significant role as nurses during the war, and a few as doctors. The Union surgeon general not only urged hospital reform but called on surgeons throughout the army to gather interesting cases, send in their reports, and perform research on promising new therapies. He likewise supported those surgeons who set up specialty wards for the study of cardiac, neurological, and ophthalmologic diseases so that research could be done on such maladies. The war allowed unprecedented government funding for these and other reform impulses.

So, then, what innovations did the Civil War bring to American medicine? World War II resulted in penicillin and DDT, for example; did similar medically related innovations emerge from the Civil War? Historian Ross Thomson has asked this question about the war's impact on technological innovation, exploring products as disparate as mass-produced firearms, shoes, and petroleum. He draws on the historiography of industrial productivity to analyze disputes about whether war accelerates industrial innovation or retards it, highlighting the problems of labor lost as men entered the army and the rechanneling of industrial capacity toward a few products (armaments, uniforms) and away from broader innovations. He argues that the Civil War maintained dynamics and paths of innovation that had prevailed (or not) before the war. In particular, preexisting networks of information diffusion persisted and were amplified by the war, with the Union benefiting particularly from its antebellum industrial health, while the South continued to fall behind.[25] In the field of medicine, the war brought no major discoveries or techniques but instead the widespread dissemination of ideas and skills that can be seen as progressive in the mid-nineteenth century, including the extensive use of disinfectants that paved the way for acceptance of the germ theory.

In a period when anatomical knowledge was central to the physician's craft, the war brought surgeons and bodies together in unprecedented numbers. No surgeon on active duty could complain about a lack of anatomical exposure during the war years. Through the mentorship systems that evolved in the ranks and general hospitals of the war, doctors learned the most effective surgical techniques of their day and the proper administration of anesthesia. The ideas of expectant medicine, which limited the use of heroic therapies like bloodletting and calomel and allowed the body's own healing powers to act, were widely preached, even if not necessarily universally adopted. The importance of hygiene and maintenance of camp cleanliness was drilled home, and the use of disinfectants to counter disease-causing substances became well known.

Above all, the proper design and organization of hospitals was made concrete and common to every surgeon's experience. Hospitals were the center of medical care in the Civil War and made that encounter novel for both patient and practitioner. In the antebellum era, hospitals were reserved for the poor, the traveling, the slave, or the seaman; few middle- or upper-class people would have been patients there.[26] Yet any army man, officer or private, might find himself in a Civil War hospital. Those hospitals were not all the same: some were quite comfortable while others were filthy, vermin-ridden pestholes.

More important in determining quality were the aspects of care that women typically delivered in the home setting. There were no intravenous lines in the Civil War; all fluid entering the patient had to come via the mouth, or occasionally via rectal enemas. The modern drinking straw was not invented until 1888, although reeds or literal straws could be used for this purpose.[27] Cups with narrow spouts were available for feeding infants and invalids. The patient prone in bed needed help in drinking—someone to fetch the fluid, someone to help hold the cup if necessary, and so on. Likewise, he needed help with a bedpan or assistance to a commode. If the bed became soiled, skin breakdown would likely follow from exposure to moisture and microorganisms. The caregiver prepared food to tempt the invalid, often soups, simple crackers, or bread with jam, or foods of liquid consistency such as cream of wheat or pureed vegetables. Getting food into a petulant feverish man took time and patience. All of these actions would have been well known to the women of the mid-nineteenth century, since sitting with the sick and assisting in their recovery was a central part of the woman's domestic role. So were maintaining clothing, cooking food, seeing to household cleanliness, and urging family members to maintain personal hygiene.

For those women who moved into the hospitals as nurses during the war, this gendered role was redefined and justified by seeing the men they served as family members—if not actual, then virtual ones. There was a sense of exchange, that while a woman cared for the men right here (where she was serving), some other woman was caring for the first one's actual son or brother or husband at a distant location. Grown men became children in their eyes, and it was appropriate for children to be seen naked by their mothers, bathed, chided, and spoon-fed, all according to traditional female roles. White physician Esther Hawks even nursed black soldiers in South Carolina, a breach of propriety hard to imagine before the war; she claimed they were children and thus expanded her domestic sphere to their care as if they were four-year-olds.[28] Civil War letters and diaries are full of reports from war nurses that a dying man whispered "mother" to her, and usually the nurse would write a letter to the actual mother to deliver the bleak news and final prayers.

This expansion of the domestic sphere was necessary because enlistment in Civil War regiments plunged the new soldier into a world without women.[29] The prewar military had learned to muddle along, accomplishing the necessary tasks on the small scale of frontier outposts. The ninety-day volunteers who rushed to sign up in April 1861 little questioned where their food would come from or what would happen to them when they became sick or wounded. It was all fun and games until someone got sick, and the men quickly began to spread microbes in the newly created cities that were military camps. The women of the country, many spurred on by the words of

Florence Nightingale, who said "every woman is a nurse," realized early on that they were needed to fill their traditional sickroom roles since men were incompetent to do so.[30] They volunteered as nurses, and through the U.S. Sanitary Commission moved to "feminize" the military experience by promoting cleanliness, nutrition, adequate clothing, and proper medical care and transport. The hospital became a liminal zone between home and camp, where women did all they could to duplicate the home atmosphere in order to duplicate the healing power of the home environment.

Although the U.S. Sanitary Commission was run at the top by a male hierarchy, it amalgamated the work of thousands of local women's groups that sought to support the war effort. Although the South had no central organizing agency, women's groups there directed similar energies toward caring for men in hospital and field. Many southern women opened their homes as temporary hospitals or organized more formal hospitals such as that run by Sally Tompkins, who privately sponsored a hospital in Richmond, Virginia, to treat wounded soldiers. The Sanitary Commission sent male inspectors into the field, but they carried a female message for the "boys" in camp—stay clean, eat your vegetables, and do your business in the sinks (latrines) and not behind a tree. The sanitary decrees of military life brought the feminine perspective into the rough-and-tumble of the military camp and were accordingly resisted, but as the war went on, it became increasingly clear to officers and surgeons that the female perspective decreased illness and increased that all-important count, the number of effectives (healthy men) available for combat. Historian Bobby Wintermute has emphasized that the antebellum military physician was a somewhat "feminized" officer, not fully accorded the privileges of his titular rank and holding a position analogous in many ways to the women left behind.[31] This feminization was a pervasive and necessary component of Civil War medicine, brought either by the women themselves or the men deputized to carry the feminine messages of cleanliness, nutrition, and the proper care of the sick.

This volume follows the stories of hospitals, women, and the U.S. Sanitary Commission through the four long years of the Civil War, using a topical rather than a chronological mode of organization. The first chapter introduces the various kinds of institutions that came under the umbrella label of "hospital" in the Civil War; it also describes the basics of contemporary medical and surgical therapeutics. Chapter 2 turns explicitly to the role of women in the war and asks how the war changed attitudes toward the formal participation of women in medicine. While every chapter is imbued with my vision of the hospital as a gendered space requiring tasks that were seen as feminized, even when performed by men, chapter 2 describes the work done explicitly by women. Since two men died from disease for every man stricken down in combat, chapter 3 describes the understanding of infectious diseases in the mid-nineteenth century and how the war contributed to debates about causation and prevention. Chapters 4 and 5 explore the history of the U.S. Sanitary Commission and the bases for opposition to this benevolent organization, which in many ways exemplified the expansion and institutionalization of the domestic feminine sphere.

Subsequent chapters take a close look at the general hospitals of both sides of the conflict. Their medical departments built massive general hospitals in cities away from the active battle zones, such as Richmond; Raleigh; St. Louis; Washington, D.C.; and Philadelphia. Chapter 6 examines Satterlee Hospital in Philadelphia, an institution with more than four thousand beds and its own newspaper, and offers insight into its operation and culture. In many ways Satterlee stands in here for the ideal general hospital, its story similar to that of many others established in northern cities. Chapter 7 looks at the Confederate medical system and the aims of its surgeon general, Samuel Preston Moore, while chapter 8 details the failures of Moore's vision and of the hospital system at the heart of Confederate medicine.

The last section of the book considers the legacies of Civil War medicine and the physicians who served in the conflict. Chapter 9 explores the unsettling behavior of the Sanitary Commission with regard to the prisoner-of-war camps and its postwar attempts, perhaps in mitigation, to get the United States to sign on to the Geneva Convention. Chapter 10 focuses particularly on the impact of the war on public health and the ways it changed discussions about the major epidemics of the nineteenth century—cholera, yellow fever, and smallpox. Chapter 11 considers the return to the strange new country that emerged from the intense war experience and the impact of the war on some of the key debates within medicine in the postwar era, including the understanding of the germ theory and the role of elite ideas in medical practice.

When I began this project in the late 1990s, I had no idea it would consume so many years of my life, or that I would find in my research material for multiple books instead of the one volume initially envisioned. The choice of foci in this volume has been heavily influenced by the richness of archival sources found on these topics. The first reel of the U.S. Sanitary Commission's microfilm papers, for example, introduced me to the fascinating topic of the health of black troops, which was intended to occupy a chapter of this project but instead grew into its own monograph.[32] A chapter on the ways in which the war did or did not spread disease, especially smallpox but also malaria, typhoid, yellow fever, and typhus, has spun off one separate paper (on typhus) and an imagined separate volume on smallpox.[33] Since this project was initially envisioned, new books have appeared on women and medicine in the war, pharmacy during the conflict, and the importance of malaria and yellow fever.[34] Jim Downs has completed a dissertation and a book on the freed people's experiences, and several edited diaries of Civil War surgeons are newly in print.[35] Historians are coming to realize the rich potential of this topic, and the present volume will contribute to this literature without by any means exhausting it.

These works build on an older literature that describes the organization of the Union and Confederate medical departments in detail. One begins with the works published near the war's centennial, such as H. H. Cunningham's *Doctors in Gray* and G. W. Adams's *Doctors in Blue,* as well as Richard Shryock's prescient essay on the impact of the war on American medicine, published in 1962.[36] The Adams and Cunningham works can be frustrating for their inadequate citations, but they contain valuable detail that in many cases I have not repeated here. More recent volumes offer synoptic accounts of the war, albeit with little attention to the surrounding context of American medicine in the nineteenth century.[37] Other works are cited as appropriate in the notes.

One theme that emerges frequently in this volume is the ways in which surgeons grasped the chance to pursue research during the war, given their extensive access to bodies. Physicians such as J. H. Salisbury studied the cause of disease; he argued that measles was caused by mold.[38] Ira Russell in St. Louis performed more than eight hundred autopsies on black soldiers, seeking answers to how their medical problems differed because of their race.[39] Burt Wilder used his time in the South to explore its natural history; he collected specimens for Harvard's Museum of Comparative Zoology and explored his pet hobby of harvesting spider silk. He never did capture the bull alligator he sought in Florida.[40] The Union Army created the Army Medical Museum and launched a massive research enterprise that asked surgeons throughout the army to send reports of interesting cases. "Specimens" of body parts thought to have value were preserved. The southern surgeon general attempted a similar system of data gathering, though on a smaller scale. From this perspective the war can be seen as a vast medical research enterprise that reveals much about the meaning and limitations of medical research in mid-nineteenth century America.[41]

Many topics of importance to the war are given short shrift here, either because of space limitations or because other historians have explored them in detail. The materials on the Civil War are so vast and varied that there are many stories to tell; the archival ocean is so broad that no one can hope to explore its full extent. It is difficult to balance the Union and Confederate sides of the story, and here I have often let the Union story stand for both, since there was little difference in medical theory or practice between the two sides. The first six chapters mainly concern the Union story, although the Confederacy is not entirely absent; chapters 7 and 8 are devoted to the Confederate medical story. While the famous "evacuation fire" of Richmond burned the records of the Confederate surgeon general's office in April 1865, there are still many extant documents to support the history of southern medicine. The letters and documents of the war became sacred texts to many once the war ended, and finding and reading all of the material that pertains to medicine within those millions of words is, well, impossible. Everywhere I have given lectures on this topic, I have met local historians who had interesting stories to tell about medical activities in their neighborhood. The medical histories of Louisville and St. Louis, both major hospital centers, as well as the Western Sanitary Commission, all deserve careful consideration. The Appalachian mountain chain divided medical care as well as armies, and it did so for both North and South.

My point of view on the war is influenced by the tenets of the "new military history," which seeks among other goals to address the broad category of "war and society." In a thoughtful review article published in 2007, historian Wayne Lee noted that the "new military history" had become the "now-old new military history" as this trend has been vibrant for the past half century. Lee defines "war and society studies" as "those that emphasize the connections between social organizations, political institutions, and military activity . . . [which include, among other topics] the social and economic background of soldiers . . . the marginalized within the military . . . the experience of the home front, and the impact of the war on society and the state."[42] This study sees the "home front," embodied in the women who were not allowed to enlist in the armies formally, as encroaching into the military space through the medium of hospitals, relief work, and other forms of medical care. It emphasizes the impact of previous controversies within medicine—about sectarianism, the role of women, the cause and control of epidemics, and the proper legitimation of medical professionals—as intruding into the newly created military medical structures and being transformed by them. And it recognizes the fundamental impact of proper medical care, including nutrition, on the fighting strength of Civil War armies.

The Civil War remains one of the most divisive topics in American history; in many real ways, the war is still being fought.[43] One can join associations, such as the Society for Civil War Surgeons, where one is asked on the registration form: Union? Confederate? Historian? Their membership includes reenactors who indeed choose which color uniform to wear, but I resolutely check the box "historian." Gary Gallagher once felt he had to defend himself, in the introduction to his history of the Confederacy, against charges of being a "neo-Confederate."[44] He was born in Los Angeles. I was born in New Orleans "early on one frosty morning," grew up amid roadside markers recalling raids by Nathan Bedford Forrest, and read *Gone with the Wind* cover to cover at age ten. I have long been fascinated by the distinctiveness of the South and keenly aware of the injustices done to its black citizens. My Irish Catholic mother from Minnesota instilled liberal values into my brother and me and exemplified what might be called the "small civil rights movement" in her efforts to teach segregated black middle-school children to read. She always regretted that she was not on that bridge in Selma. Still, the South is my home, and I look at the foolhardy actions of the men who led the secession as ignorant and destructive, as well as evil. Their actions destroyed the world they sought to preserve. Perhaps the war was necessary for abolition, although I'll leave this much-debated topic to other historians. But as a mother, sister, wife, daughter, and medical historian, I see the war mainly as a foolish, overwhelming cataclysm. One might say that all those soldiers would be dead now anyway, but the graves of my ancestors Hugh Donnelly and William Humphreys might—*might*—have been those of old men.

CHAPTER ONE

Understanding Civil War Medicine

For the men and boys who signed on to fight in the Civil War, the world changed remarkably. No longer at work in their usual trade or at school studying under the teacher's rule, they were instead cast into a new world of marching, tenting, and shooting. They left a domestic world where each of the sexes had its assigned tasks, to enter a newly ordered one in which women no longer took care of cooking, cleaning, nursing, and clothing. This was a strange new world without women, and that adjustment was one of the major ones the soldier had to make. To be sure, there were women on the fringes—some women followed their officer husbands and lived near the camp; other women filled the beds of brothels that blossomed near every major military encampment. But this was a society without a formal place for women, at least at the beginning, and the lack was felt almost immediately when it came to the care of the sick and wounded.[1]

Civil War health care occurred in some sort of hospital; for the typical Civil War soldier, this startling new experience was quite unwelcome. When sick in civilian life, men were mostly cared for at home, by the women in their families. Certainly a male doctor might be called for advice, but it was the women who provided the hour-by-hour nursing care, the care that predominated in the illness experience. Most men would never before have been in a hospital; the hospitals that existed served the poor or the transient. This alteration in health care was part of a broader reassignment in gender wars brought on by the war. In the volunteer army, both northern and southern, men had to learn new roles and set up new forms of organization to replace the work typically done by women. They had to learn to cook or to set up kitchens to provide the army with meals. Men had to learn to sew to keep their clothing in repair; small sewing kits called "housewifes" included the needle and thread for such tasks. Men had to learn to see to their own health, for example by setting up tents with dry bedding and proper drainage.[2] And men had to learn how to care for other men when they were wounded and ill. Male surgeons might know how to perform amputations or dose illnesses, but even they had scant experience with organizing a sickroom or providing the sort of care that healed when the man was at home. In her classic work *The Bonds of Womanhood,* historian Nancy Cott described in depth the gendered world of nineteenth-century America, where women's sphere was clearly based in the home, a space men returned to from the outside world for succor, comfort, and sustenance. Health care was part of the domestic sphere, and this nurturing environment had far more power to heal than the sometimes dangerous remedies offered by physicians. The medical hierarchy had an urgent need to create an effective substitute for the home sickroom from the beginning days of the war.[3]

This chapter introduces the response of the Union Army medical leadership to the health crises of the early war. The "calls" from Shiloh, the Peninsula Campaign, Antietam, and Gettysburg are foregrounded because they assumed dominance in the public criticism of the Union war department's actions; these responses will be echoed in the later account of the U.S. Sanitary Commission's interventions into these same terrible events (see chapter 3). The current chapter also introduces basic ideas about medical and surgical therapeutics that were prevalent during the 1860s, and it sketches the modes of hospital care constructed to deal with the flood of sick and wounded men. Union health care did improve over the course of the war, but it involved a steep learning curve.

The ineptitude of the medical officers in the first year of the war was quickly recognized. Not only after the first battle of Bull Run (July 1861), when all was chaos, but into spring and summer 1862, the Union medical department scrambled to provide adequate care. The U.S. Sanitary Commission and other female-dominated charities scolded the medical establishment for its cruelty in not providing adequately for the sick and wounded. Women agitated for a role as nurses in hospitals, knowing that men had no experience in providing the sort of care that they delivered in their usual role. If the army took the sick man out of his accustomed place, then the women pushed to extend the domestic sphere into the army's space to recreate the home sickroom in the military domain. The male medical officer resisted the penetration of women into the health care of the army, but women pushed hard, and they succeeded in forcing a place, especially in the more stable general hospitals behind the lines. They did all they could to recreate the domestic sickroom in this strange new place. Thus they not only provided essentials but also offered entertainments such as singing, reading, and other distractions; and they wrote letters for the men. Given the assumptions among Civil War thinkers about the importance of "will" and mental health in encouraging healing, these women saw their actions as having a direct salubrious effect.[4]

If the proper place for health care was the home, why not just send the men there? Harriet Eaton spoke for many women when she cried, "Our sick men! Why can't they be discharged?"[5] A Confederate similarly reported to a female friend at home about an acquaintance who was down with mumps on New Year's Eve, 1861: "John Mc I. Brown has them and is in the hospital at Smithville. *He wants* to go home *as bad as* he did to come. . . . Tell Sallie that she ought to come and see John Mc I—nurse him so that he will get well soon."[6] Why build a huge, expensive general hospital system at all? Of course the man who had been wounded or who was seriously ill needed immediate, proximate care. But if he could be transported to a general hospital, he could also presumably be sent home, where his care would cost the army nothing and, by most contemporary accounts, would be more attentive than care in the army hospital.

The army leadership had several aims in constructing hospitals and harboring the ill and wounded within them. In the case of battlefield injuries, moving a man any distance might well kill him. As Union surgeon Jonathan Letterman noted, "It was impossible to convince them [relatives and friends] that the removal of a dangerously wounded man would be made at the risk of his life—that risk they were perfectly willing to take, if he could only (at the end perhaps of a long and painful journey) be placed in a house."[7] A second military goal was to return the soldier to his unit as soon as possible, and when those with minor illnesses or wounds were kept close to their regiments, this was easier. If the soldier was able to travel and was sent home, it would be harder to get him back than if he was under guard within the hospital. Such care near camp was often not possible, however, especially when the regimental surgeons had to follow when the army was on the move. So regimental hospitals were broken up and the ill and wounded shipped to more distant care centers. Otherwise the sick and wounded would slow the regiment down. And of course the final goal of the army's leadership was that the men be healed and returned to service. The men's families, in contrast, wanted them home whenever possible. And because the army preferred not to feed and care for a man while he was useless to the service,

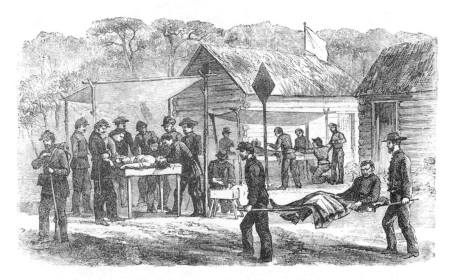

Hospital Scene—Bringing In the Wounded after the Battle. From a Sketch by Edwin Forbes. In this sketch, drawn during the war for *Leslie's Illustrated Weekly* and later reproduced in a memorial volume, stretcher-bearers bring in a wounded man and operations are under way. The patient under the tent fly has an anesthesia mask applied to his face, while the other surgeons seem to be conferring about his case. Why the man in front of the operating table on the right is kneeling is unclear; perhaps an attendant is holding him back from the table, or perhaps his job is to hold the leg steady for amputation. Paul F. Mottelay and T. Campbell-Copeland, eds., *Frank Leslie's "The Soldier in Our Civil War": A Pictorial History of the Conflict, 1861–1865, Illustrating the Valor of the Soldier as Displayed on the Battle-field*, Columbian Memorial edition, 2 vols. (New York : S. Bradley, 1893), 2:165.

furlough systems evolved, wherein a surgeon would declare that a man's wounds or illness meant he should not return to service for some given period, say, thirty or ninety days; he could go home for that period to heal. In the Confederate ranks, this course was particularly tempting as money for hospital care became scarcer and scarcer, but there was a countervailing urgent demand for men to return to their regiments, which argued against granting furloughs.

Thus the hospital, especially the general hospital, represented a compromise between allowing men to return home for care and keeping them under military discipline. Hospitals were a common locus for desertion, even so, but presumably there would have been more desertion by men who had been released to their families. As a response, women did what they could to make the hospital as homelike as possible. Unlike the prewar hospital, which was little more than an almshouse, the Civil War hospital had women attendants of various sorts who sought to bring the healing environment of the home into this barren government structure. The general hospital existed mainly to promote military discipline, not to supply some aspect of care not available on the home front. Accordingly, once the war was over, the general hospitals dissolved. While the war created a new sort of hospital, one in which the affluent classes received health care, there was little need, yet, for such a structure in the absence of military missions. Not until the 1880s, when new surgical techniques required dedicated specialized operating spaces for optimal surgical care, did the affluent return to the hospital in any numbers.

These facts should not obscure the truth that, properly constructed and staffed, the Civil War hospital could be a place for effective healing. There were a few drugs that made a difference—quinine, opium, chloroform, ether, bromine—but more important was the provision of fluids, nourishing food, clean bed linens, and wound care. Florence Nightingale recognized this principle in describing the horrible hospitals for the wounded in the Crimea, and it is in comparison with that war that the Civil War, oddly enough, shines. Progress had been made. Whereas in the Mexican-American War, seven men died from disease for every man killed in battle, and in the Crimea this ratio was four to one for the English and the French armies, in the Civil War only two Union men died of disease for each one who died of battle wounds. In an evaluation of war mortality written in 1865, U.S. Army surgeon Edward Dunster concluded: "The ratio, then, observable in our own recent war, is not only a cause of congratulation to ourselves . . . but it also shows conclusively the efficiency of the sanitary regulations enforced in our army, the skill and adequate numbers of the medical staff, and the perfection of that vast hospital system for the care of the sick and wounded." He also credited the USSC for its efforts and noted the importance of mental strength among the Union troops, their "springiness and buoyancy of spirits."[8]

Dunster's appreciation of "mental strength" illustrates an important aspect of medical care as it was understood in the Civil War. When Union physician Roberts Bartholow summarized the "various influences affecting the physical endurance, the power of resisting disease etc.," he included "moral causes, as malingering, desertion, nostalgia, etc., in operation during the whole period."[9] The word "moral" used in this context referred to mental function, although it could include the modern meaning of adherence to an ethical or religious code. When Enlightenment proponents of humane treatment of the insane instituted a regimen that involved talking, kindness, and structural routines, they called it moral therapy, not because it was ethical but because its target was to restore order to the chaotic mind. Bartholow used the label to refer to any influence on a man's physical condition that was attributable to his mental functions. "Malingering" meant faking illness or disability to avoid service, while "desertion" speaks for itself. Perhaps there were some echoes of Samuel Cartwright's theory that runaway slaves obviously suffered from a mental disorder—something he called "drapetomania"—because no sane man would want to escape slavery.[10] Was desertion perhaps a similar mental illness? "Nostalgia" was a broad term for the condition of being mentally "down," often due to homesickness, which fell into the broader category of melancholia.[11]

Nostalgia could be deadly. Bartholow reports 2,588 cases of it in the first two years of the war, and 13 deaths, but advises, "These numbers scarcely express the full extent to which nostalgia influenced the sickness and mortality of the army." A low mental state could likewise depress one's vital forces, making the difference between recovery and death from wounds or illness. Physical and mental symptoms were often intertwined. "The mental despondency and the exaltation of the imaginative faculty increase with the decline of mental strength," causing the ill body to sink further, Bartholow concluded.[12] The best treatment, not surprisingly, was to send a man home. But lacking that possibility, substitutions were widely promoted that simulated the cheering effects of the home environment. Letters from home were cherished, and men frequently thanked their letter writers for the uplift and whined when the interval since the last missive was too long. In the general hospitals, women volunteers read to men, wrote letters for them, played music in the wards, sang, and staged entertainments. Sustaining a man's will for recovery was not trivial; one surgeon caring for black troops in 1865 noted that as a rule, "when taken ill they turn[ed] away from all hope of recovery and die[d] speedily and without appreciable cause."[13] Nineteenth-century physicians were certain that strength of will was important in recovery; the same attitude survives in the modern era with the belief that a person's determination in fighting cancer, for example, makes a difference in the outcome. Whites were seen as having greater will and fortitude than blacks, and this was one explanation offered for the higher black death rates from disease.[14]

As the chapters that follow demonstrate, the hospital experience could vary widely among Civil War troops. The field hospital was a much different place from the relative haven of the general hospital safely behind the lines. Northern hospitals were more successful in approximating the well-provisioned, comfortable, home healing environment than were those in the South, and they were more successful in returning patients to health. Officers received better care than enlisted men; white Union troops fared better than their black comrades.[15] For the wounded, the initial challenge was to survive the first hours after battle, to survive surgery if necessary, and to make it to the calm of a hospital bed. The same could apply to the sick, since surviving the first few days of an illness was itself a predictor of longer-term survival. Much of the introductory material in this chapter can apply to both North and South, although information is often more plentiful on the northern experience. After a look at the therapeutic philosophy that guided Civil War surgeons, we will turn to the history of the field hospitals and then to the creation of the general hospitals.

Medical and Surgical Therapy

The Civil War occurred during a tumultuous era of therapeutic change. In the previous two decades, ether and chloroform anesthesia had revolution-ized surgery. If physicians were unfamiliar with the use of anesthetics at the beginning of the war, they became experts by war's end.[16] Bloodletting, the mainstay of therapy during the early nineteenth century, had fallen into dis-favor as progressive physicians, many of them trained in France, argued that it was not effective in treating fevers, especially pneumonia.[17] Some physi-cians claimed that the types of diseases had changed, that in earlier decades diseases had demanded the sort of bodily depletion brought about by blood-letting, but that mid-nineteenth-century febrile diseases required the stimu-lation and support of a different kind of therapy. Physicians used the terms *asthenic* and *sthenic* (from the Greek for "strength") to refer to diseases re-quiring stimulation and depletion, respectively; remedies that depleted, such as bloodletting and calomel, were known as antiphlogistics. Luckily for the Civil War soldier, most fevers of the time were seen as needing stimulation. The physician accomplished stimulation by prescribing tonics such as qui-nine and alcohol, as well as rest and appropriate foods. Whiskey and brandy were among the most commonly prescribed treatments of the war, and the control (and abuse) of the liquor supply was often an issue. Some 131 million ounces of alcoholic beverages were supplied as medicine by the Union Army, although how much made it to the sick is anyone's guess.[18]

Common diseases called forth common responses. The civilian physician had been trained to adjust his therapies according to the patient's age, sex, geographical location, and occupational style.[19] But in the Civil War, patients were mostly seen as much the same, since the soldiers were all men, were close in age, and were situated in the same environment doing the same kinds of physical activity. The physician might accommodate his therapy to locale, especially when it featured the dreaded "malaria" from swampy terrain, but the patients who appeared before him by the hundreds on busy sick-call days were more similar than different. Many of them suffered from the various forms of diarrhea and dysentery that were the number-one cause of nonbattlefield morbidity and mortality.[20] Although surgeons might distinguish acute from chronic diarrhea and measure degrees of severity according to the presence of blood in the stool or painful rectal cramping, they had no knowledge of the specific organisms involved. Physicians prescribed emetics and purgatives to clear out the offending material they felt was poisoning the gut, or opiates to slow the tumultuous bowel.

New York infantryman Constant Hanks attested to these measures (and their lack of efficacy) in a letter written to his mother soon after the battle of Antietam. It seemed that everything went through his digestive tract like water through a sieve. "Within the last few days I have passed everything through me except my hat. You write that Elisha deserted from the hospital if he had the diarrhea I don't blame him at all for deserting for they cant cure it in the hospital they give something that will stop one day then I bloat up like a drowned horse then the next day bad as ever." Hanks goes on to tell her how much letters from home matter to those in the hospital, where they are entirely among strangers. "A letter from a mother or sister or any true friend [brings] visions of home of its comforts, joys pleasures [which] come clustering around the heart like doves [and] cheer the despairing heart."[21] Hanks clearly implied that maternal attention would have more efficacy than the doctor's medicines.

Aside from purgatives, emetics, and opiates to slow the bowel, mercury compounds were among the most common medicines used in the war, especially for gastrointestinal complaints.[22] The connection of bile to the function of the liver was well known. Therefore, since a vomiting man might well produce bilious discharges or pass greenish stool, malfunction of the liver was a likely diagnosis. Physicians universally believed that calomel (mercurous chloride) and other mercury compounds could improve liver function. The drug acted as a laxative and caused excessive salivation, signaling to doctor and patient that it was active in the body. It is possible that the compound had some antibiotic effect. Tartar emetic, an antimony compound, likewise helped clear the gastrointestinal tract by promoting vomiting. These compounds were handed out with a liberal hand, so liberal that Union surgeon general William Hammond remonstrated against their overuse and removed calomel and tartar emetic from the Union supply table. Surgeons could still requisition them, but it took a special order. Calomel was one of the mainstays of mid-nineteenth-century medicine and was a flashpoint for critics of regular practice. Botanic practitioners distinguished themselves from the "regulars" by rejecting the drug and trumpeting their alternative herbal remedies (many of which were powerful purgatives). Accordingly, Hammond's order caused an uproar and contributed to his ultimate court-martial. Calomel was the partner of bloodletting; these were the two practices that typified the regular medical profession in its use of heroic therapeutics. It is interesting that such a commonly used drug was not included in the table of common remedies listed in the official history of the war; its omission may reflect either that Hammond's order had an effect in decreasing use of the drug or that its use had become so politically charged that it was best omitted from the accounting.[23]

So were calomel and other mercury compounds overused during the war? It's hard to tell. Mercury was also used to treat sexually transmitted diseases, sometimes as an oral preparation and often topically. And while the anger occasioned by Hammond's order is well documented, it turns out that ridicule of the use of calomel existed alongside it, for example in the form of a satirical poem published in a hospital newspaper in 1864, a year after Hammond's pronouncement. The poem was supposedly written by a patient, and it lambastes the drug for loosening teeth, causing skin sores, generating seizures, and other dire side effects. It tells of a man dosed disastrously with calomel who came to a sad end:

> The man grows worse, quite fast indeed—"Go call for counsel—ride with speed"—The counsel comes, like post with mail Doubling the dose of calomel.
> The man in death begins to groan—The fatal job for him is done; His soul is winged for heaven or hell—A sacrifice to calomel.

Although the poem was reproduced from an 1840s journal promoting herbal remedies, it is telling that the editors of the hospital paper (which apparently included one regular physician) published it with favor.[24] While the actual degree to which calomel persisted in use past this time is contested, it is evident that this drug was, at the least, the topic of much debate during the war.[25]

Alongside alcohol and calomel, quinine was among the most-used drugs of the war. It was seen as a general tonic and fever-lowering agent, and its special power in malarial fevers was widely recognized. During the peninsular campaign of 1862, along the seacoasts from the Potomac to the Rio Grande and deep within the Mississippi River Valley, malarial fevers struck down Union troops, and the military medical establishment responded with more than a million ounces of quinine and cinchona bark for its treatment (cinchona bark being the source of quinine and other similar compounds).[26] As described in chapter 3, the army even used daily quinine rations as a means of preventing the fevers that could so thoroughly and quickly depress troop strength. That the Confederacy suffered from a shortage of quinine in the war may have made a significant difference in the number of men able to render military service. The war not only put men in the sort of swampy environments where the malaria mosquito (unknown to them) thrived, but it amplified the number of malaria cases, because the proximity of infected men to mosquitoes and susceptible new hosts meant that what might have been a low level of infection in a sparsely settled country became a major outbreak in the "cities" that were military camps.[27]

In addition to oral medications, topical applications were common. Physicians ordered mustard plasters—irritating skin poultices—to be placed over chests harboring pneumonia and pleurisy and the disturbed abdomens of people with chronic diarrhea. This cuticle stimulation was thought to draw out poisons by inflaming the skin and also to help relieve pain. Cupping likewise promoted this outcome. Dry cups were applied by creating suction in small glass jars with a burning flame and then applying the suction to the skin. Wet cupping involved making small incisions, often with an instrument called a scarificator, and then using the cup to suck out blood as well as create the counterirritation. Leeches might also be used to suck blood from an inflamed body part, although they were hard to come by in the field environment.[28]

Such common medical problems and therapies were the daily fare of physicians treating patients at home. But the war brought them many new kinds of patients, who tried the physicians' skills and expanded their knowledge. Surgeons saw epidemics of meningitis and smallpox, diseases that would have been unfamiliar to all but a few, especially those from small towns.

Northern physicians dealt with the diseases of African Americans when the army began to enroll black troops in large numbers in 1863; such patients were so unfamiliar that the Sanitary Commission distributed a survey seeking information about how the black patients responded to remedies or experienced certain diseases.[29] Union physicians practicing in the South during the war may have made their first acquaintance with yellow fever and falciparum malaria, diseases broadly feared as particularly dangerous for northerners unadapted to the climate.[30] They could not avoid becoming well aware of the signs of scurvy and the necessity of certain foods to remedy it. Called "anti-scorbutics," these foods included potatoes, onions, pickles, and sauerkraut (all of which could be stored in barrels and survive shipping without spoilage), as well as fresh fruits such as lemons and the much-disliked dehydrated vegetables that men were supposed to mix in soup. The Sanitary Commission often supplied the Union with the necessary foods to stave off this disease, and it appeared among Union troops only when they were cut off from such supplements.

Treating gunshot wounds was the newest aspect of army surgeons' service. The kinds of injuries that might cross the path of the civilian surgeon a few times a year came in the hundreds after a major battle. The medical leadership of both the Union and the Confederacy realized that the surgeons who joined the volunteer regiments needed a crash course in military medicine, so they distributed books that outlined both camp hygiene and surgical guidelines. The books explained when amputation was necessary, how and when to control hemorrhage, and the proper care of wounds. Even if the surgeon entered the service ignorant of such matters, he quickly became exposed and, one hopes, educated in the techniques for dealing with the sequelae of the minié ball, the bayonet thrust, or the general crushing caused by a cannonball or a falling horse.[31]

Amputation has become the iconic act of the Civil War surgeon. The pile of severed limbs illustrated in Ken Burns's documentary *Civil War* is particularly memorable, and in the movie *Gone with the Wind,* an off-screen soldier faces amputation without anesthesia in the horrible Atlanta hospital where Scarlett sought a doctor for the laboring Melanie.[32] Amputation was commonly practiced, as illustrated by the large number of men who sought prosthetic limbs after the war, although amputation without the merciful use of ether or chloroform was rare. The many applications for wooden legs also illustrate a second fact—men often survived the surgery. Surgeons amputated when they believed the man would otherwise die from his wound.

What factors did they consider in coming to that conclusion? First, was the skin broken? A simple fracture, one in which the skin was intact and the bone broken only in one spot, usually required nothing more than splinting and time. Such fractures might result from falls or other trauma. But most common was the bone fractured by a minié ball, the bullet that was famous for shattering bone and macerating tissue. Such wounds, with broken skin and bone in multiple fragments, rarely healed successfully. Given a choice between amputation and life with a disability versus death from wound suppuration, surgeons chose amputation. Surgeons also saw wounds of the head, neck, chest, abdomen, and genitals. Such wounded men received pain control and sympathy, since the surgeon had little to offer those who suffered more than superficial damage to these areas.[33]

The choice to amputate was never welcome, and surgeons debated whether it was better to try to save the limb through "conservative" measures or sever it immediately. If a surgeon chose a conservative course in treatment, say, of an open tibial fracture, the plan would require immobilization, using some sort of appliance to hold the bone fragments in alignment during healing. The surgeon might need to excise some bony fragments or clean out infected bone during the healing process. Statistics gathered after the war supported the choice of immediate amputation over such limb-sparing methods.[34] They also demonstrated that quick amputations, within forty-eight to seventy-two hours of wounding, had better outcomes than waiting longer, when inflammation, as evident from swelling and redness, had set in. Mortality after primary amputations was 27 percent; later surgery resulted in a 38 percent case mortality rate.[35] The conservative approach to shot fractures overall resulted in fewer deaths when compared to amputation (18% versus 26%), but as noted in the official Union medical history of the war, those men treated with amputation most likely had more significant wounds than those treated conservatively. "It must be borne in mind that the cases reported as treated conservatively were cases selected as specially adapted to this mode of treatment, and probably were the least serious," the work's author concluded. He further pointed out that "many operations were performed in the field, where the appliances and necessary rest of conservative surgery could not be had, and where frequent transportation for considerable distances was unavoidable."[36] He also noted that as surgeons gained skill during the war, they learned techniques for conservative treatment that were unfamiliar to them earlier in the conflict.

Surgical intervention had several goals. The first was to control bleeding, from the point of initial contact in the field, during the operation, and afterward as the wound healed. Litter carriers attempted to splint fractures to prevent further damage during transport, but often this proved impossible. Upon initial examination by the surgeon, the next goal was to remove as much debris from the wound as possible, since the bullet may have carried cloth and dirt into the wound, and loose spicules of bone needed to be extracted to avoid further mechanical injury. Then the decision about amputation had to be made. If amputation was chosen, the next step was to remove as much of the limb as was necessary to clean out the destruction caused by ball or fracture, but to leave a usable stump. Surgeons tried to stay as far away from the trunk as they could, because a revision might be necessary and it could not go any farther than the hip or the shoulder. They also knew, as was demonstrated in later statistics, that the closer to the trunk they had to cut, the greater the likelihood of death. Gunshot wounds that affected the hip joint led to 83 percent mortality after amputation, whereas for amputation at the shin, the rate was 33 percent. A final surgical goal was to create a stump that allowed for the fitting of a prosthesis once the man had recovered. This entailed creating flaps of skin that were brought together with silk or silver-wire sutures. The wound was left only loosely closed, in the hope that if infection followed, there would be ample tracks for drainage. The Civil War surgeon still debated whether pus was laudable, but he knew for certain that if a wound drained adequately, it had a good chance at healing. A completely closed wound would promote septicemia, or generalized blood infection, a deadly outcome to be avoided at all costs.[37]

Surgical therapeutics began with the operation, but its subsequent course was similar to that targeting the medical patient. Surgical patients were often "low" after the operation and had to be shored up with punch made of whiskey plus condensed milk or whiskey in lemonade. Surgical wounds were treated with various forms of debridement when they became infected (debridement is the process of removing dead or infected matter in the hope that healthy tissue will emerge beneath it and promote healing). Debridement could be done mechanically with surgical tools or by using a dressing that, when removed, would pull tissue away. Surgical texts generally recommended plain cloth dressings, kept moist with cold water and changed frequently. The surgeon also used topical disinfectants, and from 1862 it became common to clean infected wounds with bromine, iodine, or nitric acid.

The surgical process usually began in the field hospital, as the prevailing wisdom held that the wounded man must be treated as soon as possible after injury, optimally no later than twenty-four hours after the skin was broken. This meant that an effective system of field hospitals—hospitals located in the immediate vicinity of the battlefield or regiment—was a key element in the medical care of both armies.

The Field Hospital

The term *field hospital* had several meanings during the Civil War.[38] It could be applied to just about any institution of healing that was not a general hospital, that is, not a large institution for patients far behind the lines. As historian John Fahey has noted, "In the official reports and correspondence of the participants, the term 'field hospital' was applied equally to austere regimental units containing a few cots and to huge divisional and corps level establishments capable of providing inpatient care to thousands of patients."[39] The defining feature of the field hospital was that it treated the sick or wounded near their regiments. Care was provided by the surgeons assigned to the soldier's regiment, or sometimes assigned to regiments that surrounded his. The army's goal was always to keep a sick or wounded man as close to his regiment as possible, in the hope that he would be cured and speedily returned to it. However, the army's leadership also knew that when that regiment was on the move, the sick and wounded were difficult to transport and in that case they would be left behind under another surgeon's care, or shipped off to a general hospital. In some circumstances, when many regiments were stationed close together, there might be a post hospital or other group facility in the vicinity to care for the sick from various regiments.

The sick soldier's first encounter with the army's medical staff was at the morning "sick call." There he presented his complaints to one of the two or three surgeons assigned to his regiment and was perhaps dosed; a decision would be made about whether he should rest for that day or go about his regular duties. Surgeon Burt Wilder described seeing sometimes 140 men at sick call, all before breakfast, so the examinations must have been cursory. The most common complaint in almost every regiment was diarrhea.[40] Men too ill for their own tents went to the regiment's hospital, a tent set up with eight cots and supplied with some sort of nursing oversight by fellow soldiers, often convalescent patients. Many illnesses were short-lived, responding to the regimen of rest, hydration, perhaps a special diet, and medication. While the man was in the hospital, his friends could visit, and they could at times help make him more comfortable by entertaining him or bringing him treats bought from the sutler (a camp peddler). The advantages of such care included the attention of friends and of a doctor who was familiar with the patients and reported, at least socially, to the man's officers and fellow soldiers. The disadvantages were the dependence on one or few surgeons, whose capabilities might or might not be helpful, and the necessarily temporary status of such units, which could be closed down whenever the regiment was ordered to march. The sick soldier might then be shipped to another hospital, where he would be among strangers.

More often *field hospital* was applied to those temporary outposts of medical care set up in the vicinity of the battlefield. One of the most important aspects of battlefield medical care was the problem of dealing with a huge surge of patients all at once. At the battle of Shiloh, for example (April 6–7, 1862), there were eighty-four hundred Union wounded and another eight thousand injured Confederates, many of the latter left behind to be taken prisoner as the Confederate army escaped to the South. And this all happened in a place where the Union Army had just arrived, where no hospitals existed, and where care occurred amid the chaos of battle. Aid stations set up to provide initial care for the wounded were frequently overrun by troops as the battle lines shifted, with the Tennessee River a definitive barrier at the back end of the field. Surgeons began to care for the wounded aboard ships, as there was no stable ground on which to set up hospitals. Many wounded lay untended on the battlefield overnight, until the tide turned and newly arrived Union troops helped those who survived the first day to push the rebels back. Through it all, surgeons established work stations, acted to control bleeding, amputated shattered limbs, and attempted to get the wounded on transports that could take them to care in general hospitals upriver in St. Louis, Cincinnati, and Louisville.

Upon hearing of the distress at Shiloh, the Sanitary Commission and various state-identified care associations chartered boats to evacuate the wounded in a disorganized sort of mini-Dunkirk. Care at Shiloh was an improvement over the even worse chaos that had followed the battle of Fort Donelson two months earlier. There, wounded men lay in the snow because sufficient tents had not been distributed for hospital use and there were not enough buildings in the sparsely occupied countryside to serve as temporary shelter. At Shiloh surgeon Bernard Dowling appropriated tents captured from the enemy and set up a hospital camp, with the tents set in orderly rows and surgeons assigned to provide both emergent care for the wounded and medical

Field Hospital at Savage Station, Virginia, after the battle of June 27, 1862. Photograph by James F. Gibson. The chaos of this scene would have been typical in the wake of battle. In the foreground a surgeon examines a man's leg to assess his injury. This image, often reproduced, is the only one I have found that depicts the medical disorder immediately after a battle. Library of Congress, Prints and Photographs Collection, LC-DIG-cwpb 01063.

care for soldiers who contracted illness in the aftermath of battle. Dowling argued that it might be better for a man to recover from surgery there than to be jostled aboard a transport ship for the trip to a distant hospital, and with his actions was born the concept of the established field hospital that was more than just a temporary site of emergency care.[41]

The Sanitary Commission highlighted the chaos surrounding the battles on the James River peninsula in spring and summer 1862, and lauded its own role in rescuing wounded men and putting them on humane hospital transports to comfortable hospitals in the North (more about this in chapter 3). Charles Tripler, medical director of the Army of the Potomac, told his side of the story in a report dated July 3, 1862. Tripler's account emphasized his inability to acquire full control of the medical needs of the men under his charge. The hospitals that could receive the wounded were first located near Yorktown but later moved closer to the fighting at Harrison's Landing on the James River and White House on the Pamunkey. As the Union admitted failure in the attempt to take Richmond and General George McClellan retreated to Fortress Monroe, some sick and wounded who could not be moved were left, with their attendant surgeons, to be captured by the enemy. Tripler was annoyed at the efforts of the Sanitary Commission, who challenged his command and at one point took a boatload of patients off to Boston against his express wishes. Tripler recognized that the army's delivery of medical care was not as it should be, but he blamed the lack of staff, and especially the lack of surgeons imbued with military discipline and skilled in camp hygiene and surgical techniques.[42] He hired some one hundred temporary contract surgeons for the army, including medical men sent by particular states to care for their own men, when there were insufficient commissioned medical officers.

Tripler had his doubts about the wisdom of sending men to the North. While he acknowledged that the James River peninsula was a swampy locale whose malarious emanations would sicken the men and further weaken those already ill or wounded, he also worried that men sent north would be lost to the army. In late May 1862, he reported that the Union troops were in much disorder, with stragglers hiding in the woods (and at times becoming ill there). Further, he complained that well or only slightly wounded men were pushing onto the transports and then escaping military observation once they arrived at the northern ports. Many more were malingering in northern hospitals. "I feel confident that more than 1,000 men perfectly fit to join their regiments are now idle in the general hospitals," Tripler wrote. His orders to keep men at the hospital stations near Richmond, instead of shipping them out, meant many were captured by the enemy when McClellan retreated. Tripler's laments about malingering, voiced early in the war and about men who were all volunteers, reveal that already the military medical establishment was at least as concerned with controlling the sick and preventing their inappropriate absence from duty as they were healing the men under their care.[43]

In June Surgeon General Hammond replaced Tripler with Jonathan Letterman, a surgeon who approached the members of the Sanitary Commission with a spirit of cooperation and brought greater order to the medical management of the department.[44] After Letterman took over and reformed the evacuation process, the USSC left the transport of men off the James River peninsula to the military and turned their attention to other fields.[45] Historian Alfred Bollet calls Letterman one of the heroes of the war and "the father of both the modern military and the emergency medical systems."[46]

There can be no doubt that Letterman succeeded in implementing plans rather than just calling for them (as Tripler had done); he did so by convincing General McClellan that reform was necessary. The horrors that followed the fruitless attempts to take Richmond in spring and summer 1862 may have convinced the commander that a better system for caring for the wounded was necessary; the political fallout of the USSC publications decrying the military's ineptitude in providing for the injured after the battles may have likewise hit home. In any event, Letterman not only devised a better system to provide for the wounded but gained the commanding officer's support in implementing it all down the line.

Letterman called for a dedicated ambulance crew to be assigned to each regiment, along with ambulance wagons for transporting the wounded. This strategy avoided the problem of a man's comrades stepping away from the battle to transport a wounded friend, and it also allowed for training in the proper way to load a wounded man on a stretcher or place him in the ambulance. Letterman also brought order to the setting up of field hospitals and the transport of men on hospital ships or trains as needed. Like Tripler, Letterman hoped for the creation of proper field hospitals, composed of tents dedicated for that purpose, because they would "be especially needed to shelter the wounded and sick, whom it would be desirable to keep with the army." Unlike Tripler, who emphasized that transported men would be lost to their regiments, Letterman justified the plan on medical grounds, saying, "No one thing so much disheartens troops and causes homesickness among those who are well as sending the sick to hospitals outside of the army to which they belong."[47] But Letterman did not achieve all this by himself. Supporting him was a growing federal infrastructure of general hospitals, the maturation of supply lines that brought food and medical supplies to the army, and a medical corps that was increasingly seasoned, with the worst incompetents weeded out. William A. Hammond was in the surgeon general's chair, directing the building of new hospitals and supporting medical directors such as Letterman in their reform efforts. By the time Hammond was ousted in August 1863, the Union military establishment had become far better organized and less in need of volunteer agency help.

Letterman did not limit his interventions to the creation of the ambulance corps and field hospitals. He inspected the camps and found a surprising amount of scurvy among Union troops camped in Virginia.[48] He also discovered that the camps were unhygienic and the men filthy. His orders had a feminine tone, revealing the need for medical leadership to take over the mother's role to ensure the health of the men. He ordered the delivery of fresh vegetables, including potatoes and onions, to the camps, "cost what they may," and said the men should be forced to eat the desiccated vegetables supplied in their ration, as well as dried apples and pickles. Fifteen hundred boxes of fresh lemons arrived shortly, and the signs of scurvy quickly disappeared. He wanted the units set up to bake fresh bread, to relieve the men of the tedium of hardtack, the thick, dry crackers that were their default carbohydrate source. Men were to be allowed appropriate time for sleep, and tents were to be pitched in healthful, dry locations. "I would also recommend that the strictest attention be paid to policing, general and special; that all the troops be compelled to bathe once a week, a regiment at a time . . . being marched to the river . . . to remain in the water fifteen minutes," Letterman wrote in mid-July 1862. He gave equally precise instructions on the construction of latrine areas and the disposal of animal manure and offal from slaughtering. Policing meant that the men would have to clean up any fecal deposits made in inappropriate spots, for "no nuisance whatever [was to] be allowed anywhere within the limits of the army." Regimental commanders were ordered to make sure that all this would be done, and Letterman had the support of the commander.[49] You can almost hear him saying, "I mean it this time!" Eating vegetables, bathing, keeping the campsite clean—such behaviors evoked a mother's command over unruly boys, but military officers were learning that the health of their camps and the strength of their regiments depended on following such feminized behaviors.

These developments all took time, however. Letterman himself did not see the system in full operation until the Battle of Fredericksburg in December 1862. While he praised the smooth operation of the care for the wounded after that battle, others wrote of persistent horrors, especially in the transport of wounded to the North.[50] And Letterman's ambulance system at first pertained only to the Army of the Potomac, although it was taken up slowly by other commands. As described later, the USSC again intervened after the battle of Antietam (September 1862), where their supplies reached field hospitals in the wake of battle before the official army wagons did. A state-based sanitary commission in Maine sent relief agents to assist Maine regiments when they heard that men were suffering for want of care after the battle. Harriet Eaton, one of the agents, found men left in houses and barns dotting the rural Maryland landscape without adequate provision of even basics like food and drink, much less medical care.[51] The battle of Perryville, Kentucky (October 1862), likewise left hordes of wounded piled into available local buildings such as courthouses and churches; there again the USSC stepped in to supplement the inadequate army care.[52]

One New York soldier sent his mother a description of the desolate neighborhood of Antietam after the battle, which was both horrible and typical of great encounters of the war. He had been all fired up on the day of the battle and almost enjoyed it. "But the day after battle when I got my pass to the rear, is *when* I see the frightful horrors of the battle, roughhewn and hardened sinner that I am, it fairly made my blood chill with horror to view the dead dying & wounded, the human woe, & suffering one short days battle had made," he told his mother. "Our Division was nearly on the extreme right so that when I left the hive of battle for the rear I came along some 4 miles, every house barn and outbuilding was full of wounded men wounded in all shapes that one could imagine hundreds lay under apple trees in orchards some groaning in pain all along surgeons were amputating legs arms, some lay as they wer[e] brought in with both legs tore off hanging by only small pieces of skin." The soldier kept walking, seeking care for his own infected knee. "Such sights as these continued until I had passed through a village called Keedysville where two churches, courthouse a school house and a great many private houses were full of wounded men used as hospitals." By the time of this letter, written a week after the battle, he knew that "there was sum 10,000 killed and wounded men." He concluded, "You can conceive what pictures of woe, pain, and destruction, one should see in coming away from the field of battle. . . . I would rather fight a battle, and run my chances, than pass, as I did, the day after amid the dying & wounded."[53]

Letterman's system received its major test at the Battle of Gettysburg (July 1863), which left more than 14,500 Union men wounded and another 7,300 Confederates left behind who were severely wounded, captured, or both. The carnage of this battle overwhelmed Letterman's careful plans. Union lieutenant Frank Haskell wrote his brother, "The whole neighborhood in rear of the field became one vast hospital, of miles in extent. . . . At every house, and barn, and shed the wounded were . . . and there they gathered, in numbers a great army, a mutilated, bruised mass of humanity."[54] Haskell saw the men with bloodied bandages who had gone under the surgeon's saw and others "like men at a ticket office, await impatiently their turn,—to have an arm or a leg cut off." Of the surgeons he wrote, "They look weary and tired . . . [and] their faces and clothes are spattered with blood. . . . How much and how long they have worked, the piles of legs, arms, feet, hands, finger, about partially tell."[55] One historian has identified 175 field hospitals serving the

Union Hospital in a Barn near Antietam Creek (ca. September 17, 1862). Although the patients have the shelter of the barn's roof, the surgeons have little natural light, and wounded men lie on straw most likely contaminated with animal dung. Note the wagons being used as ambulances. Reproduction of wood engraving after a sketch by Edwin Forbes. Library of Congress, LC-DIG cph.3b06630.

wounded of both sides at Gettysburg—field hospitals that included first-response dressing stations, locations for surgical amputations, receiving areas for those in transit, and other functions.[56]

The first problem for the ambulance service was the simple fact that day one of that battle was followed by day two and then day three; those wounded early on might lie under continued firing for hours or days before it was safe to remove them. Many of the wounded were removed at night, when the ambulance corpsmen had to find and transport them in the darkness; on the second night rain added to the nightmare scene. Still, the ambulance corps labored mightily, hauling the wounded off the battlefield on stretchers to a place out of firing range, where the injured were deposited; the stretcher bearers then returned for more wounded. From there the injured men went first to dressing stations for immediate efforts to control bleeding and pain, and then on to the appropriate field hospital for their regiments; medical care was organized into higher-level field hospitals by brigades, divisions, or corps. These stations were set up hastily in spots thought safe from the battle, usually in the vicinity of farmhouses or barns, which provided some shelter and a working well. General George G. Meade had ordered the surgeons forward, with the supplies and instruments they could carry on their own horses, but he kept the medical supply wagons back so that the roads would remain open for retreat. Thus, the surgeons were hampered from the beginning by a shortage of tents, bandages, cooking equipment, and all the basics of hospital service. Some hospitals subsequently had to be moved; one flooded in a downpour, and others were overrun by rebel troops. Through it all, surgeons labored night and day to treat the wounded, creating grisly piles of arms and legs in an attempt to stave off deadly infections in shattered limbs.[57]

Every surgeon was needed to deal with the flood of Union and Confederate wounded. A new assignment system was in place at Gettysburg, one that designated the most experienced surgeons as the principal cutters, while others were dressers or assistants. But Meade was on the move, chasing Lee's army as it retreated south toward Virginia. On July 5 Meade conveyed orders via Jonathan Letterman that he was taking the army south and that the bulk of the surgical staff needed to go with them to provide necessary care after the battles to come. Two-thirds of the medical staff walked away from the "vast sea of misery" that was the Gettysburg battlefield in the aftermath of July 2–3, 1863, and followed the army as commanded, and a disproportionate number of them were the most skilled in surgery. For the Second Army Corps this meant fourteen surgeons remained to care for more than thirty-two hundred wounded men.[58] It took a week for all the wounded to receive initial care, during which time some died who would have survived even with the primitive care available. Adding to the chaos following the battle was that the train line leading into Gettysburg had been damaged, so men waited for several days at the depot for transport trains to arrive to take them to outside hospitals, and supplies piled up on the other side of the broken rail that were destined for the wounded. The Sanitary Commission, the Christian Commission, and independent volunteers poured into the breach, bringing supplies by wagon and otherwise trying to fill the gap in care.[59]

Letterman studied the deficits of army medical care after Gettysburg and made renewed efforts to reform hospital supply and ambulance protocol. In March 1864 Senator Henry Wilson (MA), at the prompting of Boston physician Henry I. Bowditch, pushed a reform bill through Congress that codified Letterman's system into law and applied it to all of the Union armies.[60] Bowditch had become rabid on the subject after his son Nathaniel died of wounds sustained during a minor skirmish in Virginia in March 1863; it was a death hastened by inappropriate care in the wake of the injury.[61] Never again would volunteer aid be required in such a dramatic fashion as at Gettysburg—or at least it would not be allowed to assist the troops. When Harriet Eaton returned from Maine in summer 1864, determined to once again provide "relief" to the Maine troops, she was told that volunteers such as her were only permitted to work in the receiving hospitals at City Point, southeast of Richmond.[62] The Sanitary Commission has little to say about immediate relief after the major battles of summer 1864. Its members served as a source of supplies rather than the providers of care. This did not mean that all went smoothly—it was impossible for the ambulance men and attendant surgeons to keep up with the care of men wounded in the battles of the Wilderness, Spotsylvania Courthouse, and Cold Harbor—but the assumption had grown that the army was able to care efficiently and humanely for its wounded and no longer needed to depend on charitable organizations to take up the slack.

Even though the field-hospital system was not perfect in summer 1864, it had evolved perhaps as far as possible given the means and personnel of the Civil War army. Further refinements came when medics were able to respond quickly to shock in World War I and World War II with the provision of blood transfusion and intravenous fluid resuscitation. In the Korean War, mobile surgical hospitals brought sophisticated care close to the battle lines, and general progress in surgery and medicine meant that the wounded were more likely to survive.[63] But it was not a story of even progress over time. During the Spanish-American War, outbreaks of typhoid illustrated that the army had forgotten the hygienic lessons of the Civil War. Once again line officers had to be schooled in the necessity of "policing" their camps and commanding men to use sinks properly and keep themselves clean. After that war the army instituted camp hygiene training at West Point and established a regular female nursing corps to amend some of the worst problems in hospital care that plagued the Spanish-American War. Perhaps most tellingly, in the struggle to make cleanliness and control of bathroom habits a manly endeavor as important as learning to manage weapons or attack the enemy, wily army doctors took to calling such endeavors "sanitary tactics," putting camp hygiene on an equal footing with battle strategy, at least in semantics.[64]

The General Hospital

For the Civil War soldier who survived the first days after being wounded and the process of surgery, if it was necessary, there was a healing haven waiting at the end of a rough trip by boat or train. Called "general" hospitals because they were not assigned to particular regiments, divisions, or corps, they marked a new departure in American medical history. While the Union Army did not invent the hospital—hospitals had existed for millennia—they perfected it on a large scale. Their Confederate colleagues, by and large educated in the same schools and reading the same books, did much the same, albeit with fewer resources. The general hospital was vastly different from the chaotic scenes that defined the field hospitals after a battle. The general hospital was deliberately located safely behind the lines, in such large cities as Washington, Philadelphia, Louisville, and St. Louis, cities in contact with battlefields by rail or water. Some, such as hospitals in Annapolis or Point Lookout, Maryland, were built to be particularly close to water transport from the James River peninsula. Many hospitals occupied large buildings that had previous been almshouses, schools, and warehouses. But as the war went on, both sides began erecting new, massive hospitals that were built on the pavilion plan and could hold thousands of patients. Tents might be pitched around the building when demand swelled beyond the indoor bed spaces. These were huge institutions—Chimborazo Hospital in Richmond could house more than five thousand patients, and Philadelphia's Satterlee Hospital approached that capacity when its grounds filled with tents.

While these institutions shared some features with antebellum hospitals, in other ways they were wholly new inventions. The antebellum hospital was a place for the poor, for the insane, or for the improvident traveler who was far from home. Its condition might be epitomized by a famous drawing of a Bellevue Hospital ward, published in *Harper's Weekly* in 1860, which showed a sleeping woman with rats crawling over her bed and table.[65] In such hospitals discipline, minimal care, and perhaps cure were the goals, but there was a wide gap between the status of doctors and that of patients. In Europe the gap even larger, as medical students walked the wards of charity hospitals with famous physician teachers and openly claimed potential dissection rights by tying toe tags on patients who still lived. Patients went there when the only other choice was to lie (and die) in the gutter. Anyone with means stayed home during sickness and was tended by family members.

The Civil War armies were composed almost entirely of volunteer troops, including men from across the social strata—mechanics, farmers, clerks, businessmen, and scholars all donned the blue and the gray. There was some truth to the adage that it was a "rich man's war and a poor man's fight," for particularly after the draft came, and a wealthy man could buy a substitute, the class differences were amplified. But even so, the Civil War hospitals were the first in the nation's history to see large numbers of affluent patients, men who would have been cared for at home before the war. On December 17, 1864, 83,409 men occupied Union hospital beds; there was room for 34,648 more.[66] Just as the war brought a new style of death to the country, it also brought a new vision of the sickroom, a tableau repeated for hundreds of thousands of patients. The war put the middle class in the hospital, and their presence made the Civil War hospital much different in philosophy and design from the civilian counterpart.[67]

Antebellum fascination with hospital design had primarily targeted the mental asylum. While the affluent could fairly easily care for the physically sick patient at home, given the limited requirements of nursing care in the mid-nineteenth century, the situation for the mentally ill was different. Even the affluent could go crazy, need restraint, and require twenty-four-hour attention that families could not provide. The antebellum asylum offered not just care, but cure, thought to be generated by the creation of architectural order and the soothing solace of routine. It is no accident that when affluent patients were under consideration, hospital designers were more likely to focus on comfort as well as restraint. General medical hospitals might have private rooms for paying customers, such as businessmen who took ill in their travels, but only mental hospitals had large numbers of affluent ill in the prewar era.[68]

The war vastly changed this calculation. The Civil War hospital was careful to maintain class structure—there were wards for officers that were separate from wards for privates even in field hospitals—but the officers were, nonetheless, often in the hospital. The general hospitals were also under constant inspection by families, visiting dignitaries, chaplains, and convalescent soldiers who could write letters of complaint home. While initial efforts were marred by poor organization, as 1862 progressed the general hospitals became increasingly pleasant, orderly, and curative places in which to recuperate. Hospitals in more distant cities like Nashville or Corpus Christi might escape close scrutiny, but the general hospitals within the northeastern and midwestern states had an audience that demanded respect and proper care for the men. And the men demanded it for themselves.

One such was insurance agent James O. Churchill of Freeport, Illinois, who entered the army as a twenty-six-year-old second lieutenant and left it a brevet lieutenant colonel. He was wounded in the assault on Fort Donelson, suffering a compound fracture to his right femur and injury to the other thigh as well.[69] Churchill refused bilateral amputation for his thigh wounds in the field, so instead he was bound into a lower-body casing that failed to support the fracture through the rough trip to a hospital in St. Louis. Once there, Churchill noted multiple problems with the traction apparatus that kept his legs elevated, not to mention bed sores and an infected finger. So great was his misery that Churchill slipped his ward nurse a dollar to fetch him extra morphine at night. He complained about the depressing effect of hearing men screaming in the "amputation room" or seeing coffins being carried down the hall. He was no passive patient. Churchill admired the skill of his surgeon but also saw his flaws. In an April 1862 letter to his parents, he reported that the man was "wholly bound up in his profession. He looks upon a wounded man as a piece of mutilated flesh and bone, and his duty is, with nature's assistance, to place it back in its normal condition." Although Churchill appreciated such dedication, he continued, "The individuality of the subject is wholly lost sight of."[70]

Churchill illustrated this lack of sensitivity with an anecdote about his own care. "Frequently surgeons are here from Paris, London and other large European cities to examine into our methods of treatment. [Dr. Hodgen] would often bring them to my room, throw down the sheet, explain the character of my wounds, and his treatment, without speaking or paying any attention to me whatever." Churchill found such behavior intolerable, no doubt in part because revealing his thighs also displayed his genitalia. "I felt very much humiliated, and made up my mind to devise some means to stop it." The next day the doctor came by with two surgeons from London. "As they turned to leave, I said 'Stop gentlemen,' and holding out my hand, said, 'Fifty cents a piece, if you please.' The visiting surgeons looked surprised, but the Doctor seemed to take in the situation." The local surgeon, at least, understood this was a different creature from the usual impoverished hospital patient. "All right," he said. "Every time since then he has always asked my permission, which of course is granted with the greatest pleasure."[71] Churchill survived his wounds and the war, although he walked afterward with a cane. Over time, middle-class patients like Churchill came to demand a new sort of hospital care, care that respected patient rights, provided a pleasant, clean environment, and moved from caretaking to comfort and cure. Accompanying this transformation in attitude was a new architectural shape for the modern hospital, one that began in the Civil War and directed hospital design well into the twentieth century.

The Pavilion Hospital

"The pavilion . . . is a sanitary code embodied in a building."[72] By the end of the eighteenth century, French physicians began imagining hospitals having long, slender wards with airy spaces between them, and English planners of the early nineteenth century likewise envisioned an ideal hospital that emphasized light, air, and ventilation in its design. By the mid-1850s, at least one hospital in each country had been built according to these specifications. But it was Florence Nightingale, and her little book *Notes on Hospitals,* who brought the pavilion plan of hospital architecture to the forefront of Civil War medicine. Nightingale had seen the deadly effects of poorly designed hospitals in the Crimea, where wounded and sick British soldiers housed in a converted Turkish barrack suffered nearly 50 percent mortality. She toured hospitals in Britain and France, formed her own opinions about the best hospital architecture, and described the ideal hospital as one that had ample air per bed, ventilation to prevent the accumulation of deadly foul odors, an adequate water supply, and abundant sunlight. She recommended that wards be 30 feet wide and around 130 feet long, with a ceiling 16 feet high. Windows should reach near the top of the ceiling to prevent the accumulation of fetid hot air. Civil War designers added ridge vents to keep the air circulating and further reduce the ward's smells.[73]

Nightingale's works on hospitals and nursing were best sellers in the United States and widely read among educated physicians. Her hospital designs emphasized order and fresh air. As Charles Rosenberg has recognized, Nightingale's "hospital seemed to her quite literally a microcosm of society, every part inter-related and all reflecting a particular moral order. Just as order in the body and an appropriate physical and psychological equilibrium constituted health for the individual, so order in the hospital implied a low incidence of fever and wound infection for its inhabitants."[74] The symmetry of the hospitals created according to the pavilion plan likewise echoed the concomitant concern about building asylums for the insane, whose architecture would supply the order found missing in their troubled minds.[75] Orderly pavilions embodied a vision of therapeutic efficacy for the most forward-thinking of nineteenth-century medical leaders.

Members of the U.S. Sanitary Commission were particularly well aware of Nightingale's dictates for hospitals and critical of the initial use of existing buildings to house the sick and wounded Union soldiers. After inspecting hospitals in Washington, D.C., many of which had formerly been schools of various sorts, the committee concluded in summer 1861: "Old buildings do not make good Hospitals." They recommended that "hereafter instead of hiring old buildings for General Hospitals they should order the erection of a sufficient number of wooden shanties or pavilions of appropriate construction, and fully provided with water for bathing, washing, and water-closets, and ample arrangements for ventilation." Their judgment could have been copied word for word from Nightingale: "This suggestion embodies the latest and best views as to the construction of hospitals, and its adoption would save both lives and money."[76] When the USSC succeeded in ousting the old medical guard and putting their man, William Hammond, into the surgeon general's office (see chapter 3), this plan went forward quickly. Hammond devoted one-fifth of his massive tome on military hygiene to the proper construction of hospitals.[77] The Confederates were reading from the same Bible, and they built in Richmond the largest pavilion hospital of the war, called Chimborazo.[78]

Conclusion

Hospital care was central to the experience of the ill and wounded in the Civil War. Military leaders learned the hard way that "nothing so much embarrasses operations in a campaign, as a large sick report." Military and medical leaders had to discover and devise ways to hold the army together and maximize the number of effective troops even though, as Charles Tripler wrote, "however scientific, zealous, intelligent and active the medical officer may be, men *will* get sick, and in a fight, they *will* get wounded."[79] The system of ambulances, field hospitals, and general hospitals effected by both sides was conceived to replace the typical flow of health care delivered in the home. The military provision of camp hygiene regulations and nutritious food likewise replaced the female role; it took on women's function as the supervisors and enforcers of cleanliness and the source of food and clothing. As later chapters demonstrate, the northern general hospital evolved into a healing environment that mimicked as much as possible the comforts of home, while the southern medical system shared the same aspirations but failed in their execution. The southern soldier was accordingly less likely to recover successfully and return to the army that so needed him in the ranks.

CHAPTER TWO

Women, War, and Medicine

At the top of anyone's list of "Famous Women in the Civil War" is Clara Barton, who organized battlefield relief stations, visited Andersonville after the war to identify the dead and notify families, and in the 1880s established the American Red Cross. So when she told a Memorial Day audience in 1888 that at the end of the Civil War, "Woman was at least fifty years in advance of the normal position which continued peace . . . would have assigned her," she had ample personal experience upon which to draw.[1] Barton's term "position" referred to the women's movement in general: the drive for suffrage, equal rights in the workplace, political voice, and gender neutrality in the courtroom. Was this assessment of the war's impact on women's rights an accurate one? Historian Mary Elizabeth Massey certainly thought so, according to her pathbreaking book on women and the war, published in 1966. The female presence in the public arena expanded markedly after the war, and there is no doubt that, as Massey concluded, "the Civil War compelled women to become more active, self-reliant, and resourceful, and this ultimately contributed to their economic, social and intellectual advancement."[2]

With millions of working-age men engaged in combat, the two nations required women's labor to maintain the functions of everyday life. And with thousands of those men in hospitals, the nations required an influx of hospital workers as never before. While in the modern era we see precise distinctions between roles such as doctor, nurse, social worker, and charity agent, the situation was much more fluid in the mid-nineteenth century. This chapter considers the stories of women who filled these various roles. Antebellum charity had emphasized the care of the worthy poor and of people unavoidably away from home and in desperate circumstances. Because of the war, vast numbers of men were away from home, beyond the reach of family support, and often disabled, and these men became the proper subject of organized national charity on a scale hard to imagine previously. How women would have fared if the war had never happened is anyone's guess, but the conflict was clearly a vehicle that carried women from the home into the public sphere, bringing undeniable changes in their "position."[3] That the war also may have worsened their aggregate health, temporarily, was one side effect of this transformation.

The "position" of women related not just to socially assigned roles but also to geography. The border between the private and public spheres—as well as between the home front and war areas—was stretched, transformed, and in places ruptured during the war. Historian Judith Giesberg has especially emphasized how the war displaced women from their usual places. "As they moved into new spaces, or expanded to fill the void left in old ones, women redrew the lines that separated home from war and mapped an alternate wartime geography dictated by the material conditions of war rather than the ideological constraints of gender." Giesberg's lens takes in not only the middle-class northern and southern women who have received the predominant attention from historians, but also the ways in which the war disrupted women's lives on the farm and pushed working-class women into factories and other public work. Disruptions in the Confederacy were even more severe than in the Union, especially for black slaves and poor white women who struggled to survive in a world with many sources of danger and few sources of support.[4] A simplistic narrative about the progress of medical women during the war that ignores these complexities is insufficient. In June 2012, historian Stephanie McCurry spoke to an audience of Civil War historians regarding such an approach—the emphasis on progress—as something "we used to do" in her plea for, instead, a gendered approach to Civil War history. And Thavolia Glymph told that same audience to be wary of searching for women heroes in the war, especially African American ones. For many women, the war was simply about survival.[5]

But historians of medicine have not completed the first lap in this historiographic race. We need to better recognize how the war altered the course of the feminist revolution in medicine that began when America's first female physician, Elizabeth Blackwell, took her MD degree in 1849. Women have served as domestic healers for as far back as we have records, but it was not until the mid-nineteenth century in the United States, England, and continental Europe that they were allowed access to formal medical training that led to the MD degree. Women who had sought training and work as physicians during the 1850s (and there were some three hundred trained female doctors at the end of that decade, according to Elizabeth Blackwell's estimate) found their world and opportunities changed by the war's disruptions. Mary Putnam Jacobi, writing the history of women in medicine in 1891, noted that "it was amidst the exigencies of a great war that their opportunities opened, their sphere enlarged, and they 'emerged from obscurity' into the responsibilities of recognized public function." Jacobi felt the war ushered in a new stage of development for women physicians, one that was directly tied to the expansion of hospitals, such as a hospital founded in Chicago to care for the women and children (including black refugee women and children) who had been left destitute and ill by the war's depredations.[6] Putnam herself pushed the boundaries of the possible during the war, taking degrees in pharmacy and medicine, interning at the New England Hospital for Women and Children (Boston) in 1864, and traveling (alone, no less) to Louisiana and Virginia to care for sick family members engaged in war work.[7]

While Jacobi saw a new stage of women in medicine beginning with the war, it was mainly in the decades that followed that women made significant strides as physicians. She found 2,432 women physicians listed in the 1880 census, and most of them would have trained in medical institutions that opened their doors to women after 1865.[8] Even though the war encouraged the career opportunities of female physicians, they were largely stymied in their attempts to take an active role in military medicine. A few were so bold, and their stories follow shortly. Some women with medical degrees served as nurses instead of gaining positions as physicians, given the strength of the barriers against them. Where tolerated at all, women physicians were seen as appropriate caregivers for women and children, not the male bodies integral to warfare.

Prevailing attitudes pronounced middle-class women too sensitive and refined to perform in the public sphere of hospitals, since the exposure to male bodies would both challenge their modesty and assault their sensitive souls by exposure to nasty sights and smells. Yet such exposure in the home with regard to relatives, including dead relatives, was common; it was the matter of pushing such experiences into the public arena, not the experiences themselves, that was novel. Furthermore, women had worked as servants in hospitals since the Middle Ages, either as nursing sisters or as hired hands from the servant class. What was new in the Civil War was the presence of middle- and upper-class laywomen performing the work of nurses for men who were not family members. They justified this move by creating an imaginary extended family—the soldiers were all their children, even soldiers who were older or of a different race.

The experiences of Harriet Eaton, a Maine woman who was a forty-three-year-old widow with three children as the war began, illustrate the complicated identities and attitudes of women who took up war work. After her oldest child, Frank, joined a Maine regiment, she found employment with the Maine Camp Hospital Association, working as a field agent in Virginia. She was a paid employee who distributed the food, clothing, and bedding gathered by the Maine group to Maine troops, especially the sick and wounded, in fall 1862. She went to Virginia in part to be near her son, although she saw him only a couple of times. So she came to refer to the Maine troops as "our sons."[9] In January 1863 she visited the Seventeenth Maine regiment, and "carried with [her] a supply of shirts, drawers, stockings, &tc., also a supply of condensed milk, pepper, ginger, preserves, farina, and passing through the quarters gave each man who was sick a paper of tea, a cracker and some dried apples."[10] The grateful men to whom she brought items and cheer in turn saw her as a substitute for the family members they longed to see. "One poor young boy said he thought of his mother most of the time," she reported, "only sometimes when he thought of me. Another in the same tent said he should call me 'Aunt' and I willingly accepted the title."[11] Eaton saw all the Maine men as her relatives, a common conception for women extending their household world onto a broader stage.

Historian Jane Schultz has found evidence of more than twenty-two thousand female hospital workers in the pension records in the National Archives and from various southern sources.[12] Some provided direct patient care, while others worked as hospital administrators, laundresses, or cooks. Some women worked only briefly—Louisa May Alcott nursed in Washington, D.C., for only three weeks before she became sick and was sent home—while others labored throughout the war. Some volunteered, and some worked for pay. Most probably shared Alcott's desire to serve just like the "boys" who joined the army, although for some women paid employment was necessary for survival. Black women in the South worked for wages if they were free or on consignment from their owners if they were slaves. Throughout, they were performing home tasks of nursing, cleaning, washing, sewing, cooking, and feeding, doing so in a public place for strangers only because the world had been turned topsy-turvy by the war. These women, by and large, did not go into the war seeking to further women's "position," as the female physicians did. They entered wanting to serve their country, to serve their newly envisioned extended family, or merely to survive. Yet the nursing role did lead to new professional opportunities for women through the subsequent creation of formal nursing schools, a transformation that built on the evident success of female nurses during the war.[13]

Another category of women involved in health work during the war were those associated with the various relief and aid societies, although many of these women engaged in work that could also be called nursing. Throughout the North and the South, women responded to the call to arms by organizing societies whose initial functions included sewing, bandage rolling, and the preparation of food for their soldiers. Aid societies were clustered into state units, and in the North some of these were in turn united under the aegis of the U.S. Sanitary Commission or the Western Sanitary Commission. Although these societies were often headed by men, the bulk of the work was done by women, often volunteer women who donated their time and food. Such women served in organizational and leadership roles that built on traditional charity activity but translated into more active political expression at war's end.

This chapter considers each variety of female health worker—physicians, nurses, and relief workers—in turn, while recognizing that the categories often overlapped. Historians have begun to realize that women medical activists in the postwar period were unavoidably influenced by their wartime experiences. Carla Bittel, who wrote a biography of Mary Putnam Jacobi, is a prime example; she recognized that "the war served as a major catalyst for engaging middle-class women in the healing professions" and had an important impact on her subject.[14] Another historian of women physicians, however, blithely told me when I inquired about her subject's wartime experience, "Oh, I leave the war to the re-enactors."[15] Historians of the postwar flowering of feminism and women's access to education and the professions need to carefully consider how the war changed the lives and careers of these women, who were as profoundly influenced as their men.

It is easiest to discern the impact of the war on the position of black women. Before the war many were slaves; with the Emancipation Proclamation (1863) and the passage of the Thirteenth Amendment to the Constitution (1865), they became free. But the transition to freedom and postwar society was rough. Slave women were particularly vulnerable during the war and bore the heaviest burdens of food shortages, the increase of work intensity that followed the conscription of male slaves by both sides for war work, epidemic disease, and the dislocations of liberation. There are glimmers of improvement in "position" in the aftermath of Union victory, however. Slave children, including female slave children, learned to read in the schools set up by the Freedmen's Bureau. In St. Louis and elsewhere, black women learned to be hospital nurses, and former slaves turned cooks earned their first paychecks working for Union establishments. Wives of returning black soldiers started Reconstruction life with a nest egg of military back pay. For most black women who survived the war, life was better in 1870 than it had been in 1850, but it took a century and more for black women to approach the desired status of equality with other members of society.[16]

The literature on the subject of women and the Civil War is both large and growing. Women made up roughly half the population of the United States when the shots fell on Fort Sumter, so to reserve only one chapter for them in a book on Civil War medicine is, on the face of it, absurd.[17] All of the war's history was, in a way, women's history. Women obviously did not live through the war tidily isolated from men and separate from their concerns. Historians always have to narrow their focus; choosing gender as a lens is one way to carve out a useful narrative from limited resources. If there is an area in which historians have particularly answered Maris Vinovskis's call for a social history of the Civil War, gender history may be the best populated.[18] This chapter takes on the large subset of this social-history literature that concerns women, war, and medicine. It was within the field of medical care, broadly conceived, that women were most active in the war and made the mark that allowed Barton to assert their progress.

Women Physicians in the Civil War

Before the war those who sought to expand the role of women in the public sphere of medicine argued that women should be trained as doctors. One might imagine that this was too ambitious a goal, that women should have sought training and positions as nurses first, as an entering wedge into the hospital hierarchy. But this step had already been taken: female nurses were at work in hospitals, albeit with the social status of servants. Nursing was not a profession before the Civil War; it had no standards, no formal training programs, no journals, no learned-vocation insignia. Within the home, female family members cared for the sick; if a nurse was hired, she was of the servant class. As noted previously, the principal occupants of the public hospitals were the poor (deserving or not); those serving them were either convalescent patients or hired servants, all of them of approximately the same class as the patients, in most cases. There was no glory in nursing, no image of it as a profession worthy of a middle-class woman. Many women had also worked as midwives and at times had been called "female doctresses" for filling roles that included the care of women in childbirth as well at the dosing of ill children and adults. One of those, Martha Ballard of Bath, Maine, is particularly well known to us from the record of her diary.[19] Midwifery was a traditional role and a position held by women throughout the country and across ethnic groups. Being a physician, though, had the cachet of a learned profession, involved an advanced degree, and was a suitable target for reformers hoping to open the professions to women. Acquiring this position, especially with the added prestige of an army commission, was a lofty goal worth the attention of feminist reformers.

By the start of the Civil War, there were more than two hundred women in the United States with a diploma that said "M.D." Elizabeth Blackwell, the first such graduate, completed her studies at New York's Geneva Medical College in 1849. During the 1850s Geneva enrolled no further female students, but a scattering of other regular medical schools admitted women, including the New England Female Medical College (Boston), the Female Medical College (Philadelphia), and Western Reserve Medical School (Cleveland).[20] Women were even more successful in gaining admission to sectarian schools, and indeed some historians have estimated that about half of the female medical graduates of the 1850s had attended eclectic, homeopathic, or hydropathic schools.[21] Supporters of women physicians emphasized that the doctor's role was an extension of the woman's role as caretaker in the home and that women had a natural sympathy for the sick that made them ideal healers. Some went further, such as Samuel Gregory, who founded the female medical college in Boston on the premise that it was immoral for men to attend women in childbirth, so women physicians must be trained for that role. Most of the women physicians trained during the 1850s limited their practice to women and children, and those schools that admitted only women focused their education along these lines. Opponents charged that women were too delicate for medical training, that the sights and smells of medical practice would coarsen the feminine mind, and that the whole enterprise was improper and scandalous.

The U.S. Army had no expectation of enrolling women among the ranks of commissioned officers, medical or otherwise, and the women who applied were turned down. But while the army medical system was officially rigid, there were gray areas on its edges where protocol broke down, so a few women found opportunities for practice. None of these women ever achieved the status of a formal commission, but at least three of them could claim with some evidence to have served as physicians to the troops. Other women physicians served the army as nurses, accepting medical service of that kind in lieu of recognition for their more advanced training. On the Confederate side, there is evidence that one female physician practiced in military hospitals, for a brief time.

The best-known female physician who served in the Civil War—a controversial one—was Mary Edwards Walker. Walker had grown up in Oswego, New York, with parents who encouraged women's education and their liberation from unhealthy corsets and physically limiting clothing. Walker wore trousers on her wedding day, remained a lifelong advocate of dress reform, and often appeared in the attire made famous by Amelia Bloomer in the 1850s.[22] She attended Syracuse Medical College, an eclectic school, where she earned the medical degree in 1855 and also, as it happened, met her future husband. By the time of the war, thirty-year-old Walker had separated from her husband and had failed to establish a medical practice. She was unfettered by career or family and ready for the new opportunities presented by the war.[23]

In fall 1861, Walker went to Washington, hoping to gain a commission as a surgeon in the army. The army examining boards refused to consider her application, on the grounds both of her sex and of her training at an "irregular" medical school. Undaunted, Walker found an overworked physician at one of the Washington military hospitals and volunteered her services. She helped him with the ninety or so patients under his care and in return received food and a place to sleep. She failed to persuade authorities to give her a regular appointment there, and spring 1862 found her in New York, studying for a second medical degree from a hydropathic medical college. She went home to Oswego for a while, but by fall 1862 she was following the army in Virginia, volunteering her services after battles. In the wake of the second battle of Bull Run and Antietam, when the army struggled to care for thousands of casualties, Walker's attentions were apparently welcome. In mid-November General Ambrose Burnside signed an order that allowed her to accompany the sick and wounded to hospitals in Washington, D.C. After delivering her charges, she returned to perform similar duties after the battle of Fredericksburg. Throughout the bloody fall, her only compensation consisted of rations and a tent.[24] In spite of testimonials from physicians who observed her work, government authorities persisted in denying her either pay or a formal appointment.

After some months spent in various projects in Washington, Walker headed south again in fall 1863, this time in the wake of the battle of Chicka-mauga. The surgeon-in-charge there disdained her medical credentials and would only accept her employment as a nurse. While in eastern Tennessee, she made friends with General George H. Thomas and other men in power in the western theater. Continually frustrated by her lack of recognition, Walker wrote directly to President Lincoln in early January 1864, asking for a position in a female ward, if she could not receive a surgeon's appointment. He deferred to the medical department, which ignored her request. But somehow, via General Thomas, she received an appointment as a contract assistant surgeon to the Fifty-Second Ohio Regiment, replacing a physician who had died. She finally had the legitimate paying appointment as a surgeon that she sought (although not the commission that would have made her an army officer).

Before she could take up the position, however, she had to pass an army board examination. Walker's account of her examination describes it as a farce, an exercise in which all the male protagonists entered with the conviction that she could not possibly succeed.[25] One member of the board was Roberts Bartholow, a well-known physician who had penned the manual for physicians examining military recruits and was a prominent member of the Cincinnati medical establishment after the war.[26] Bartholow said of Walker, "She betrayed such utter ignorance of any subject in the whole range of medical science, that we found it a difficult matter to conduct an examination . . . [and that] she had no more medical knowledge than any ordinary housewife, that she was, of course, entirely unfit for the position of medical officer." In a letter written to a medical journal after the war, he jeered not only at her "notorious" clothing but at her medical school, which he said was "a 'hydropathic institution' at Geneva, N.Y." (although Walker had failed to provide a copy of even this school's diploma).[27] She received the appointment in spite of this negative report and took up her position with the Fifty-Second Ohio in eastern Tennessee.[28]

It is unclear what Walker's duties were upon joining her regiment in fall 1864. She apparently got on well with the commanding colonel, but one observer later reported that "she did little or nothing for the sick of the regiment" and that "the men seemed to hate her."[29] As a recent biography by Sharon Harris reveals, most events of Walker's life appear contested in the

Photograph of Dr. Mary Walker. She is wearing her signature pants and the medal of honor granted her on November 11, 1865, by President Andrew Johnson. National Library of Medicine, Prints and Photograph Collection, # B010947.

historical record. She repeatedly aroused both warm admiration and ardent censure in her encounters during the war. Walker arrived at camp in February, months after the last engagement of her unit, so there would have been little surgery to do, and the troops were in fairly good health. Instead she began to nurse the civilians in the surrounding countryside. She later said this was a deliberate step to win the hearts and minds of the people and bring them over to the Union cause. On April 10, 1864, she rode too far and was captured by the Confederates.[30] So ended her two-month stint as a contract surgeon in the U.S. Army. On the way to Castle Thunder in Richmond, Walker was mocked by one Confederate officer as "a *thing* that nothing but the debased & depraved Yankee Nation could produce—'a female doctor.'" He also commented, "She says she came to our pickets to bring some letters for her friends in the South expecting we would let her go back, but we all think she came with some design for mischief."[31] She was, they thought, a Union spy.

After four months Walker was exchanged, but the poor conditions typical of Confederate prison camps in 1864 had sapped her health. She returned briefly to her regiment but requested a transfer to a hospital in Louisville that cared for female Confederate prisoners of war. She served there for a few months and then in April 1865 moved to her last wartime post, a federal orphanage in Clarksville, Tennessee. Descriptions of her behavior in Louisville vary from exemplary to cruel, and it is at any rate evident that she could not work amicably with the other (male) surgeons there. After the war Walker apparently never maintained a medical practice, instead building a career as a lecturer, reformer, and author. She received the Congressional Medal of Honor for her wartime work in November 1865, the only woman ever to be so honored.[32]

Was Walker a skilled physician whose detractors were just typical bigoted men who could not acknowledge the skills of a woman? There is no doubt that she met significant prejudice, yet it also seems likely that she had little training in the key skill of the Civil War practitioner, surgery. She had trained at an eclectic school that emphasized herbal remedies, and it is unclear how much anatomical or hands-on clinical training she received. Her subsequent limited practice would not have offered many opportunities for surgical experience. She does not describe surgical procedures or her response to them in her memoirs, and it seems most probable that her care of the sick involved mainly nursing care and the choice of medicines. In this she did not differ from many surgeons at the start of the war, but by the time she received her appointment in 1864, her lack of the intense field-hospital experience common to surgeons associated with regiments worked against her. She seems to have been torn between wanting to establish her status by caring for men and feeling more comfortable dispensing care to women and children. The sources we have about her wartime career are heavily tinged with self-promotion (her own memoir and requested letters of recommendation) and were proffered to acquire either an appointment or financial reward. She seems to have been capable of charming authority figures, but it is less clear whether she would have met the standards of Elizabeth Blackwell or Marie Zakrzewska as a model of her profession. While Blackwell was famous as the first female MD in the United States, Zakrzewska (trained in Europe) likewise rose to prominence as a woman physician and medical leader by creating a modern hospital and training program for women, the New England Hospital for Women and Children, founded in 1862 in Boston.[33]

Walker has become something of an icon for feminists and those seeking female heroes in the history of medicine. In 2010 I purchased a magnet with her face on it as one of the noted "women of the war." Walker's Congressional Medal of Honor was one of hundreds rescinded during World War I, in a legislative act declaring that only those who had shown valor in *combat* could receive it, but Congress restored the medal in 1977 after being lobbied by Walker's great-grandniece. The U.S. Postal Service issued a Mary E. Walker stamp in 1982, and she remains prominently displayed on websites as a pioneering woman physician.[34] In 2003 John Raves, a trauma surgeon from Pittsburgh, gave a talk to the Southern Association for the History of Medicine and Science titled "A Woman Surgeon: Eclecticism or Oxymoron?"[35] In it he recounted Walker's story and then told us that he used her biography as inspiration for his own female surgical residents. Biographer Harris presents Walker as an embattled feminist fighting for women's rights, although her admiration is tempered by recognition that Walker's style stirred animosity as well as support. Walker persists as a legendary, if quirky, figure.[36]

A second female physician who served during the Civil War did so less formally. Esther Hill Hawks trained at the New England Female Medical College, graduating in 1857. After practicing in Manchester, New Hampshire, for a few years, she traveled to Washington in 1861 to volunteer as a war surgeon or, failing that, as a nurse. Not surprisingly, the medical bureau turned down her application for service as a doctor; apparently Dorothea Dix also found her background (and perhaps her beauty) inappropriate for her nascent nursing corps. Hawks's husband, a sectarian physician trained at an eclectic medical college, was likewise turned down for military service. After federal forces had captured the Sea Islands of South Carolina, the Hawks couple found new fields for their talents. Husband Milton received an appointment as manager of one of the government plantations there, to be worked by freedmen, part of a scheme one historian has called a "rehearsal for reconstruction." Esther Hawks followed him, having been hired as a teacher of the freedmen.[37]

When General Rufus Saxton began organizing black troop units in the Sea Islands, he accepted Milton Hawks as a contract surgeon for a black regiment in 1863. By that point in the war, the army was desperate to find surgeons, especially for the black regiments, and standards had dropped. An eclectic degree was better than no degree at all.[38] By April there were enough black troops in the area to generate the need for a general hospital, and a once-grand house in Beaufort was commandeered to meet the need, with Milton Hawks in charge. Initially Esther Hawks's main task was to help clean and prepare the building for its new function. But when her husband had "been taken from the hospital detailed by Gen. Hunter, to accompany a secret expedition to the coast of Florida," no surgeon was "sent to take charge, of the hospital." She was "left manager of not only the affairs of the Hospital, but . . . to attend Surgeons' call for the 2nd [S.C. Volunteers]." She reported in her diary, "Every morning at 9 o'clock the disabled are marched down to the hospital in charge of a Sergent and I hold surgens call, for hospital and Regt. and with great success." Hawks was proud to note that "Hd. Qtrs." did not discover this situation, and she managed it all until her husband returned. Her brother's appointment as hospital steward no doubt made the situation easier to manage.[39]

Hawks does not write or speak as a physician in this diary. She does not lament, for example, that she has skills that are not being put to use. In indicating that she ran the hospital and organized sick call, there is no language to suggest that she "showed them" or that they were blind not to see that she could perform the task just as well as a man. Instead, she adopts an attitude that reveals her own understanding of the reality of her situation and her view of her limitations as a physician. "I suppose I could not have done this if my brother had not been hospital steward," she admitted. "Or if the patients had been white men—but these negroes are so like children that I feel no hesitancy in serving them!"[40] Her training was in the care of women and children; by relegating the black soldiers to the latter category, she maintained propriety.

Elsewhere in the diary, it is clear that Hawks was content in carrying out a nursing role when her skills were called for in emergencies. The small general hospital in Beaufort received five hundred wounded men at once in the aftermath of the bloody engagement by black troops of the Fifty-Fourth Massachusetts regiment at Battery Wagner. Hawks pitched in to create as comfortable a setting as possible for this tide of wounded humanity, a work done, as she says, by "two surgeons and one sick woman." She was the sick female. The men called her Mrs. Hawks, and she said, "[I] endeavored with my whole heart, to make this dreary hospital life, as home-like as possible— and I was richly rewarded by their grateful thanks." She seemed to apologize at times for even acting as a nurse, commenting about one surgical case,

"There was no one to assist the Dr. but me." Throughout her narrative, one hears the voice of a woman who interpreted her medical training and duties through the lens of domesticity, who saw her tasks as extensions of home-making, mothering, and bedside nursing.[41]

Less is known about the third female physician who provided care to Union troops during the war. Born Sarah Chadwick in 1824 and married to Henry Clapp in 1866, she appears in the records as either Dr. Chadwick or Dr. Clapp. Chadwick trained at Cleveland Medical College in the 1850s and was a classmate of Marie Zakrzewska. She returned to her family's home in Lee County, Illinois, after a short period of practice in Cleveland. When the war erupted, she offered her medical services to the Seventh Illinois Cavalry, then stationed in Cairo, Illinois, a notoriously pestilential setting.[42] She may also have worked on the hospital transports that brought the wounded and sick from the western battles in spring 1862, most notably the encounters at Fort Donelson and Shiloh. Dr. Chadwick resigned from her unpaid post in August 1862 and did not practice medicine again after her marriage in 1866. In 1890 she petitioned the U.S. Congress for a pension in recognition of her wartime service, but the private bill authorizing this payment was not passed until 1907, one year before her death at eighty-three.[43]

Only one female physician who attended Confederate troops has been identified, a woman named Orianna Moon Andrews. Born and raised in Virginia, she attended a female seminary in Troy, New York, and Female Medical College in Philadelphia, where she graduated in 1857. She followed this formal education with a year of medical study in Europe and spent time with her uncle in Jerusalem before returning home to Virginia. In the first hectic days after the battle of Bull Run, Dr. Moon traveled to Charlottesville, where university buildings were rapidly being converted to hospital wards. She may have been formally commissioned a surgeon in the Confederate army at that time, as family legend attests, but no paperwork survives. At Charlottesville she met Dr. John Andrews, whom she married in November. For the rest of the war, they worked together, although whether the female Dr. Andrews acted more as a nurse than a doctor is not recorded. After the war she had twelve children, one of whom became a senator from Virginia.[44]

It was far more common for women with medical training to serve in nursing positions during the war than as formal surgeons. It is likely that they were able to function as surgeons in all but name, particularly in hospitals away from the front lines, where the patients were in immediate need not of surgery but rather of medicines, nourishing food, and attention to wound healing. Those nurses who worked in hospitals with their physician husbands probably had even greater freedom of action. One woman remembered that her mother, Mary Thomas, who had received medical training in Philadelphia in the 1850s, first helped the U.S. Sanitary Commission transport sick and wounded soldiers back to hospitals in Indiana. When her physician father returned from another assignment, "he served as Surgeon in charge of the Refugee Hospital at Nashville, and my mother as Matron,—a woman could not hold the position of Surgeon."[45]

The impact of the war on the position of women as medical practitioners was mixed. Mary Walker likely did not help the cause, because her emphasis on "physiological" dress (a knee-high skirt worn over trousers) attracted ridicule to her profession. The war brought her fame, but the fame often amounted to contempt more than admiration. The *New York Medical Journal* reported on a speech she made in London in 1867, for example, and seized the opportunity to sneer at her costume as well as her rhetoric.[46] Other women physicians appear only in private memoirs, not in the public press of the day or the sentimental accounts of "women of the war" that appeared within a few years of its end.[47]

It may well be that the war promoted the cause of women physicians in less direct ways. At a time when male doctors were in short supply, more women may have turned to female physicians on the home front. If these encounters were satisfactory, the female doctor's practice would have grown, and her reputation with it. After the war, opportunities for women physicians expanded. Mary Thomas, for example, found that the Wayne County, Indiana, Medical Association, which had rejected her before the war, accepted her as a member in 1870. Medical schools established in the decades after the war, including the University of Michigan in the late 1860s and Johns Hopkins in 1893, began to accept women. Women's hospitals in Philadelphia, New York, and Boston grew dramatically, as female patients became more accustomed to female doctors.[48]

It is certain that the war put paid to the argument that women were not sturdy enough for medical practice or could not stand the sights and smells of the hospital, for thousands of women nurses proved their mettle during the war. Some women fainted and ran away, but many did not. If a woman could do all that nurses did during the war and not flinch or lose her respectability, there was no moral or character barrier between women and medicine. At least a few of those nurses went on to medical school, providing a direct link to their wartime experiences.[49]

Nursing and the War

In contrast to the few women doctors chronicled in the preceding section, thousands of women became nurses during the war, thanks in part to the example set by English women in the Crimea a few years before. Florence Nightingale was the single greatest influence on Civil War hospitals and the people who worked in them. Nightingale, an English woman who had fought British bureaucracy to bring order to hospital care in the Crimea from 1854 to 1856, was a master publicist whose fame reached America through newspaper accounts in the late 1850s. American poet Henry Wadsworth Longfellow sanctified her in his poem "Santa Filomena," published in the first issue of the *Atlantic Monthly* in 1857:

> The wounded from the battle-plain,
> In dreary hospitals of pain,
>   The cheerless corridors,
>   The cold and stony floors.
> Lo! in that house of misery
> A lady with a lamp I see
>   Pass through the glimmering gloom
>   And flit from room to room. . . .
> A Lady with a Lamp shall Stand
> In the great history of the land,
>   A noble type of good,
>   Heroic womanhood.[50]

Nightingale's influence grew with the publication of her books *Notes on Hospitals* (1859; 1860 in the United States) and *Notes on Nursing* (1860), both widely available in the United States.[51] In the months following the surrender of Fort Sumter, when patriotic fervor was at its height, women on both sides of the conflict sought ways to achieve their own "heroic womanhood." As Elizabeth Blackwell told her sister, the war had created a "mania of women to 'act F.N.' "[52]

Nightingale is undoubtedly the most famous figure in the history of nursing and is regarded by many as the founder of nursing as a profession. But her initial impact was ambiguous. She called upon *all* women to be nurses, and her *Notes on Nursing* is mostly about what the conscientious woman should do in her own home to mind the sick properly. It was only one step for the aspiring young heroine to move from home to hospital, if the figure in the hospital bed is imagined as her brother, husband, son, or father. But it is quite a different thing to imagine training for an occupation as nurse, to take on the role as a lifelong way to make a living. Plenty of women already worked as nurses in 1860, but they were not starry-eyed heroines. Their representative, in caricature, was Sairey Gamp, Dickens's drunken, dilapidated character who attended laying outs with as much gusto as lying ins.[53] Even if sober and attentive, nurses in antebellum hospitals were of the servant class and were no more heroic than a laundress or a housekeeper. The inmates of hospitals were likewise of the impoverished classes, and although an affluent woman might visit there to dispense charity, she would be unlikely to aspire to work in such a place.

What changed in the Civil War? The volunteer soldier was a hero, and the lady who waited on him basked in that glory and served her country in a time of heightened patriotic zeal. Many women who went into hospitals longed to achieve this "heroic womanhood"; others merely wanted to make a living. Few thought they were entering a profession, one defined by special training, licensure, learned knowledge, and specialized societies.[54] Some found the harsh reality of the hospital more than they could bear, but others reconfigured the strange environment into a home—and many knew exactly how to "keep the keys" and run a household. The hospital, after all, had beds to make, floors to wash, food to prepare, children (the patients) to tend, and men (the surgeons) to manage. Middle- and upper-class women were at home in such managerial positions.[55] Because they were the ones most likely to write letters and memoirs, from them we learn of the working-class and slave nurses and all the barriers to keeping order within the hospital world.[56]

There were no nurses in the Civil War, if by "nurse" we understand the job in anything like its modern designation. The historian needs to be careful not to retrofit the modern concept onto the positions held by women during the Civil War. *Hospital workers* is a better generic term, as historian Jane Schultz has argued, one that encompasses the many jobs that women (and men) did, which included the modern occupations of hospital administrator, chain-of-supply clerk, nurse, nurse's aide, laundry worker, housekeeper, emergency medical technician, and even candy striper. These jobs were filled by affluent women, working-class women, convalescent male soldiers, male and female slaves, and black contrabands.[57]

There were few jobs in the hospital that were specifically gendered, with perhaps the exception of laundress. The custom when the war started was that convalescent male patients would wait on their brethren who were still bed-ridden, and the hospital steward would supervise medication and perhaps the production of food. This system worked fine in the small post hospitals that preceded the war, when there might be fifteen or twenty patients at a time. But the war produced thousands of sick and wounded men, and as hospitals became larger and more complex, a new system was needed. Both sides cobbled together hospitals on the run, and women found jobs there almost from the beginning.

In the North, the fledging Sanitary Commission at first offered to screen, train, and supply female nurses. Dorothea Dix, a well-known reformer and advocate for the insane, volunteered her services as head of nurses and received the appointment in May 1861.[58] At first she worked with the Sanitary Commission, but her prickly personality alienated many, and she soon had a falling-out with USSC leaders. Dix continued in her post until fall 1863; after that, surgeons assigned to hospitals were allowed to hire their own nurses. Dix shared the concerns of many that hospitals would challenge the delicacy and respectability of women. She wanted women who were matronly, serious, and not likely to view the hospital sphere as one for courting and silliness. Nurse candidates were to be over thirty and dressed in plain clothes without "ornaments of any sort." This caused a flurry of aging, as young women aspiring to "heroic womanhood" suddenly wished to be plain old maids. Dix eventually hired more than three thousand women who received a salary from the U.S. Government and mostly worked in the general hospitals far from battle lines.[59] Other northern women bypassed the Dix system and went straight to the battlefront to help or worked for the Sanitary Commission.[60]

We know less about how nurses were recruited in the South. There was no central office or single coordinator. True to their commitment to states' rights, the Confederate states avoided setting up a centralized hospital system, choosing instead to establish hospitals by state, located wherever battles were waged. Even when the government organized large general hospitals such as Chimborazo in Richmond, the men remained segregated by state, and nurses were likewise recruited by state. Ada Bacot attended a church service in Marietta, Georgia, where the Reverend Mr. Robert Barnwell, a representative to the Confederate Congress, recruited nurses to go to hospitals in Virginia. Kate Cumming heard a lecture in Mobile calling for volunteer nurses. The men issuing these calls saw to it that women accepting the plea had an appropriate escort and transportation to the hospital location. These southern women were called to manage the hospitals as they managed plantations—directing the work of slaves or servants, organizing the kitchens and laundries, and carrying the keys to the storehouses of food and, particularly, whiskey. Women were hired from the local working class around each hospital to do the laundry, and free blacks were hired or conscripted as well. Some hospitals had slaves attached to them, but more often the labor was done by either working-class locals or convalescent soldiers. Slaves cost money, and the hospitals were often short of funds. The hospitals could also be evanescent entities—set up here this month and moved nearer the battle lines in the next. It would have been hard to maintain slave discipline with such mobile entities, and slaves near the Union lines were prone to disappear behind them.[61] Southern hospitals are described in more detail in chapter 8.

Southern women were more likely than northern ones to see the immediate aftermath of battle and feel called to deal with it. Some northern women went as near the front as they could get and ministered to men who had come straight from the battlefield. But more often they saw the men in general hospitals at a distance from the fighting. Many southern women, in contrast, lived where the fighting was happening; they found wounded soldiers in the barn, by the road, and in the town square. Think of Scarlett O'Hara picking her way across the downtown Atlanta plaza littered with wounded soldiers. Southern women saw that scene repeatedly, at every train depot a few stops down the track from the latest battle. Some women became part of an organized response team that fed, cleaned, clothed, and transported the wounded to a hospital bed. Others came out on the days of crisis, bringing whatever they could carry to feed and hydrate the wounded. Medical thought of the time claimed that wounded men needed immediate "stimulation" to counter the effects of what would now be called shock, the body's response to significant wounding, pain, and blood loss. So the preferred drinks for these angels of mercy to carry to the wounded were alcoholic stimulants, either in the form of straight whiskey or some kind of punch, such as whiskey mixed with lemonade or whiskey and sweetened milk. This sort of episodic nursing was not uncommon during the war. Women might work for a week, a month, or a half year until the immediate need in their neighborhood waned or they were called home to deal with problems in the family. Being a nurse did not mean a fixed commitment in time or space.[62]

How intimate were the Civil War nurses with the bodies of the men they cared for? This clearly varied from place to place, and fears of undue familiarity no doubt worried kin back home. Hannah Ropes, working at a hospital in Washington, D.C., in summer 1862, described stripping new patients of their bloody, muddy uniforms, washing their bodies, and dressing them in hospital garb. "Everything they had on was stripped off—and, weak, helpless as babies, they sank upon us to care for them," she wrote. "With broken arms and wounded feet, thighs, and fingers, it was no easy job to do gently. One quite old man, sick every way, and a bullet hole through his right hand, called me "good mother" when I laid his head upon his pillow." This trope of mother and child helped obviate the amazing impropriety of a twenty-five-year-old woman handling a naked, strange man.[63] Other middle-class nurses rarely touched or viewed the bare skin of their patients, leaving such work to the convalescent male nurses or the more lower-class women. With chronic diarrhea the number-one disease of the war, and many of the men too weak to ambulate, there must have been many nasty messes to clean up, but women rarely mention such commonalities of hospital life in their letters or memoirs. We do know that Ropes, and her friend Louisa May Alcott, acquired infection from those bodies. Both women contracted typhoid fever in January 1864, and Ropes died of it.[64] Those women who had hysterics at the sight of a man sitting up in only his shirt, as one novice nurse did at Chimborazo Hospital, would not have lasted long in hospital service.[65]

Women did not enter Civil War hospitals unopposed. Male surgeons rightly feared that if middle-class women were allowed to take charge of the administration of hospitals, the male prerogative would suffer. The women who served as nurses in hospitals before the war were of lower class than the physicians who walked the wards, and thus those women posed no threat to physicians' authority. But the women who volunteered to nurse during the Civil War were often of the same or even higher social class, and they carried a moral authority that occasionally bested the physicians, especially the ones who were shirking their duties. There were several orders of nuns who nursed during the war, and physicians much preferred them over laywomen, for they obeyed and never got above their station.[66] Medicinal whiskey was often the center of controversy, with matrons controlling access to it and guarding it for therapeutic use against surgeons and stewards who wanted it for themselves.[67] Some nurses were also unafraid to call malpractice and malfeasance when they saw it, and they had the political clout to see discipline enforced. Hannah Ropes in Washington, D.C., discovered that the surgeon and hospital steward with whom she worked were stealing from the hospital and starving the patients. She went outside of the hospital hierarchy to Secretary of War Edwin Stanton, who had the men arrested.[68] In spite of these sources of conflict, as the war went on, physicians came to see the value of women in the hospitals and became accustomed to their presence.

Most of the women who worked in Union Civil War hospitals were white, but in a few places black women were also hired to nurse. Whether this was common in hospitals that tended black soldiers is hard to determine. We do know that one of the best-documented training programs for black nurses was in St. Louis, where a physician sympathetic to the black cause, a nurse leader who was likewise inclined, a large black hospital population, and abundant black refugee women all came together. Emily Parsons, nursing superintendent of the Benton Barracks Hospital, organized a training program for black nurses, who were sorely needed on the extensive black wards of the hospital. White women rarely nursed black men, and in the few documented instances of their doing so, the white women probably did not have intimate knowledge of the black bodies.[69] Jane Schultz has identified more than two thousand black women among the records of hospital workers on the Union side during the war; most of these women carried labels other than "nurse" and were employed as cooks, laundresses, and "maids of all work."[70] Some African American women, such as Susie King Taylor, stayed near the regiments of their husbands, brothers, and other family members and did work as varied as cleaning guns, washing clothes, and nursing small-pox patients.[71]

There is no doubt that the experience of affluent women in Civil War relief work led directly to the founding a decade later of the first training schools for nurses in the United States. Following the model set by Nightingale at St. Thomas Hospital in London, women leaders pushed for the establishment of hospital nursing schools in New York, New Haven, and Boston in 1873. By 1890 there were thirty-five nursing schools in the country. The women most active in creating these schools had served either as army nurses or as managers for the U.S. Sanitary Commission.[72]

As mentioned before, one of the most famous of Civil War nurses was Clara Barton. Barton was aghast at the terrible conditions that followed the first battle of Bull Run and organized her own relief agency to help the wounded in battles that followed. She operated outside the established structures of the USSC or Dix's nursing corps, raising money on her own and remaining apart from other reformers. Near the war's end, she took on the task of finding missing soldiers whose families were unaware of their fates. She went to Andersonville and recorded the names of the thirty thousand Union troops who were buried there, for example. After the war she traveled to Europe, where she learned of the war relief efforts of the International Red Cross. Home again, she founded the American Chapter of the Red Cross, an organization she felt should respond not only to the needs of war relief but to the human suffering that followed natural disasters as well.[73]

Women in the Relief Organizations

The history of the U.S. Sanitary Commission is told in greater detail in chapter 4. But an account of women's wartime medical work and its influence on subsequent progress for women must include at least a brief summary of their role in relief agencies.

Elizabeth Blackwell was in New York when the Civil War began, running a small hospital for women and children and offering clinical training to new female medical graduates. She was a friend of Florence Nightingale and was deeply familiar with Nightingale's work in the Crimea. Blackwell suspected that the U.S. Army would be just as ill-prepared to deal with casualties as the French and British had been in the Crimea and that the actions of educated women could make a major difference in the health care of wounded and ill soldiers. In April 1861 she called meetings of like-minded women to organize the effort to promote women nurses; provide adequate food, clothing, and bedding to army hospitals; and raise money for other relief efforts. Out of her initial efforts were born first the Women's Central Relief Association and later the U.S. Sanitary Commission. Blackwell herself was brushed aside as others took the leadership in these efforts, in part because association with such a controversial figure as a woman physician threatened their ability to gain mainstream support.[74]

Multiple aspects of the USSC were women's work, and in many ways the organization embodied the female influence on the war effort. Inspired by Florence Nightingale, the USSC was a beacon for order, nurturance, and cleanliness. Cleanliness was traditionally the particular province of the mother inside the home; women expected tidiness, and men and boys were the sources of mud and disorder.[75] The insistence on cleanliness was a feminine characteristic of the USSC, and even when the male USSC physicians inspected camps, it was with the fussiness of a mother scolding the menfolk for their disregard of hygiene. The USSC extended the female tradition of benevolence and charity, a local phenomenon before the war, into a national organization. Thousands of women's aid societies in the North sewed quilts, made bandages and pads, boxed foods, and otherwise created what was needed to supply the hospitals and more generally the men serving in the army. The USSC sought cash as well, some from its female associates themselves and also from well-heeled donors, to buy directly the items most in need. In the last years of the war, women's aid groups organized local "sanitary fairs" to raise money, some of which went to the national organization.

Historians have argued about the impact of the USSC experience on women's rights and political activism. Historian Judith Giesberg sees the activities of women in the U.S. Sanitary Commission as the "missing link" in the history of the "women's movement" in the nineteenth century. She points to the active role of women in antebellum reform movements, such as abolitionism, dress reform, women's suffrage, and religious enthusiasm, and the surge of activism from the 1870s on, beginning with the Women's Christian Temperance Union and culminating with the 1920 passage of the Nineteenth Amendment allowing women's suffrage. Giesberg argues that the leadership roles taken by women in the Sanitary Commission's work connect the two periods of activism and that their war work trained a new generation of women leaders in political and organizing skills, skills they used to change postwar American society.[76]

Other scholars are critical of the USSC, seeing it as an organization begun by women but stolen from them by male leaders who had an agenda driven by nationalism and the desire to reclaim social status and impose social control. Historian Jeanie Attie, for example, emphasizes that women were pushed out of top leadership positions in the USSC and demonstrates their resistance to the nationalizing tendencies of the male management. Attie points out that women who sewed and canned for the USSC were not paid for their labors, whereas the men who inspected camps under its aegis received a regular salary. Women's work was devalued, and women resented and resisted the USSC as this became more and more evident. She refers to letters from 1862 and beyond in which women complain that they can no longer afford the fabric to make clothing and that inflationary food prices have made canning too expensive. The initial fervor (and access to closets with old clothes) had faded, leaving women who were now struggling to make ends meet without a male in the household unable to answer the expectations of the USSC. Attie sees resistance in these letters; it is easier to see reality and honest indications of what was possible.[77] Altogether, Giesberg's argument that the war empowered at least some women bears more weight than Attie's negative view, although more examples beyond the few prominent women Giesberg considers would strengthen her argument.

The South had no equivalent of the USSC; local aid societies in the Confederacy tended to be, at most, state-linked rather than joined in a federal web. Historian Drew Faust estimated that more than one thousand local aid societies sewed uniforms and bedding, gathered food for the troops, and supported nurses who went into southern hospitals.[78] The ladies of Mobile, for example, sent oranges to Chattanooga at the request of a nurse there who had come from and been sponsored by an aid society in Mobile.[79] Women throughout the South learned to weave, spin, and sew (or put their slaves to work doing it) as the supply of cloth dwindled with the growing effectiveness of the blockade. Some of that cloth went to the troops in the form of blankets, clothing, and bandages. As the war went on, shortages on the home front meant that there was less and less surplus to send. Southern communities were stripped of all available food and supplies by the armies of both sides, leaving little behind for civilians. Faust has emphasized that the loss of morale among southern women, and the despairing letters they wrote to their men in uniform, was one factor in the South's defeat.[80] At least as important was the absence of the kind of supplies that the USSC was able to provide for northern troops. Many southern observers commented, upon seeing the dreaded Yankees for the first time, on how well fed and clothed they looked. By 1864 southern troops were ragged and thin, and the abundance of their country had been exhausted.

Women's Health during the War

In considering the impact of the Civil War on women's "position," we must pause to ask, How did the war affect women's health? With thousands of doctors serving in the army, there were many fewer on the home front to minister to women. Given the quality of nineteenth-century medical practice, this may have been beneficial. Physicians tended to bleed women during pregnancy, for example, and a decrease in this practice could only have been to women's advantage.[81] Male doctors had the tools to help laboring women whose deliveries were not progressing, though, and the absence of this last resort, whether it was forceps or the destruction of the fetus, may have cost the lives of mothers.[82] Local studies might help elucidate how many women even noticed the absence of male physicians; how many of them, in other words, would have only called a female midwife anyway. It may be that in the South the female black midwife (who had gone to the Yankees or was otherwise displaced) was more missed than the doctors at the front.

Prissy may not have known much about birthing babies, but her mother did.[83]

A second effect of the doctor shortage may have been the promotion of sectarian medicine. It has long been argued that women were more likely than the male of the household to choose homeopathic or other sectarian healers for themselves or their children. With both the husband and the doctor away, one can imagine that the homeopaths, eclectics, and other natural healers had an open field. Homeopathy and other alternative medical systems peaked in the third quarter of the nineteenth century, and it may be that the war gave them this (unintended) boost. It is also possible that women who had seen medical care during the war on a scale previously unknown in their lives felt more empowered to make medical choices—and they chose the sectarians. It is clear that sectarians fought fairly unsuccessfully for an official place in the army's medical corps (in the North), but this lack of recognition may have allowed them to further the cause at home.[84]

Women would have been as prone as men to contract the diseases that the war amplified, such as typhoid, smallpox, and malaria. A further threat for women during the war was malnutrition, if not out-and-out starvation. In the North poor women had trouble buying food as inflation drove prices up for all goods and foodstuffs. Working-class women whose men were at war were particularly suffering, and some women's aid societies in the North took up their cause particularly, as being worth of charitable support.[85] As Judith Giesberg has shown in her history of northern women on the home front, working-class women often failed to receive relief, and their men could not help them, being dead or far away with the army, and unpaid at that.[86] In the South, areas ravaged by war were notoriously short of food. Armies moving through much of Virginia, Kentucky, Tennessee, and Georgia destroyed agriculture and its products. They took stored food and farm animals and trampled crops in the field. Slaves departed from the plantations when Union troops were near, leaving too few people to harvest crops. Southerners sold corn and pigs to the Confederate army, only to find that the Confederate money received in return bought less and less food for the farmer's table. Observers in Tennessee, western North Carolina, and northern Georgia described a farm population that was gaunt and starving.[87] Drew Faust describes the desperation of southern women who begged their men to come home from the war and feed their families. Women and men rioted in Richmond over high food prices in 1863, smashing store windows and taking food by force. Whether these deprivations translated into higher disease rates or shortened lives has not been analyzed in detail, and perhaps the question cannot be so studied, given the inadequacy of southern vital statistics.[88]

One survival strategy of these needy women was to turn to prostitution, a trade much in demand by soldiers in both armies. While there are no hard figures on visits to prostitutes, the resulting venereal disease does appear in the medical records. Among white Union troops, there were almost two hundred thousand cases of venereal disease recorded by the military doctors (how many went unrecorded is anyone's guess). This amounted to a disease rate of about eighty-two cases per one thousand men. The incidence among black troops was lower, and apparently southern soldiers also had fewer cases, perhaps because they were less likely to be garrisoned in cities and towns where prostitution flourished. The prostitutes themselves were affected by the diseases, as were the women back home with whom the men had subsequent sexual encounters. Syphilis can cause insanity, cardiac failure, and bony deformities in its later stages. Both syphilis and gonorrhea contribute to infertility and stillbirths. To treat these diseases, army surgeons used mercury, which had its own side effects and was not universally effective. Whether the spread of venereal disease had a postwar footprint of decreased fertility or increased mental illness has not been determined.[89]

The most vulnerable female population during the war and its immediate aftermath were the recently liberated slaves.[90] Women who remained in slavery after their men were recruited into the Union Army (in the border states) or impressed into Confederate service had to work all the harder on the shortened rations that characterized the southern food supply, especially after 1863. Those black women who fled to freedom or were liberated by Union armies gathered in refugee camps or on the fringes of army camps, suffering extreme deprivation and dying of exposure, disease, and maltreatment. Government policies on both sides that favored putting black men to work while ignoring their families created what historian Jim Downs has called a "crisis of dependency."[91] Women, often with children and the aged in tow, lacked food, clothing, housing, and bedding. The winters of 1863 and 1864, in particular, were bitterly cold, and smallpox rampaged through refugee camps and impromptu black encampments. Conditions were particularly harsh in the lower Mississippi Valley. One observer described a black refugee camp housing 2,199 people near Vicksburg. "There appears to be more squalid misery and destitution here than in any place I have visited. The sickness and deaths were most frightful. During the summer from thirty to fifty died in a day, and some days as many as seventy-five," wrote Western Sanitary Commission head James Yeatman in 1864.[92] Others put the mortality rate in contraband camps in the Mississippi Valley at 25 percent.[93] The Freedmen's Bureau made some attempt to help people in these camps, but its efforts were often inadequate. The war may ultimately have improved the health of African American women by the ending of slavery, but they paid a very high price during the process of emancipation.[94]

Conclusion

Did the Civil War promote the "position" of women? If this question is asked of the position of women in nursing, then the answer is undoubtedly yes. The war established nursing as a respectable field for middle-class women to enter and encouraged the growth of educational institutions to train them. The war answered the question of whether women had the mental fortitude to witness the bloody and nasty scenes of hospital work without fainting, and it demonstrated how useful women could be in hospital administration and nursing practice. This consensus emerged more slowly in southern states than in the North, just as the development of hospitals lagged in the South, until spurred by the Hill-Burton Act a century later.[95]

The answer is less clear for women who sought to become physicians. There were few locales in the war that matched the talents of "specialists" in the care of women and children, even when the medical authorities allowed women physicians access. Mary Walker's final appointments were to care for refugee women in Louisville or orphaned children in Clarksville. Esther Hawks cared for black male soldiers for three weeks but justified it by calling them "so like children." Mary Thomas served as matron for a refugee hospital in Nashville, one where many of the patients would have been women and children. So the women physicians' role was limited, but they may have benefited from the widespread experience of women nurses just the same. The hospital professional was no longer solely male, and many male surgeons would have found that women worked well in the setting. As thousands of women made the geographic shift from home to hospital, from the care of family members to that of strangers, restrictions on the places that women could work in the health care field crumbled before the reality of wartime experience.

Women who worked for the U.S. Sanitary Commission and the thousands of women's aid societies that undergirded it undoubtedly learned powerful lessons about political activism and organizing. Southern women may have acquired skills as well, although the societal disruption that resulted from the war overwhelmed this effect. It is certainly clear that women did not take an active role in public health or public sanitation after the war, leaving this sort of reform to the men who were connected to the USSC. Instead their most prominent reform issue, at least in the 1870s, was the Women's Christian Temperance Union. Did their experience with the control of the "medical whiskey" in Civil War hospitals fuel their desire to see men imbibe less of it? The indications are mixed. Women seemed to have no doubt that the physicians were right in dispensing whiskey to the recently wounded—but they were scornful of drunken doctors and the mistakes they made.

So had the position of women and medicine been put forward fifty years, as Clara Barton claimed? Certainly she had done well by the war, learning organization skills and honing the chutzpah that led to the founding of the American Red Cross in 1881. By the 1880s American women had the choice of a new, respectable career in nursing, and many more women were attending medical school than three decades before. Some of those nurses in training were African American, and many of them would have been in bondage if not for the war's outcome. As an engine for change, the war, by forcing women into new positions, for better or worse, very likely did radically change the position of women in the United States, especially in medicine. Although this change met resistance from men, that resistance was blunted by women who recast the public space of the hospital as the home, their patients as family, and their roles as nurses and even doctors as endorsed by a moral perspective shaped by traditional domestic expectations.

Chapter 3 steps back from the feminine focus on the war's medical history to explore more thoroughly the investigation of infectious disease and its prevention in the mid-nineteenth century. Even here there are gendered aspects, since purity and cleanliness were so central to societal understandings of what it meant to be civilized, and women were the acknowledged guardians of such aspects in society. More broadly, however, the arguments that had raged over contagious disease, and the meanings of such debates for governmental action in public health, move us into the world of medical men who saw in sanitation and disinfection a path to improving the lives of ill and wounded soldiers and the societies to which they would hopefully return.

CHAPTER THREE

Infectious Disease in the Civil War

One of the most remarkable changes in medical thought in the nineteenth century was in the understanding of infectious diseases. At the start of the century, most physicians in Europe and America would have agreed that the fevers (a nebulous category marked by the symptom of high temperature) were caused by inhaling the foul odors that arose from various forms of filth. A few diseases were in a special category of contagious—smallpox, plague, and syphilis most prominently—while others became the subject of heated debates because they appeared to move across the globe by ship and train, yet were not obviously transferred person to person. By 1900 American and European physicians had largely abandoned the foul odor as an icon of danger and now adhered to the idea that tiny life-forms specific to each disease were the causative agents in infections. Most of the dramatic work in this transformation was done in Europe, where Louis Pasteur and Robert Koch argued from the 1870s that microorganisms caused disease. Koch invented a method to establish the connection between specific organism and specific disease. Pasteur promoted his own research methods, including techniques for creating vaccines. And Joseph Lister, an English surgeon, began the conversation that eventually led to sterile operating theaters and the modern surgical encounter.

In America, the Civil War ignited new debates about contagion and brought disinfectants into the forefront of discussions about controlling the spread of disease. When historians claim that the Civil War came at the end of the Middle Ages as far as medicine was concerned, it is lack of knowledge of the germ theory of disease that is often mentioned. Yet 1861 was not 1800, much less 1348, when bubonic plague ravaged the European population. For decades American physicians had been actively debating whether certain diseases were contagious, and they continued that debate with the new experiences offered in hospitals and barracks during the war. Increasingly they considered contagion as a serious possibility. Conditions during the war were optimal for the amplification of contagious diseases. Civil War camps were instant slums, bringing thousands of men into close proximity with one another and with various insect vectors of disease.[1] The omnipresence of smallpox on both sides during the war prompted both awareness that the cause of disease could be carried in an envelope (the dried vaccine scab) and knowledge that contact brought new infections. Physicians who had not seen much smallpox before, physicians from rural areas of the North and the South where it was rare, saw it now. Measles was even more of a problem early in the war. It was obviously a disease that swept through the troops, and its cause was suddenly the subject of much interest. The research on measles may not have led to a new discovery by modern standards, but it put the animalcular theory of disease, the idea that tiny living things cause infection, into the medical press and into doctors' minds. Other diseases, such as meningitis, pneumonia, dysentery, and wound infections, spread through wards and camps and prompted questions about causation and spread. Physicians left the Civil War more familiar with concepts of contagion and the possible role of microorganisms in causing it than they were on entering the conflict.

Historian Charles Rosenberg has identified the use of disinfectants in the 1866 cholera epidemic in New York City as a turning point in American public health.[2] Here a newly created board of health stopped a cholera epidemic, the first time such a major epidemic outbreak had been contained by scientific methods. But where did this idea of disinfectants come from? Were New York City physicians familiar with the work of English physicians such as John Snow and William Budd, who had targeted the water supply as the source of poison? There is little evidence of such a connection. Rather, I would argue, it was the Civil War experience with disinfectants that led to the 1866 cholera triumph. It led also to an emphasis on disinfectants in the 1870s war against yellow fever. Paradoxically, the use of disinfectants at midcentury was almost entirely unrelated to the idea that microorganisms cause disease. But their use did pave the way for accepting bacteria as causative agents in later decades. The Civil War familiarized America's doctors with the power of disinfectants, which readied their minds for the idea that the chemicals succeeded because of their bactericidal properties. This development is explored further in chapter 10.

Diseases killed two soldiers for every man who died of wounds during the Civil War. Most of those diseases were infectious, and many occurred in outbreaks that brought units to a standstill. Physicians who had argued for public health reform before the war, who emphasized the need for public cleanliness and the keeping of vital statistics to provide information about such diseases, were now ready to answer the call of officers and families to do something about these unexpected foes that were weakening the armies. Suddenly debates about contagion were no longer confined to discussions of quarantine in ports or street-sweeping budgets. If before the war the principal victims of infectious outbreaks were the teeming poor of city slums, the target now was every military man camped in unhygienic camaraderie with the men around him. Questions about infectious diseases assumed new importance during the war, and with that attention came government funding for research and prevention. There was a new urgency to solve the health crisis that now affected not just the least among them but the millions of men at risk who went to war perhaps to die in glory, but little thinking they would be mowed down by measles.[3]

Contagion, Anti-contagion, and Infection in 1860

If one word could describe attitudes toward the causation of fevers in the United States and Europe in the mid-nineteenth century, it would be *confusion*. Any attempt to find consensus runs immediately into an opposing view that does not fit the historian's faux consensus at all. Physicians used terms like *contagion, infection, miasm* (or *miasma*), *malaria, virus,* and *poison* to mean what they wanted in a particular setting; the next article in the same medical journal may or may not have used the words synonymously. There was no uniform theory of disease causation, since each disease had its own set of predisposing, exciting, and proximate causes. Many of these causes overlapped from one disease to the next. Exposure to excess cold or moisture or spoiled food or undue exertion could exacerbate the symptoms of just about any disease. Yet a new spirit of empiricism was abroad in the land. With obvious roots in Renaissance and Enlightenment tenets of epistemology, physicians increasingly argued, in print, that they had seen a disease pattern with their own eyes and had drawn their own conclusions. "We are aware that the views we have expressed are counter to those of many, perhaps a majority of our professional brethren," wrote two Mississippi physicians when arguing that yellow fever could be transported even if it was not contagious. "These convictions, however, have been forced on us from observation and reflection, in opposition to early imbibed impressions, and views of those in whose opinions we were very thoroughly indoctrinated."[4]

There is no doubt that disease theory was in flux at midcentury. For decades physicians had been debating whether diseases like yellow fever and cholera were contagious. Lacking conclusive evidence for either position, they chose up sides based on political or commercial interests.[5] Politicians continued to debate the proper response to epidemics, with one side arguing for urban sanitation and the other advocating quarantine. Neither method saw much success in limiting epidemics of cholera and yellow fever, for reasons evident from a modern understanding of their modes of transmission. European and American physicians were aware of the failure of medical science to solve either of these disease problems, and by the mid-nineteenth century they were looking for new tools in their struggles against epidemic disease. In putting forward new speculations, they tended to avoid hot-button words like *contagion* or *contagious,* as those terms had acquired so many negative connotations. Some modified the word *contagion* with *contingent* to emphasize that this was not the contagion of old. Others used new labels, like *transportable* or *communicable,* to talk about how disease could travel, even when the means by which the disease moved from one place to another remained in doubt. Some diseases were riper for new explanations than others, and these differences are analyzed later in the chapter.[6]

Early in the nineteenth century, along with general notions that "bad smells cause disease," a concomitant view held that fevers should not be differentiated into separate diseases.[7] The foul smells were labeled miasm, or miasms, or miasma, and they arose from rotting vegetable or animal matter, including dead animals and feces. Fever was fever, caused by these foul miasms and modified by the season of year, the susceptibility of the patient, gender, diet, race, age, and so on. One disease could segue into another, if proper care was not taken. So a cold could turn into pneumonia, which in turn could become tuberculosis. When cholera first arrived in the United States in 1832, physicians argued that it was merely a severe version of the usual dysentery made worse by environmental or social factors. Yellow fever might be seen as remittent fever (malaria) made more virulent by the climate or other peculiar circumstances of an epidemic year.

With such malleability, causation was likewise protean.[8] But by midcentury physicians were coming to see more and more diseases, especially the ones that occurred in major epidemics, as specific entities. Sufficient personal experience with cholera, or yellow fever, or tuberculosis led them to see these diseases as distinct syndromes. The idea of transmutation was not entirely gone, but the process was advanced enough that physicians could begin to ask what the specific cause of each specific entity might be. There was filth and heat and moisture every year in New Orleans, for example. Why did yellow fever appear in one year and not another?

One answer to the question was that each disease had a specific cause, or poison, that was necessary for its inception. Physicians discussed predisposing causes, including the various influences that could debilitate a patient's body, but they reserved for the category "exciting cause" that factor which immediately sparked disease symptoms. While those who argued that a pile of rotting coffee or a sudden exposure to cold was enough to set off a fever were still vocal, more and more physicians began to speak in terms of a specific poison causing a specific disease. The war, with its networks of information dispersal, accelerated the exposure to this idea and standardized its place in medical thought. It might be necessary for the special poison to join with certain predisposing causes in the patient, such as dietary insufficiency, excessive fatigue, or exposure to bad air, but still the presence of the specific poison was necessary for the evolution of a specific disease. Perhaps a particularly foul environment set the stage for a specific disease poison to thrive, but both were necessary.

Poison was a useful word. Its precise nature did not need to be identified for the concept to have value. Physician Elisha Harris described "specific infectious poisons" in a pamphlet titled *Control and Prevention of Infectious Diseases* that was widely distributed by the U.S. Sanitary Commission to the northern army medical corps. He explained, "The diseases that owe their origin and diffusion to specific infectious poisons, though comparatively few in number, are the most troublesome and fatal that can afflict an army." Further, "the specific poisons upon which [these infectious maladies] . . . respectively depend are unquestionably capable of being communicated from the persons sick to persons uninfected. And in the exceptional group, in which we place certain very fatal maladies, the infecting poison is, in some instances, susceptible of transportation by porous substances, or vessels with contaminated air, acting as the media or vehicles of such transmission or transportation."[9] So the specific poison might move person to person or be carried from place to place in a trunk or a ship. It might emanate from a foul chamber pot full of diarrheal discharges. But if it was not present, the disease would not occur. And if it could somehow be neutralized, the disease would cease to spread.

The Power of Disinfectants

The promotion of disinfectants and their powers to neutralize disease poisons was one of the most dramatic medical outcomes of the war. Harris highlighted disinfectants in his pamphlet; hospitals north and south made disinfection part of their routine. Even if they missed the fine points of explanation that elite physicians could offer, the dullest doctors came home from the war knowing that disinfectants could block the generation of some infectious diseases by dissipating the dangerous smells of putrefaction. But what did medical researchers think the disinfectants were doing when these chemicals limited the spread of disease? The disinfectants somehow neutralized the poisons that caused infections. During the 1860s well-read physicians would have encountered the hypothesis that the poisons were organic chemicals of some sort, rather than living microorganisms. Many believed that the specific poison might work as a sort of catalyst, in a way analogous to how Justus Von Liebig had explained putrefaction and fermentation. The poison was an organic chemical that acted through molecular excitation on some component in blood or tissue. This action either destroyed or changed the tissue, while at the same amplifying the power of the chemical poison. These organic particles were self-replicating, without being alive, rather like crystals forming in the appropriate mother liquor. Disinfectants, it was believed, interrupted this chemical process.[10]

Although these views of disinfectant action were innovative at midcentury, the use of such agents themselves was hardly novel. Various compounds had been used to purify the environment from at least the Middle Ages.[11] One chemist, discussing the use of burning sulfur to fumigate a ship or building thought to be contaminated with cholera poison, said in 1867, "The vapors of burning sulphur have been used for such purposes from immemorial antiquity."[12] Captain William Bligh had the *Bounty* scrubbed weekly with vinegar to cleanse it of impurities, following a common practice of the British Navy.[13] Quicklime was spread over corpses in graves to limit odors and promote dissolution. Charcoal's ability to absorb odors was also well known. Vinegar preserved fruits and vegetables in the pickling process, and of course salt was a common way to prevent meat spoilage. But the mid-nineteenth century saw a new emphasis on the wielding of disinfectants to stop disease, to make the air of the sickroom safe, and to prevent the transmission of disease from one sick body to another.

Civil War physicians were less concerned that foul water—as opposed to air—could cause disease, although they did understand generally that some water sources were unhealthy. Rather than locating specific poisons of, say, cholera, within water, however, they focused on the general constituents of the water. Water that was loaded with organic material or with certain minerals (especially salt) or demonstrated excessive alkalinity was not healthy.[14] For example, William A. Hammond, the Union surgeon general in the early years of the war, made such claims about water in a pamphlet circulated throughout the army. He said: "A water to be suitable for this purpose [drinking] should be free from any considerable quantity of organic or mineral constituents, and consequently colorless, and without any peculiar odor or taste."[15] Others pointed out that the addition of certain chemicals, such as permanganate of potassa, to the water could render such murky water potable.[16] In areas with particularly unhealthy water, hospitals used stills to purify water for patient use.[17]

Hammond explained the dangers lurking in dirty water. "The earthy matters which are so abundant in some of our western river waters almost invariably cause diarrhoea in those who are unaccustomed to their use, though this effect gradually ceases to be produced if the drinking of the water is persisted in." He added, "Organic matters are frequently present in water, and give it qualities which render it deleterious. They may be either gaseous or morphological, as portions of decomposing vegetable or animal remains, infusoria, algae, fungi, etc. Water in which such matters are found readily becomes putrescent, and is most noxious to the health of those who use it as a drink, producing diarrhoea and fever."[18] It was the products of putrefaction in the water that made it unhealthy, not the organic material itself. In another Union government pamphlet, Alfred Stillé wrote: "If the drinking water is not pure, it should, as already remarked, be boiled before being used; and this precaution against the mischievous effects of vegetable and mineral impurities, is of the first necessity in the treatment of the sick."[19] For both men, the concern is not about specific microorganisms, but rather that the putrefaction of organic material would in itself give off poisons that would make the water unhealthy. Treatment of water with disinfectants or boiling the water acted to stop this process of putrefaction.

Nevertheless, the major concern of Civil War physicians was air impregnated with disease-causing properties. Many disinfectants were recognized and used during the Civil War to counter the effect of the supposed putrefactive poisons connected to a diseased person or other rotting organic material. In his pamphlet Harris listed the kinds of disinfectants, according to their actions:

1. absorb and retain noxious effluvia, like charcoal and porous clay.
2. absorbents of moisture; chemical agents that act on organic matter, and recombine some elements of noxious effluvia (quicklime, sulfuric acid, HCl acid, nitric acid)
3. soluble salts that arrest process of decomposition, and control production of phosphuretted and sulphuretted gases—nitrate of lead, chloride of zinc. These stop process of putrefaction by chemical reaction.
4. antiseptics that act diffusively and rapidly—chlorine gas, hypochlorite of soda, chloride of lime
5. the most prompt and efficient antiseptic known—bromine
6. antiseptic oxidizer, produces ozone—permanganate of potassa
7. antiseptic and deodorant, capable of great variety of applications, carbolic acid and coal tar compounds
8. destructive of contagious virus and all transportable infections—heat
9. destructive of yellow fever miasma and the malaria that produces paludal fevers—frost.[20]

The first item in his list acts directly to absorb the bad air that causes disease by taking it out of circulation. Numbers two through seven disrupt the chemical processes that destroy living tissue, the processes of putrefaction and decomposition. The last two recognize the climate as a factor in the spread of disease and also the role of temperature extremes in decreasing infections.

Harris allowed for the possibility that some diseases might be spread by a "contagious virus," but is he talking about microorganisms or some other variety of small life-forms? Not very likely. Elsewhere in the pamphlet, he speaks of the smallpox virus as "the most strongly marked type of specific poisons" and comments that "the certainty of its contagious quality and the liability of its infectious diffusion made it the scourge of the civilized world until Jenner" introduced vaccination. Harris does mention that some of the chemical agents "form indestructible compounds with putrescent materials; or . . . destroy cell-life and the cryptogamic and infusorial organisms," but such living tiny agents were very low on his list of disease culprits. Like others of his time, he used the word *virus* in a nonspecific way. Louis Pasteur changed notions of putrefaction and fermentation by saying they were due to living, multiplying microorganisms, but this idea had not made much progress in the United States by the time of the Civil War. Theories of animalcules had been entertained but scoffed at by and large. They might be on the list as "also possible," but they rarely headed it. J. K. Mitchell had suggested in 1849 a theory that animalcules floating in air caused disease, and he was widely cited.[21] But as Michael Worboys has noted for germ theories in the 1860s United Kingdom, there was no cumulative effect of these animalcular hypotheses.[22]

It is essential to separate the idea of contagion from its later explanation, the transmission of germs. To say a disease was contagious in 1860 was to say it acted like smallpox. Transmission was from person to person and happened easily through the air after near contact. The concept of a specific poison that arose from one person and wafted to another explained this adequately; there was no need to postulate tiny life-forms as the transmitted material. Indeed, the idea that animalcules of some sort caused disease could fit either with a contagion model or with the spontaneous generation of disease from piles of rotting filth. In an era when some scientists still believed the theory of spontaneous generation of microorganisms, it was possible to argue that such creatures arose in the proper conditions of filth and moisture, escaped into the air, and caused disease. Person-to-person transmission and the germ theory of disease remained separate concepts until united by developments in bacteriology in the 1880s.

In reading a mid-nineteenth-century text on the spread of disease, one must pay special attention to how terms such as *virus, contagion,* and *infection* are being used, particularly since they may have different meanings from their use in modern medicine. Florence Nightingale, for example, defined *contagion* as the spread of animalcules from person to person causing disease—and she found no evidence for that. She called the transfer of a poison through the air from a sick person to a well person "infection."[23] Her emphasis was on the possibility of preventing such a transfer with adequate ventilation, since infection could be prevented with fresh air, whereas "contagion" implied inevitability. Others used the phrase *contingent contagion* or said a disease was "transportable" to avoid implying that tiny living life-forms were transmitted. This allowed them to emphasize the transmission while not having to prove exactly what passed from one who was sick to one who was well.

To avoid person-to-person transmission of disease in the Civil War hospital, disinfectants were key. The one "disinfectant" that Harris failed to mention was the most touted: abundant fresh air in hospital wards to blow away the noxious gases arising from ill men. Florence Nightingale had provided directions for building a healthful hospital, directions that were widely acknowledged during the American Civil War. She argued that the footprint of a man's bed and the free space around it should be no less than one hundred square feet and that altogether he should lay claim to fifteen hundred cubic feet of air.[24] Wards should be long and narrow, with abundant windows and roofline vents to provide adequate exchange of air. (As described in chapter 1, many Civil War hospitals were built according to this pavilion design).

The great Chimborazo Hospital in Richmond was one of the largest hospitals to be constructed using the pavilion system. At one point it consisted of some one hundred small buildings, long narrow rectangles of rough wood eighty feet long by twenty-eight feet wide with seven-foot-high walls. Each long wall had ten windows, and there were doors at both ends and in the middle.[25] These were ideal structures for ventilation (although difficult to heat in winter), and the air space for patients was jealously guarded. One nurse supervisor at Chimborazo, Phoebe Pember, reported disciplining a nurse who decided to divide off a section at the end of one building for her private quarters. Pember explained to her defiant subordinate that, "each patient being allowed, by law, a certain number of feet, every inch taken therefrom was so much ventilation lost, and the abstraction of as much space as she had taken for illegal purposes was a serious matter, and conflicted with the rules that governed the hospital." Besides, Pember continued, "no woman was allowed to stay in the wards [overnight], for obvious reasons."[26] Indeed, Confederate medical regulations dictated that the physician was to "enforce the proper hospital regulations to promote health and prevent contagion, by ventilated and not crowded rooms [and] scrupulous cleanliness."[27] Pember had trouble getting help from the hospital physicians in disciplining her rogue nurse, mainly because they and the patients were enjoying the dustup so much. Finally she enjoined the hospital's carpenter to remove the barrier, and the nurse was sent packing.

If the air in a hospital ward became foul in spite of the best efforts at ventilation, or where ventilation was inadequate, surgeons also resorted to the application of disinfectants. Walls and floors could be washed down with disinfectant substances. J. J. Woodward, in a book on army diseases widely circulated during the war, advised the generation of chlorine gas within wards. It worked, he believed, by "actually combining with the noxious ingredients of the atmosphere, and producing inoxious compounds."[28] He provided formulas for the generation of chlorine gas and also advised that bromine gas could be similarly efficacious. How the patients on the ward tolerated the toxic atmosphere is not recorded; perhaps they were removed during the fumigation process. Milder versions of such a process were common. An article in the *Confederate States Medical Journal* cites a French physician's advice with approval: "The author next alludes to disinfectants, recommended as a means of remedying in some measure, the defects of ventilation, and gives the preference to permanent fumigations of chloride of lime diluted with water, by means of vessels scattered through wards and passages, the chloride being removed every three to four days."[29] Such practices were common in hospitals, north and south, during the Civil War.

Controlling Wound Infections

Florence Nightingale had argued that the best gauge of the healthfulness of hospitals was the presence or absence of wound infections.[30] She believed the poison that created the infected wound moved from bed to bed if the hospital was dirty, if it was poorly ventilated, or if the men were poorly fed and cared for. Civil War surgeons were likewise intensely aware that a seriously infected wound indicated a failed therapeutic intervention. And the appearance of gangrene within a ward was cause for particular alarm. Surgeons actively sought new ways to deal with these sequelae of operations and wounds and took the opportunity the war offered to experiment with new techniques.

There is a difference between the infected wound and the gangrenous wound, a difference well known to Civil War surgeons. Skin infections are caused by a variety of microorganisms, including streptococcal, staphylococcal, and clostridial species. Physicians recognized erysipelas as an inflammation of the skin that could occur in the absence of skin perforation or in the skin adjacent to any sort of wound. The modern distinction between erysipelas (infection of the superficial skin layers) and cellulitis (infection of deeper skin layers and underlying tissue) was not sharply made in the 1860s. A variety of terms referred to these degrees of skin infection, and if the case was severe and resulted in systemic disease, surgeons used terms like *pyaemia* or *septicemia,* understood as the effect of pus entering the general circulation. Gangrene was different. Here the wound became acutely painful, changed color to a dark purple or even black, and the skin around the wound began to collapse as tunnels of infection undermined and dissolved the adjacent tissues. Probably every Civil War wound became infected to some extent, but surgeons were precise in noting when gangrene first appeared in their wards.

A complicated science of wound healing underlay surgeons' understanding of these processes. Cellular pathology was in its infancy (Rudolf Virchow's seminal work on the subject had not appeared until 1858 [in German; published in English in 1860]) and was not widely cited during the Civil War.[31] Nevertheless, a few American physicians were familiar with the microscope and could appreciate European writings on inflammation and wound processes. To some extent they communicated this knowledge to their medical peers. Although microscopes were not widely available to American doctors in the 1860s, many would have heard that pus, the product of inflammation, consisted of small colorless cells. Whether these cells formed spontaneously at a wound site or migrated there from the bloodstream was a matter of debate. Many thought a thin layer of pus, so called laudable pus, was necessary for wound healing to occur. According to this theory, the process of healing went something like this: (1) the wound occurs; (2) inflammation follows, generating pus cells; (3) this inflammatory state, which is one of hypernutrition, conveys a formative force; (4) the formative force leads to granulation and wound healing. The pus thus calls forth the healing force, often reified in that era as the "healing power of nature" or the "vis medicatrix naturae."[32]

If the process of inflammation got out of hand and proceeded in too vigorous a manner, the wound filled with dead and dying cells. These cells rotted and in turn irritated living tissue. This damaging putrefaction in turn stimulated more inflammation and the appearance of harmful pus. The catalysts that turned on putrefaction could also spread to other wounds of patients in neighboring beds, causing them to putrefy. Physicians argued that if there were microorganisms visible within the wound, they were attacking these dead cells, not leading the process of inflammation. Some thought that microorganisms arose spontaneously in such cases. But according to most theorists, this sort of putrefaction was seen as analogous to fermentation, which meant that both were caused by Liebig's active organic molecules, not microorganisms. It was Lister's contribution to claim that wound infections were caused by living microorganisms floating in the air, an idea he credited to Pasteur. But he did not publish this proposal until 1867.[33]

How widespread was this understanding of wound inflammation among American physicians? It depended on their previous level of education and how avidly they read government educational documents or learned from their better-trained colleagues. Probably all expected to see some pus in a wound, and they certainly understood that a wound needed to drain in order to heal. Premature closure was dangerous. What is also clear is the universal acceptance of the idea that one infected wound, be it cellulitic or gangrenous, could affect other wounds in a hospital if care was not taken. Accordingly, the appearance of wound infections within a ward was an emergency that led to immediate action. The affected patient or patients were moved to isolation, either a separate ward or even a hospital tent where ventilation and isolation could be maximized. If more cases appeared in the ward, it would be emptied, scoured, whitewashed, disinfected, and prepared anew with new mattresses and washed bedding. When Confederate surgeon general Samuel Moore offered the following advice in his handbook for surgeons, he was expressing common wisdom: "When the disease [gangrene] makes its appearance, *nothing short of complete segregation of each case . . . with a liberal use of disinfectants, can give a promise of security against its further spread.*"[34] This did not imply any understanding that germs moved from one patient to another. Rather, the common idea was that a poison arose from the infected wound, traveled through the air, and landed on another wound, poisoning it in turn.

Disinfectants were not only commonly used to prevent wound infection; they became critical in the management of such wounds, especially gangrenous ones. Surgeons recognized that wound healing required the removal of dead tissue, the draining of abscesses, and the reduction of wounds to a floor of healthy flesh. This might be accomplished with a scalpel or with dressing changes that brought layers of flesh away with the bandage. It was particularly difficult with gangrene, because the disease process burrowed tunnels into surrounding areas, and it became a challenge to clear out all of these infected paths while conserving healthy tissue. By 1864 it had become standard practice to debride these wounds with caustic disinfectants, clearing out nasty and dead material and, hopefully, leaving behind a clean wound that could now heal. Surgeon General Samuel Moore recommended nitric acid as the agent of choice.[35] Middleton Goldsmith, a Union surgeon, had such good results using bromine in cases of gangrene that almost all of his patients survived.[36] Such procedures were very painful, and most surgeons employed anesthetics to make the process tolerable. Surgeons reported high rates of success using these procedures: patient survival rates reached 80 percent and 90 percent.

Surgeons in the Civil War preferred preventing wound infections over treating gangrenous ones, and they viewed disinfectants as key to achieving a healthy ward. This approach reflected the assumption that it was the air of the hospital ward contacting the wound that was dangerous, not the environment or tools of the surgical encounter itself. For example, John H. Packard wrote a manual for surgeons that the U.S. surgeon general had distributed widely during the war. Packard described the various surgical tools in detail and cautioned the prudent surgeon to have his implements, including those used for unlikely exigencies, assembled before the operation began. He said nothing here about cleanliness of the surgical space or how the instruments were to be treated between patients. A chapter on disinfectants is all about how they were to be used in the ward, where "an atmospheric influence [due to] . . . organic particles or *fomites* floating in the air," or possibly some sort of chemical, could poison all the wounds, no matter how minor. He also mentions the "articles employed for changing the character of the surface in foul or sloughing sores," such as silver nitrate or nitric acid.[37]

As discussed in chapter 1, manuals that gave instruction on the immediate response to gunshot wounds emphasized the dangers in delay and the need for speedy surgery following the injury. Surgeons worked throughout the war to accelerate the transport of the wounded from battlefield to surgeon's care. Field hospitals were set up as close to the front lines as practicable, with speed and accessibility valued over such refinements as cleanliness or comfort. The shattered arm or leg had to come off, and the sooner the bleeding was definitively controlled and the ragged wound closed up, the better for the patient. Wounds were often contaminated by the dirty uniform of the soldier, fragments of shot and shell, and the soil on which he had lain. The influence of the surgical hands and tools, which might at least have been cleaned in water before the case, must have seemed minor. Surgeons believed that rapid surgery saved lives and that surgical infections, if they occurred, were the fault of badly ventilated hospitals, the depleted constitution of men deprived of adequate food, warmth, and rest, or other factors beyond their control as surgeons.[38] One surgeon summarized, in a section on the causes of gangrene: "The causes . . . are, overcrowding of patients in confined and filthy apartments, . . . bad or innutritious food, stimulating drinks, great loss of blood, the shock of a severe injury, insufficient attention to cleanliness and to the dressing of wounds, mental depression," and so on.[39]

The struggles to understand, treat, and prevent wound infections were central to the experience of physicians treating the Civil War wounded. One can see this process as a borderland between the male world of the scalpel-wielding surgeon and the female world of nurturance, cleanliness, and maintenance of morale. The widespread use of disinfectants to prevent and cure wounds draws jointly on the male world of science and the female arena of routine cleanliness and scrubbing with common household substances such as vinegar. As male surgeons gained more surgical experience in their few years of war than they would ever have had in peace, they likewise accumulated new knowledge about the importance of key factors in wound healing. Overall, the omnipresent discussions of wound healing during the war prepared these men to be especially ready for the new theories about wound infection that emerged in England after the war's end.

Infectious Diseases

There is little doubt that by the end of the Civil War most surgeons believed that the cause of hospital gangrene, if not of wound infections more generally, passed through fetid air from patient to patient. They also were confident that isolating such patients with abundant air around them, such as in free-standing tents away from the main hospital, would prevent gangrene from spreading, that disinfectants could interrupt the infection process, and that caustic disinfectants could clean wounds and allow healing. So widespread were these understandings that even nurses reported them in matter-of-fact tones as established truth. It is hard to know how many physicians changed their minds on this subject during the war, but the evidence supports a claim that by 1865, if not sooner, there was consensus.[40]

But what of other diseases? How widespread was the notion that disinfectants could interrupt their transmission? And how was this related, if at all, to the idea that they were contagious or caused by a specific poison? For those diseases seen as most strongly contagious, as stemming most directly and consistently from person-to-person contact, disinfectants could do little to stop their spread. Smallpox could be prevented only by vaccination, for example; disinfectants had no power against it. There was more hope for a second category of disease, those viewed as subject to contingent contagion. And for those not attributed to contagion at all, disinfectants were particularly powerful. The usefulness of disinfection was, in other words, almost inversely correlated to the contemporary assignment of contagiousness.

Infectious diseases during the Civil War can be roughly categorized as (1) certainly contagious, (2) showing some evidence of contingent contagion, and (3) indicating no evidence of contagion at all. Tracking with these categories were the concepts of specific poison and specific disease entities, with category one having the most specificity and category three having the least. Smallpox and measles were the clearest representatives of category one. Cholera, yellow fever, and typhus fell into the middle ground of category two. Pneumonia and meningitis were at least considered as possible candidates for category two as well. Category three included the major diseases of the war, the "continued fevers," fevers usually accompanied by diarrhea or dysentery.

Smallpox and Measles

Smallpox was not only widely recognized as contagious; it was the exemplar of contagiousness. No special knowledge was needed to see that each new case followed contact with a previous case. Furthermore, the cause of smallpox could be carried around in an envelope or vial. In the early decades of the eighteenth century, Cotton Mather and Zabdiel Boylston introduced smallpox inoculation into America, having learned of its practice in Africa, the Middle East, and South Asia. This procedure entailed taking some lymph from the sore of a patient with smallpox, scratching the skin of a person who had never had the disease, and rubbing the "smallpox matter" in. The inoculated patient would experience a mild case of real smallpox but suffered less risk of death or disfigurement than the person acquiring smallpox in the natural way. When a practitioner had no cases of active smallpox, he could acquire matter for inoculation by receiving the scabs taken from smallpox pustules as they dried and fell off patients some distance away. These scabs, sent in the mail or traveling via saddlebags, could then be reconstituted with a little water, and the resulting mixture was potent to cause new cases. It was also widely believed in the eighteenth century that the poison of smallpox could persist on blankets or clothing. The British made use of this feature as a form of biological warfare against American Indians during the French and Indian War in North America. There was a clear consensus that smallpox was contagious and was caused by a specific poison.[41]

This consensus guided the medical response to smallpox. In the Revolutionary War, George Washington had set a precedent by having all of his troops inoculated who could not demonstrate evidence of previous infection. After the discovery of vaccination by Edward Jenner, and the 1798 publication of his method, inoculation was succeeded by the safer vaccination procedure, which initially used cowpox instead of smallpox as the source of the matter to be rubbed into the open wound.[42] By 1860 smallpox vaccination was commonly practiced in the urban areas of the North, although smallpox and subsequently its prophylaxis were rare in rural areas, especially in the South and the Midwest. Once smallpox came to prevail in the Union and Confederate armies, the surgeons general of both sides pushed for the enforcement of the rule that all soldiers must be vaccinated or show a vaccination scar. Vaccine became scarce, and what was available was not always active, but the response on both sides in the combat was the same and was dependent on the common understanding of smallpox as a specific contagious disease with a specific cause and a specific mode of prevention. Physicians also used isolation in special smallpox hospitals, or at least separated hospital tents, to control the spread of the disease.

Measles was the other disease considered contagious by almost all physicians during the Civil War. In his essay on that disease written after the war, Cincinnati physician Roberts Bartholow noted, "In some instances it was not possible to trace the source of contagion. In the absence of such information, the non-contagionists assumed that the local conditions were sufficient to develop the disease. The advocates of this view were not numerous." In fact, Bartholow reported, "the positive facts of contagion were too numerous to be overborne by such uncertain negative testimony."[43] Although measles had a steady presence in urban areas of antebellum America, most rural families had not seen it. The disease ravaged the armies north and south in the early years of the war; when black troops joined in large numbers in the winter of 1863–64, measles returned to torment this fresh crop of nonimmune men. Commanders responded by keeping fresh recruits in a seasoning camp so that they could weather the infection before joining the army in the field and by trying to isolate cases when they occurred. This latter strategy could not be effective because measles is infectious for several days before the distinctive rash occurs, making such newly infected cases unwitting and undetectable vehicles of disease.[44]

One Ohio physician, J. H. Salisbury, thought he had found an explanation for the measles epidemics, and he framed it within an animalcular theory fairly typical of those put forth in the mid-nineteenth century. He noted that soldiers often slept on straw and that straw frequently became wet, moldy, and rotten. Salisbury scraped some mold off a mesh of straw and examined it under a microscope. He identified it as a species of penicillium and proclaimed that this organism was the cause of measles. His discovery received much attention. Published in the prestigious *Boston Medical and*

Surgical Journal in 1862, it was mentioned in several subsequent medical papers about measles in the army.[45] Detractors quickly disputed the finding, by countering that the soldiers most likely to fall ill were from the country; urban recruits rarely erupted in measles. Yet surely the rural youth was at least as likely, if not more so, to make the acquaintance of moldy straw before arriving in the army. Salisbury's theory exemplified the many animalcular theories that evoked exploration and were ultimately rejected because either lack of evidence or facts argued against them.[46] Physicians were open to the possibility that an animalcule spread measles, but Salisbury's fungus did not explain the facts of the disease very well. Although it was an animalcular hypothesis, it did not explain the contagiousness of measles; instead, it tied the disease to an environmental factor, moldy straw in mattresses. A germ theory could be linked to the environment as well as to person-to-person transmission.

Salisbury built his case in the wake of another publication, which had argued for an animalcular cause of intermittent fever, again one that put the cause in the environment and not in person-to-person contagion. In 1849 J. K. Mitchell had explained why the foul air, or malaria, around swamps was able to generate intermittent fever and exert a negative impression on other disease states: it was because the air contained floating spores of a fungus that poisoned the system. Mitchell's short book *On the Cryptogamous Origin of Malarious and Epidemic Fevers* became well known and is often mentioned in American medical writing of the 1860s.[47] Yet it is usually cited only to be dismissed or at best accepted as "possible, not proven." The U.S. Sanitary Commission pamphlet on intermittent fevers noted: "Of the intimate nature of the 'paludal poison,' 'Marsh miasmata,' or 'malaria' we are in complete ignorance. Of the hypotheses thus far advanced, the most plausible are those which refer the morbific influence to the sporules of cryptogamic plants, or to the infinitesimal ova of infusoria." The article went on hopefully, "This confusion of ignorance still leaves us in possession of certain knowledge concerning malaria, from which much practical good may be derived."[48] Note that here the word *malaria,* bad air, is used as the cause of intermittent fever. Malaria did not become the name of the disease caused by "malaria" until the 1880s.

Salisbury's work was the most prominent exposition of a germ theory of disease during the American Civil War. And it was roundly dismissed. Surgeon J. J. Woodward and others used microscopes to search for animalcular causes of disease but failed in their quest. The idea was there, and it beckoned those attracted by its explanatory power. But physicians in the 1860s needed more than microscopes to make the necessary connections between germ and disease. They needed the stains that Robert Koch and others in Germany later developed for viewing microorganisms. They needed Koch's method of inoculating suspect germs into animals or growing them on agar plates. And they needed to abandon the fungi as the most probable causative agents; that transition did not happen in Europe until the 1870s. And it did not help that the most obviously contagious of diseases were caused by sub-microscopic viruses (using the term in its modern sense). Until new tools were available, physicians in the Civil War made do with the concepts of specific poisons and miasms, but the most adept among them constantly sought new information amid the wealth of misery and disease that surrounded Civil War camps.

Other diseases that fell into the "obviously contagious" category generated less debate during the war. Physicians had recognized the venereal transmission of syphilis since the Middle Ages and had urged men to be continent—but with very little effect. Particularly where armies had established camps, such as the Union garrisons at Memphis, Nashville, Louisville, and Cincinnati, bordellos flourished and syphilis spread unchecked. The other disease that was widely perceived as contagious was bubonic plague, labeled by some historians as "The Great Teacher" for its lessons about contagion. Yet the latest plague outbreak had occurred in Europe in 1722, and plague did not appear in North America until just after 1900, when cases emerged in San Francisco. So plague was rarely mentioned by Civil War medical writers. For great teachers, we must turn instead to yellow fever and cholera.

Yellow Fever and Cholera

Yellow fever and cholera were the major epidemic diseases in Europe and America during the nineteenth century. "Major" here means that these were the diseases that disrupted trade, aroused panic, prompted debate, and forced government action. Other diseases, such as tuberculosis, might kill similar numbers each year, but tuberculosis was endemic and was believed to be due to heredity, indiscretion, and multiple deficits of poverty. Yellow fever and cholera both traveled and thus inspired the use of quarantine to wall off the epidemic and keep the community safe. Yet neither disease appeared to be as contagious as smallpox. Hospital workers could tend yellow fever or cholera patients and not fall ill. The first local cases of yellow fever or cholera could occur in people who had not had contact with any foreign person arriving on a disease-carrying vessel. Because there were intermediate steps in the communication of both diseases—mosquitoes for yellow fever and contaminated water for cholera—cases seemed to arrive de novo. They were not contagious as the nineteenth-century physicians used the term, and yet they did seem to travel.[49]

While these two diseases were at the center of the controversies around contagion during the nineteenth century, they were not of major importance during the war. Yet both remained key to the contagionism debate and immediately emerged after the war to assume centrality in medical discourse.[50] Paradoxically, the war changed this discourse in irrevocable ways. The power of disinfectants to purify hospital wards suggested new ways to combat dangerous epidemics, and the use of disinfectants in combination with isolation and quarantine came to dominate the response to these two diseases. This postwar story is considered in chapter 9. Physicians also gained new experience with local epidemics during the war, as they unfortunately had plenty of opportunities to watch infections like pneumonia and meningitis move through crowded barracks. They were led to question the role of contagion in these outbreaks, even when they ultimately decided against it. As in the frequent positing of animalcular hypotheses, even when these ideas were dismissed, they gained currency by being considered seriously.

Southerners had hoped that two major southern diseases, yellow fever and malaria, would fight on their side during the Civil War by weakening the northern soldier who lacked the southerners' acquired tolerance to these diseases. Malaria was important in the early years of the war, especially during the James River peninsula campaign of 1862.[51] In 1862 northern officers ordered a daily quinine ration to battle the influence of swamps, and the situation improved. The Union blockade then caused an acute shortage of quinine to the South, leveling the playing field further.[52] Southerners placed even more reliance on yellow fever as a sure scourge of these invading strangers. One Yankee physician noted: "Our civil war had scarcely commenced when throughout both the North and the South it was prophesied that the great scourge of the tropics, yellow fever, would decimate any northern armies that might penetrate the 'Cotton States' within the 'yellow fever zone.'"[53] Yet this disease, too, was apparently tamed by public health measures taken to limit it.

Several small yellow fever epidemics occurred during the war, mainly on the Atlantic coast. The largest was at Wilmington, North Carolina; other outbreaks occurred at New Bern, Charleston, and Hilton Head. These port cities were visited by blockade runners, whose illicit actions made it difficult to track down the source of each epidemic.[54] Some continued to argue, as did Dr. William Wragg of Wilmington, that yellow fever emerged spontaneously from the filth of the city.[55] But others held that some sort of disease spark had to be imported for an insalubrious neighborhood to explode with yellow fever. The most interesting story, for sanitarians, was the epidemic that did not happen. New Orleans had been visited by yellow fever almost every year since 1840, but from 1861 to 1865 there was no major epidemic. General Benjamin Butler occupied New Orleans in 1862, and the federal blockade limited commercial traffic in the city. Butler ordered the city streets cleaned and also instituted a military quarantine against ships coming into the city. New York sanitarian Elisha Harris, who believed that yellow fever could spontaneously generate in New Orleans *or* be imported from outside, was impressed with this result. He wrote, "There can be no reasonable doubt that the restraints of quarantine prevented exotic causes of yellow fever from being disseminated, and possibly prevented epidemic outbreaks of the fever in the city, and . . . it is regarded as reasonably certain that with the aid of such artificial conditions as prevailed in the 'ram fleet,' the infective poison of the fever may be and was generated at New Orleans in 1863 and 1864, and . . . by good sanitary conditions the city was saved from the pestilence that gained foothold at the hospital landing by the riverside."[56] The fact that the hated Yankee had prevented yellow fever when the local government had been powerless to do so in preceding years was both an indictment of southern public health and a goad to reform once the Yankees went home.

Are Other Diseases Contagious?

While yellow fever and cholera had been at the center of major disputes about the causation of infectious disease, the war offered physicians more opportunities than private practice had, to see diseases in clusters. For example, Ira Russell saw more than fifty cases of cerebrospinal meningitis among the black troops gathered at Benton Barracks in St. Louis in January 1864. He found the disease to have about a 50 percent mortality rate, and few remedies helped, beyond quinine early in the course. He concluded that the causation was "miasmatic," and it appears that he surveyed his fellow physicians on the topic, for they agreed.[57] This epidemic was followed by an eruption of the same disease among white soldiers. This time he found that "local miasmatic influences, if not its sole originating cause, . . . had much to

do in producing it." Russell does not say whether contagion was also a possible cause. He admitted that he had never seen any cases of meningitis in
his medical school years or in private practice. He also reported that at first
the physicians treated the cases with an antiphlogistic course (bloodletting,
calomel) but later decided that the disease was asthenic and needed stimulants such as quinine and alcohol.[58]

Russell was equally careful in studying an epidemic of pneumonia at Benton Barracks that fatal winter. There were 784 cases and 156 deaths. As in
the study of meningitis, Russell performed autopsies, gathered data from the
various physicians working under him at the hospital, and summarized his
information in tables and text. His research plan was explicit. "I have endeav-
oured to ascertain, as far as practicable, the causes that operated in producing the disease and the most effectual mode of combating it." He attributed
importance to the very cold and wet weather that prevailed that winter, the
exposure of the men and the poor construction of the barracks, and the presence of measles, which he recognized as weakening the men and often leading into pneumonia. "One hundred men were crowded into rooms originally
meant but for fifty, necessarily rendering the air very impure. . . . The effect
of an epidemic influence is shown by the fact that physicians and nurses,
who had not been greatly exposed to the vicissitudes of the weather . . . have
suffered from it." While Russell appears confident that the main causes are
environmental, he reported that other ideas were being discussed. "The surgeons on duty with the regiments in the barracks report that men occupying
the same bunks with those affected were very much more liable to be attacked than those more remote. Some of the most intelligent surgeons were
led to believe that the disease was actually contagious."[59] Russell's papers
are important not because he came to radical new conclusions or otherwise
was highly influential on the medical practice of his time. Rather, his written work illustrates a kind of experience that must have been repeated many
times, in many places, during the war. Physicians saw diseases they had
never seen before, or they saw diseases in clusters whereas before they might
have the seen only the isolated case. Working in proximity with other physicians, they could debate causation and therapy, and given the circumstances
of military life, they could perform autopsies. Ideas such as contagion were
debated, rejected, reconsidered, and reformulated. Medical knowledge was
in a dynamic state that allowed the emergence of novelty and the new configuration of old debates.

Even when diseases were scarce during the war, that very fact could promote new thoughts about specificity of disease and its causation. As I have argued elsewhere, typhus was notable by its near absence during the Civil War.[60] At least one physician pondered why it did not occur. Joseph Jones, a well-read and educated southern physician, surveyed Confederate officers, including physicians, seeking well-documented cases of typhus. He was particularly struck that no epidemic erupted at the prison at Andersonville, where surely the men were deprived and filthy enough for an epidemic to erupt, if all that was necessary was sufficient filth and a weakened constitution. "This great experiment of Andersonville, perhaps the greatest and most remarkable of modern times, strongly sustains the view that typhus and typhoid fevers are dependent upon the action of special poisons, the conditions for the origin and action of which are as definite and as limited as in the case of the poisons of smallpox and measles," he wrote in 1867. He concluded that typhus could not "be generated by animal exhalations from putrefying excrements or bodies, but that these diseases are propagated by a special poison emitted by the living body, either directly or through the excretions and secretions."[61] He left open the question of exactly what the special poison might be.

The Continued Fevers

Gastrointestinal infections ravaged Civil War soldiers, and here physicians were sure that contagion was not an important actor, that these diseases instead arose from multiple environmental causes. Directly tied to life in camp, these causes might have been prevented if prevalent notions about camp cleanliness had been followed, argued many contemporary observers. Guidebooks for officers and army physicians emphasized finding a clean source of drinking water and locating the latrines (called "sinks") away from it. Men were ordered to use the sinks and not to relieve themselves in convenient bushes, and the sinks were to be covered daily with dirt and, if possible, a disinfectant such as lime.[62] These standards were rarely met, and the Civil War soldier lived amid a soup of fecal organisms. Soldiers with dysentery were particularly likely to soil the ground near their tents, as the urgency of their evacuations prevented travel to the distant sinks. Rains then washed fecal material from horses and humans into the water supplies. Many soldiers suffered the consequences in bouts of acute or chronic diarrhea; one study found that a quarter of the veterans sampled had persistent disabling diarrhea after the war was over.[63]

Civil War physicians tended to clump most diarrheas and dysenteries as "continued fevers," a reference to their daily fever spikes; malaria, in contrast, gave its victims days off. Typhoid was the best known of this class, and the one most commonly separated out as a distinct disease, possibly with its own specific poison. Typhoid had been distinguished from typhus in the decades just before the war, based on the microscopic appearance of the gut wall upon autopsy. Physicians' recognition of this as a key sign is evident in case reports on diarrheal illnesses that note whether this sign was or was not present.[64] William Budd's work on the contagiousness of typhoid was widely known, if not accepted.[65] Some considered contagion a part of the etiological picture, as in this physician's comments: "The fact is now well ascertained that this fever is, under certain contingencies, infectious, and communicable through the agency of the bodily excretions of the sick; but the greater truth is, that effete animal and organic matter in a state of putrescence, as in badly policed camps, barracks, and latrines, and especially the mephitic effluvia from sinks, etc., are the most powerful localizing causes of its endemic prevalence."[66]

Surgeon J. J. Woodward, whose account of camp diseases was widely distributed to northern physicians, dismissed Budd's work altogether, including the idea that typhoid fever was contagious. "In America the leading practitioners have so completely abandoned the idea that typhoid fever as manifested in civil life is a contagious affection, that the question need not be discussed in this place," he concluded, arguing not just for his own opinion but the general consensus. "The continued fevers of the army during the present war . . . have certainly presented no phenomena to justify a belief in the possibility of contagion. The whole weight of the facts is in favor of the opposite view, and the general tenor of the opinions and the practice of army surgeons is against the contagion theory." They based these views on their own experiences in the war's hospitals. "While separate hospitals have been carefully established for measles and variolous disease, the fever cases have been invariably sent indiscriminately to the general hospitals with other diseases and wounds. They have been treated side by side . . . [with] no bad effect upon the sick."[67]

If little support for the contagiousness of typhoid could be found, even less was evident for the other forms of dysentery and diarrhea. Dysentery was diagnosed in cases with fever, bloody stools, and tenesmus, colonic and rectal cramping that caused a continuous urge to defecate. Absent the blood and cramping, the case was categorized as diarrhea, which might be acute or chronic. The causes of dysentery and diarrhea were many, according to Civil War physicians. One Sanitary Commission brochure on diarrhea proclaimed: "Camp diarrhoea is a disease probably more largely occasioned by general causes than most of those which the military surgeon is called upon to treat. Let an army be fed upon improper or scanty rations, and an increase in the number of diarrhoea cases will soon follow. . . . In a word, *want*—want of food, want of clothing, want of rest, want of spirits, everything which tends to break down the vital power of the soldier, is almost certain to occasion diarrhoea, and it is produced by the same causes with equal certainty, if not equal severity, in all our armies."[68] None of these inciting factors were specific or contagious.

Another physician proclaimed, "The *predisposing* causes of diarrhoeal maladies in the army are evidently hot weather, fatigue, malarial poison, and irregular habits," where "malarial poison" referred to the air near swampy, stagnant water. He went on, "The *exciting* causes are, exposure to the cold damp ground at night, with insufficient clothing or bedding, or bivouacking in the open air, and under every variety of weather; partaking of food at irregular intervals, which is often scanty, precarious, and of bad quality." He also condemned "overloading the stomach with green fruit and other indigestible matter; indulging in large draughts of water while over-heated; resting upon the damp ground or in the heat of the sun while upon the march; in many cases debauchery and intemperance whenever opportunity presents itself." After this long list of indiscretions, he concluded, "All these are fruitful causes of the diseases of the digestive organs and alimentary canal, which are manifested by acute diarrhoeas and dysenteries."[69]

These sorts of multifactorial proclamations tended to overwhelm the few voices promoting the role of microorganisms. There were physicians who argued for a fungal cause, such as one who concluded "that camp diarrhoea is caused by cryptogamic formations generated in camps and barracks, and disseminated in the air; and that good and absolutely perfect police will insure against this form of disease."[70] Such voices were rare. Yet in spite of the unpopularity of this point of view, it was commonly argued that the diarrheal discharges of all patients should be treated with disinfectants, especially if their containers were going to be sitting around the hospital for any length of time. But, again, the disinfectants were not used to kill microorganisms. They stopped putrefaction and the formation of dangerous poisons in the decomposing mess that would float into the air and endanger others.

That Civil War physicians indicted multiple causes in explaining diarrheal

Chickahominy Swamp. *Harper's Weekly,* July 19, 1862. Such marshy land was typical of coastal and riverine military sites in Virginia, South Carolina, and North Carolina and along the Mississippi River. "Chickahominy fever," which plagued General George Mc-Clellan's army on the James River peninsula in spring and summer 1862, was likely a mix of malaria and typhoid. The standing surface water allowed for abundant mosquito repro-duction, and the high water table meant that latrine discharges would have quickly con-taminated the water supply. In the view of contemporary doctors, it was the foul marshy odors, a mix of methane and sewage, that made the men sick.

illness is not surprising given the many diseases that they lumped under this category. From a modern perspective, the historian can identify the likely presence of entamoeba histolytica, shigella, salmonella, campylobacter, giar-dia, multiple viruses, and other infectious organisms. Intestinal worms were common. And we now know that scurvy and other forms of malnutrition can weaken the intestinal wall and lead to wasting diarrheas.[71] A single soldier might endure multiple different infections, and his ravaged intestine would in turn poorly absorb nutrients, decreasing his ability to mount an immune re-sponse. Diarrheal illnesses were omnipresent in the Civil War and returned in the Spanish-American War to again demonstrate to military officers that good camp hygiene was critical to maintaining an effective fighting force. This all-too-common disease spectrum was too confusing to stimulate an awareness of the contagiousness or the transmission of waterborne organisms.

Conclusion

The Civil War brought questions of contagion and disease prevention to the forefront of physicians' minds and introduced many doctors to newer notions of specific poisons and their transmission. Physicians who had never personally handled an epidemic before, or felt responsible for stopping one, were · in charge of hospitals where disease outbreaks demanded a remedy. They learned from errors and came to respond in similar ways. Diseases thought to be contagious—smallpox, measles, yellow fever, and later cholera—were isolated in separate hospitals or "under canvas" in hospital tents located away from the other sick men. Bodily fluids given off by the patients, whether vomit, diarrhea, or pus from wounds, were disinfected. Clothing, bedding, and bandages were dipped in disinfectants before a thorough washing or were burned. Many physicians left the war with a clear notion that epidemics could be controlled or even stopped if the right steps were taken.

These concepts of disease transmission and prevention did not depend on accepting the animalcular theory of disease causation. No such theory was established until the late 1870s, when Robert Koch and Louis Pasteur offered proof, not just evidence of association, that certain microbes caused certain diseases. Physicians were right to dismiss Salisbury's theory that measles was caused by a fungus on rotten straw. And the microbes that cause measles, smallpox, and yellow fever could not be seen until the electron microscope became available to resolve viruses in the 1930s. As historian Nancy Tomes has pointed out, there was no clearly established "germ theory" to accept before the work of European bacteriologists. What was an active subject for debate during the 1870s was this central question: "Is the causal agent responsible for infectious disease a chemical substance or a living organism?"[72] Either answer fit the ruling doctrines of public health, which favored sanitation, isolation, and disinfection.

Civil War physicians were convinced that, aside from disinfectants, camp hygiene was key to maintaining a regiment's health. But officers and enlisted men had to be convinced that hygienic discipline was important and necessary. As we have seen, army surgeons such as Jonathan Letterman did their part to spread this sanitary gospel. Another force promoting camp cleanliness was the U.S. Sanitary Commission, the voice for the nation's women, who were, after all, traditionally in charge of maintaining domestic cleanliness. Chapter 4 turns to the history of the USSC and the many facets of its organization that sought to answer the health crisis brought by the war.

CHAPTER FOUR

Connecting Home to Hospital and Camp

The Work of the USSC

In the war fever that followed the firing on Fort Sumter and President Abraham Lincoln's call for troops, men rushed to volunteer while women and men not of fighting frame rushed to support the war effort in other ways. In the North, the U.S. Sanitary Commission was born amid this wave of "excited benevolence of the country towards the army."[1] Its efforts funneled disparate impulses active in spring and summer 1861: the drive of women's groups to supply the Union Army with clothing, food, and bedding; the desire of some women to serve as nurses; the determination of those familiar with the story of the Crimean war that losses from disease should be prevented; and the general wish for obvious deficiencies in the care of sick and wounded men to be remedied. The USSC was similar to the Red Cross in the way it functioned: it was independent of the government and received funding from contributors and not the public purse, but it was officially authorized to provide relief and to work in tandem with military authorities.

As noted earlier, the USSC began with a meeting called by Elizabeth Blackwell on April 29, 1861, at the Cooper Union Hall in New York City. There women were summoned to support the war effort, either by becoming nurses or by channeling food and sewn goods to the men. A group of male civic leaders, headed by Henry W. Bellows, assisted this initial effort and subsequently took command of the organization, ultimately dubbed the United States Sanitary Commission. This group of male commissioners went to Washington to gain a charter that would allow the agency to serve as the interface between women's charity groups and the sick of the Union Army. Although headed by men, the organization embodied the impulse of women to extend their knowledge of health care within the home to their men who were injured or ill within the military. It also conveyed the sanitary gospel of northern medical reformers seeking to bring the best possible conditions to

Women's Meeting at Cooper Union Hall, April 29, 1861. This large gathering marked the beginning of the U.S. Sanitary Commission. Artist unknown. Paul F. Mottelay and T. Campbell-Copeland, eds., *Frank Leslie's "The Soldier in Our Civil War": A Pictorial History of the Conflict, 1861–1865, Illustrating the Valor of the Soldier as Displayed on the Battle-field*, Columbian Memorial edition, 2 vols. (New York : S. Bradley, 1893), 1:242.

military camps and hospital. The official goal was to "assist the medical department" in whatever way a volunteer organization could contribute.[2] Much of what follows concerns the words and actions of men, but always motivating them was the feminine nurturing response to the bloody and diseased outcomes of masculine aggression.

The governing members of the USSC were well-known men in the professional and financial circles of New York, Boston, and Philadelphia. Henry Bellows, the president, was the pastor of the First Congregational (Unitarian) Church in New York City. Alexander Bache, the vice president, was a great-grandson of Benjamin Franklin who taught science at the University of Pennsylvania, organized the coastal survey of the United States, and founded the National Academy of Sciences. Elisha Harris had distinguished himself as a New York physician with strong interests in public health. William Van Buren was a well-known Philadelphia physician, and George Templeton Strong, the treasurer, was a wealthy New York lawyer (who left a memorable diary quoted frequently in Ken Burns's documentary *The Civil War*). Frederick Law Olmsted joined the organization in summer 1861. Olmsted later became famous as the designer of major urban parks, but he had already made a name for himself by the 1860s for his travelogues based on his own journeys through the South and other parts of the United States. Another member was physician and chemist Wolcott Gibbs, who was called to a chair at Harvard in 1863. These well-educated men brought to the Sanitary Commission a strong expectation that science should guide medical action.

Efficiency, organization, expertise, and order were their governing watchwords, layered on top of female expectations about the proper ways to care for the sick; and the medical department of the Union Army was in sad need of these contributions.[3] When Fort Sumter came under fire, there were 114 surgeons in the army; 24 of them resigned to join the Confederacy, and 3 more were cashiered for disloyalty. The army began the war with an aged surgeon general who died in May 1861 and was succeeded by the similarly senior Clement Finley, who was unprepared for the job. The army's medical structure had been customized for frontier service, where the men suffered from occasional infectious diseases or the casualties of encounters with American Indians. These doctors, at least, had field experience, whereas the thousands of physicians who soon joined the war effort along with their volunteer regiments had none.

These volunteer units, formed in the first frenzy of response to Lincoln's call for troops after Sumter, had officers who had been commissioned yesterday and doctors pulled from private practice by the decisions of these officers. The physicians were not screened for competence, the regiments lacked medical and hospital supplies, and the ambulance service consisted of asking the drummer boys or other supernumeraries to help bring men off the field. At the first battle of Bull Run in July 1861, wounded men lay on the field four or five days awaiting help, and the Medical Department scampered to create hospitals in Washington to care for them. The medical leadership was not up to the vast and chaotic task facing it.[4] Bull Run laid bare the need for massive reform in the management of the war, including the Medical Department. The general situation is described in chapter 1; this chapter recounts the efforts of the USSC in answering this call by encouraging reform and providing services during the period when the U.S. Army's medical management was still struggling to find its way.

Spring 1862 brought a general transformation in the War Department. Lincoln's first secretary of war, Simon Cameron, was a Pennsylvania politician who proved incapable of managing a complex enterprise. The Union Army grew from 16,000 to 670,000 men between Lincoln's inaugural and the end of 1861. While Cameron kept track of contracts and orders via notes to himself that littered his desk and pockets, subordinates took advantage by bilking the army with rotten food, blind horses, knapsacks that dissolved in the rain, and armaments that failed to function. Little wonder, then, that Cameron failed to recognize the ineptitude of the Medical Department or see to its correction.[5]

When Lincoln replaced Cameron with Edwin Stanton in January 1862, the men of the USSC saw their chance. They were disgusted with the in-efficiency and ineffectiveness of the army's Medical Department and had been trying for months to promote its reform. Their attitude toward Surgeon General Finley was reflected in a letter written by Frederick Law Olmsted to a friend in late 1861. "It is criminal weakness to entrust such important responsibilities as those resting on the surgeon general to a self-satisfied, supercilious, bigoted block-head merely because he is the oldest of old mess-room doctors of the old frontier-guard of the country," spewed Olmsted. "He knows nothing and does nothing and is capable of knowing nothing and doing nothing but quibble about matters of form and precedent and sign his name to papers."[6] First the USSC tried to get Finley dismissed and replaced with William Van Buren or his student, William A. Hammond. When that failed, Hammond helped the leaders of the USSC draft a bill that would dethrone Finley and allow his own appointment. In December 1861 Senator Henry Wilson of Massachusetts introduced this legislation. Secretary of War Simon Cameron opposed Hammond's appointment, and hence the USSC in general, because of an old family quarrel (or so said the official USSC history). But after Cameron resigned in disgrace, the bill made it through Congress.[7] Lincoln signed the medical reform act on April 16, 1862.

It is not clear exactly how the Sanitary Commission achieved this objec-tive. Olmsted told his father three days later, "Our success is suddenly won-derfully complete. The Medical Bill after having been kicked about like a football, from House to Committee & Committee to House and . . . over & over again, at each kick losing on one side & gaining on another," until the bill itself was "so thoroughly flabbergasted that nobody knew where or what it was," suddenly sailed through both houses and "before we know it is a law." Then Stanton "kick[ed] the old Surgn Genl out of his seat" and the president promised to nominate Hammond to take his place, at the USSC's urging.[8] It may be that the battles of Fort Henry, Donelson, and Shiloh, and preparations of war on the Peninsula, had awakened Congress to the neces-sity of the measure. General McClellan was a supporter of it, and his reputation as the glorious leader who would organize the army and ensure victory reached its peak in spring 1862. The USSC organized medical societies to pepper the president with letters favoring Hammond once the bill passed. Hammond and the USSC were temporarily in the favor of Secretary of War Stanton, as well, but by summer he had formed an animus against the organization that lasted throughout the rest of the war.[9]

The bill gave the USSC most of what they desired. It instructed the president to appoint a new surgeon general "from the medial corps of the army, or from the surgeons in the volunteer service, without regard to their rank when so selected, but with sole regard to qualifications." This provision was designed to circumvent the current army policy of promotion based solely on seniority in the service, so that the USSC candidate, William Hammond, could be appointed instead of Clement Finley, who had gained his position by rising to the top with years of service. The bill also created a corps of medical inspectors whose job was to report on the conditions of camps and hospitals as well as to alert the surgeon general when incompetent surgeons were identified.[10] Hammond, who had spent several years of his young career as an assistant surgeon in the army (serving mainly on military posts in the West), had some "street cred" as an experienced army physician, but even more importantly, he had earned a reputation as a scientist from his publications on aspects of human and animal physiology. Although his direct exposure to European medicine was limited to a few months of sightseeing, Hammond was well read in the best of European medical works and was all around a "coming man" in the medical world. The USSC leadership equated scientific rigor with efficiency, which they hoped Hammond would bring to the Medical Department. The senate approved Hammond's appointment on April 25, 1862.[11]

Having put a proper administration in place, the Sanitary Commission turned directly to the challenges that the spring, summer, and fall of 1862 brought, beginning with the immediate need to raise money. The commission at first appealed to the insurance companies, arguing that their interventions would save lives and thus reduce company costs.[12] They called on others with deep pockets as well, but the main font of funding was a surprising source. Far out west, the men who had made fortunes in the gold rush passed the hat, and a half million dollars came over the Rockies to fill the USSC coffers in fall 1862.[13] The commission used the funds in several ways. They paid medical inspectors to evaluate camps, and they purchased the supplies needed in hospitals, such as bedsteads and water jugs. They opened offices and depots in cities near the fighting but far enough away to be safe from enemy encroachment. And they assisted women's groups in creating a mountain of baked and sewn items to be delivered to the army. One USSC source called this "funneling the river" of women's beneficence, and it was one of the main functions of the Sanitary Commission. "The Sanitary Commission is the great artery which bears the people's love to the people's army," began Katharine Wormeley's sketch of the USSC in 1863.[14] The story of the USSC in Washington politics is well described by historians; less familiar is the more important story of what the USSC accomplished in the field, how it worked and how the health experience of the Civil War soldier was changed as a result.

The pages that follow detail these many functions of the USSC. They included

- organizing and delivering women's donations for the troops (bedding, clothing, and food),
- assisting the government in setting up hospitals with as many features of home as possible,
- assisting the Medical Department in transporting wounded and sick men to general hospitals,
- maintaining depots of supplies and creating rapid-response teams that rushed those supplies to doctors after battles,
- educating army surgeons about camp hygiene and the best modern medical practices through camp inspections and the distribution of didactic materials, and
- other "special relief" duties.

Funneling the River of Women's Efforts

The USSC originated in the impulse of northern women to assist the war effort. They were accustomed to caring for "their men"—brothers, sons, husbands, and fathers—and knew intimately all the ways that women's work made a difference in men's lives. The USSC served not only to connect the work of women to the army; it attempted to recreate the role of women in the sickroom for the wounded or ill soldier.[15] As the location of sickness was moved from private house to government hospital, Americans sought to preserve as much as possible of the customary ways of caring for sickness. Women moved directly into the hospital as nurses, cooks, laundresses, and matrons, duplicating their roles in the household. The USSC channeled the energies of women who could not be physically present at the bedside; these women provided as many components of home care for hospitalized soldiers as could be delivered under wartime conditions. Early in the war, little thought had been given by army medical men to fitting out hospitals. The sickroom was a female domain, and it took a female perspective to recognize what was needed to create the optimal recuperative environment. Although the USSC at times supplied aid to field hospitals, most of their applications were administered farther behind the lines, in general hospitals.

Henry Bellows used the simile of a river to describe the USSC's role in delivering the efforts of women all over the North to the Union soldiers who needed their ministrations. Writing in fall 1862, his main point was that the cash-strapped USSC could not supply raw materials or money for salaries to the women's aid societies that furnished the necessities distributed by the USSC. "Nothing short of the free activity and free contributions of every family, hamlet, village, church, and community throughout the loyal States, continued as long as the war continues, can avail to meet this never-ending, always-increasing drain," he wrote in the wake of the battles of Antietam and Perryville. "It is the little springs of fireside labor oozing into the rills of village industry, these again uniting in the streams of county beneficence, and these in State or larger movements, flowing together into the rivers which directly empty into our great national reservoir of supplies, which could alone render possible the vast outflow of assistance which the Sanitary Commission is lending our sick and wounded soldiers." Bellows concluded that if the women stopped producing, the river would run dry and the men would go wanting. He called on the women not to abandon the nation's troops, not to give up their virtual place at the bedside.[16]

The USSC leadership saw itself as essential to bring discipline to this female effort, however. If women sent care packages to their own family members, or the regiments that came from their communities, the packages could get lost or spoiled. Armies on the move had no need for fripperies, and the healthy fighting man was not the proper target for philanthropy. Symbolic of the silliness of undirected benevolence was an early craze to sew havelocks for the men; a havelock is a combination hat and neck cover designed to protect the soldier from sunstroke and popularized by Sir Henry Havelock of the British army during the 1857 Sepoy Rebellion in India. Thousands were made, but the soldiers found them too hot to wear and used them instead as dishrags or coffee strainers. There was a similar mania to

Our Heroines. This sketch was inspired by the many functions of women associated with the U.S. Sanitary Commission. It appeared as a double-page central spread in *Harper's Weekly*, April 9, 1864. In the center is the female nurse at the bedside, ready to sponge the man's forehead or read to distract him from his discomforts. The corners show women providing battlefield relief, sewing clothing for soldiers, and raising money at a "sanitary fair." The upper right corner shows a nurse wearing the headdress of the Sisters of Charity, the nursing order that staffed Satterlee Hospital (see chapter 6).

scrape lint for bandages, although bandages were way down the list of needs in army hospitals.[17] The USSC saw its purpose as learning what was truly needed, by visiting hospitals and talking to army physicians, and directing the women's aid societies to produce appropriate goods.[18]

Bedding

A wounded or sick man arriving at the hospital needed, first, a place to lie down. In the most primitive settings, such as a field hospital established in a barn, that place might be a bare floor or a pile of hay. Sanitary Commission women sewed bedticks, cotton sacks shaped like long pillowcases that could be stuffed with straw to make mattresses. Such mattresses were well suited to an age that lacked plastic mattress covers; the soiled mattress could be easily disassembled, the tick sent to the laundry, and the stuffing replaced with fresh straw.[19] More established hospitals in major cities might have mattresses stuffed with feathers or horsehair. But it took time for the army to realize that it had to buy and maintain hospital beds—many, many hospital beds—and the USSC often supplied the want early in the war.

Women's aid societies likewise made bedclothes, sewing sheets, pillow cases, and quilts, all to the specifications dictated by the USSC, which supplied necessary discipline and order. Commanding medical officers in the Union Army established a standard bed size for hospital beds and set the dimensions of bedclothes to fit those cots. The USSC in turn told the many sewing circles what was needed, and quilts and sheets of just the right size and composition were created. Early in the process, the USSC would buy fabric and thread for these sewing projects and forward the raw supplies to women's work groups. This practice ceased in 1862, and the central organization instead spent its cash on bedding produced in factories, telling the heads of the women's groups that they were on their own to purchase necessary supplies. Some affluent women's groups bought sewing machines and hired women to work them, creating miniature factories of their own.[20] In 1862, after the battles of Fort Donelson and Shiloh resulted in thousands of hospitalized casualties, the Chicago branch of the Sanitary Commission set to work to supply them. A large hall was fitted out with sewing machines, and women, many of them soldiers' wives, were hired to sew hospital clothing and bedding. "Never was clothing manufactured more rapidly; for the machines were run into the small hours of the morning, and there was no slacking of effort while the urgent demand lasted," recalled Mary Livermore.[21] Such bedding was stamped with the USSC brand, both to let the men know where their comforts had originated and to discourage the reselling of these items meant for the hospitalized soldier.

Clothing

The Union soldier typically had one uniform made of wool, a shirt or two, and perhaps a change of underwear, commonly made of linen or cotton. He had no pajamas in his backpack, and when he was wounded or sick in hospital, the army provided no clothing appropriate for such a situation. The USSC was the first to recognize that the hospitalized man needed different clothing in this setting, particularly if he arrived in a muddy, bloody uniform. It provided the simple comfort of cotton nightshirts and drawstring shorts to be worn in the hospital, as well as bathrobes for the ambulatory man. These too were sewn according to standard patterns, and women's aid societies shipped the completed items to central depots. From there the USSC sent the clothing to hospitals in need of it.

A physician stationed in a Nashville hospital in the winter of 1862–63 highlighted the need for the USSC's mode of supply. "As an illustration of the difficulties sometimes experienced in the effort to provide for the wants of the sick in hospital, when obliged to rely wholly upon the ordinary channels of supply let me relate an incident which occurred while I was an Act. Asst. Surgeon on duty in one of the large hospitals in Nashville, during the winter of 1862–3," he began. "My patients were insufficiently provided with underclothing. I applied to the surgeon in charge of the hospital—an excellent man who was really anxious to promote the welfare of his patients, but who was very properly desirous to act in accordance with the regulations. He promised to draw clothing from the Quartermaster as soon as a list of articles could be prepared." So, the surgeon prepared the list. "It was then decided to delay the promised requisition till the whole hospital could be canvassed. I had this done. Then the necessary blanks could not be procured, and as a supply was soon expected from the printer, another delay was ordered." But having the blanks did not resolve the matter, since in the interval the hospital's population had shifted considerably, with many of the initial patients sent to a convalescent camp. "The surgeon now decided that the amount of clothing required was not sufficient to justify the trouble of a requisition. I proposed to him to draw a quantity of clothing which might be kept in the hospital and issued to the patients as they needed, but he knew a surgeon in charge of another hospital who had lost a considerable amount of clothing taken from his storeroom after it had been drawn from the quartermaster." The surgeon feared being accused of fraud, so he was reluctant to institute this plan. "In this way I was put off from day to day, during the cold damp weather of December, though the warehouses of the quartermaster were overflowing with warm clothing of every description. One excuse followed another for more than a month till an order of some sort from the Med. Director's Office removed the difficulties and provided us with all the clothing desired."[22] He would have called on the Sanitary Commission, but its Nashville depot was insufficiently stocked at the time, fall 1862 being a low point in USSC financial health.

In contrast, as a general rule, an army surgeon had only to send a written request to the USSC depot to receive needed goods. The USSC employed on average three hundred male and female agents, although at times of peak activity the number reached seven hundred. One of their jobs was to staff the depots, where they would see that goods were received, sorted, stored, and sent out at surgeons' requests. Agents often anticipated the impending medical needs in the field even as battles were ongoing, ready with wagon loads of emergency supplies. But they also maintained supplies well behind the lines and were (usually) ready on a day's notice to supply underwear, or bathrobes, or bedticks for the hospitals.[23]

Food and Other Supplies

The army diet was not sufficient to support health and often was inappropriate for hospitalized patients. The standard ration per man per day included salt or fresh pork or beef, soft bread or crackers, beans, rice, coffee, sugar, and salt. (Later in the war the army added potatoes and dried vegetables to the list).[24] This was understood only as a basic diet, and the expectation was that meals would be supplemented by local produce, which found its way to the men's mess kits in several ways. The soldiers foraged on the march, picking blackberries, grabbing apples from trees, or stealing from farms. Soldiers with enough money in their pockets could buy fruits, vegetables, or baked goods from camp peddlers, called sutlers. Sutlers also sold the men alcohol, in some cases hidden in pies to elude officers who were trying to enforce rules against outside alcohol.

The official method of dietary supplementation was in the hands of the regiment's officers. They could sell any unused part of the regiment's food allowance back to the subsistence department and build a company fund. With that money they bought local produce for the men. This system worked well when there was local produce to buy and when the officers were efficient in implementing the system. But when an army had been in an area for any length of time, the markets were stripped bare, and scurvy threatened the troops. Armies also disrupted agriculture, by burning the fences that kept animals out of crops, stripping the fields of growing plants, or destroying crops by burning, trampling on, or fighting on fields. And when armies were on the move, particularly when thrusting deep into hostile territory, it was often difficult for supply trains to maintain even minimal food supplies.[25]

This sort of shortage was well documented during the war. Virginia was hard hit by the presence of both armies, and food prices in Richmond soared. By spring 1865, scurvy was common among Union troops in Virginia, especially the African American regiments. Even though food might be abundant farther north, interstate traffic in produce did not move fruits and vegetables quickly enough to the Union armies in Tennessee or on the Mississippi River outside Vicksburg. Such insufficiencies were made worse by corrupt and inefficient officers. During the winter of 1862–63, scurvy appeared among the men of the Army of the Cumberland. General William Rosecrans was perplexed, because he knew that a hundred barrels of vegetables were daily delivered to his command. His investigation discovered that most of the vegetables had been consumed by officers and their families, such that the privates had received barely three rations of vegetables in the twelve months ending in April 1863.[26]

The USSC responded vigorously to such needs—including the needs of Rosencrans's men in spring 1863—when they recognized them. Not only did its depots stockpile canned foods and barrels of vegetables; it responded to emergency needs with its network of women's aid societies. J. S. Newberry, the USSC secretary for the western states, remembered one particularly heroic response in spring 1863, when scurvy (aka scorbutis) threatened Grant's troops. "Early in March the Chicago Commission issued an appeal to the North-west for anti-scorbutics to be used in the army of General Grant. It was dated March 4, 1863, was short and very urgent. In addition to this little circular, which was scattered broadcast throughout the North-west, and to articles inserted in the Chicago daily journals, the Commission telegraphed concerning the emergency to many of its larger auxiliaries." The telegrams said simply, "Rush forward anti-scorbutics for General Grant's army," and were sent to cities and towns throughout Wisconsin, Michigan, and Indiana.[27]

Newberry described the organizational scheme that governed the food gathering. "The towns were divided into districts, and every house was visited; a central depot of deposit was appointed. . . . In the country, committees went in wagons, begging as they went . . . from house to house. This was done day after day, first in one direction, then in another, through mud and rain, by men and women." This great harvest was phenomenally successful. "All through the month, potatoes and onions, sourkraut and pickles, rolled across the Central Railroad, and sailed down the Mississippi. A line of vegetables connected Chicago and Vicksburg. Not less than a hundred barrels a day were shipped, and generally the average was more." Newberry reported proudly, "This movement is more striking from the fact that the Government had endeavored to obtain these articles and had failed." The Sanitary Commission workers in the upper Midwest were far more determined to feed their sons, and far more aware of the importance of proper diet, than were the officers entrusted with the task. "No exertion, no sacrifice, was considered too great to be made in behalf of the army investing Vicksburg. Nothing could have been devised better calculated to arouse and unite the West, than the claim of the South to the exclusive control of the Mississippi," Newberry triumphantly concluded.[28] On July 4, 1863, Grant's well-fed army took Vicksburg.

In April 1863 the USSC learned that the Army of the Cumberland, then bogged down in Tennessee, was equally in need, and Newberry again directed needed foods to stave off scurvy. His agent in Murfreesboro, Tennessee, went further, starting a vegetable garden with the specific purpose of growing needed foods for the army and the hospitals that were multiplying in Nashville.[29] General William Rosecrans and his army pushed through the mountains at the edge of the Cumberland Plateau and into the Tennessee River valley of eastern Tennessee. On September 20, 1863, he met Confederate general Braxton Bragg at the battle of Chickamauga and lost, retreating to Chattanooga. There the Confederate forces effectively besieged Rosecrans's army, frequently interrupting the supply lines by rail from Nashville and by river from points north. In Chattanooga eighty thousand men and sixty-one thousand horses and mules were on half and then quarter rations, and the civilian population suffered along with them. The ten thousand work animals that died of starvation in turn became food for hungry men. Scurvy blossomed as fresh produce disappeared from the markets.

Unable to break the siege, Rosecrans was relieved from command, and Ulysses S. Grant took over. His troops broke through in late November, defeating the Confederates at the battles of Lookout Mountain, Missionary Ridge, and Chattanooga and driving Bragg's army into Georgia. The flow of food and medical supplies quickly resumed to Union forces. The Sanitary Commission was ready with fresh fruits and vegetables (many of them grown on USSC farms in middle Tennessee) for the starving and scurvy-wracked men in Chattanooga. And USSC workers helped care for the wounded and sick who were transported out to the waiting hospitals in Nashville.[30]

While the USSC was always ready to supplement the general diet of the soldier in times of need, the hospitalized soldier was its special target. The standard army-issue diet was even less appropriate for hospitalized patients than for men in camp. This was obvious even to Union Army officers, and a scheme was put in place under which the hospitalized soldier's standard ration was sold, the money put into a hospital fund, and that fund used to buy foods appropriate for invalids. Puddings and toddies made from fresh or condensed milk, soups, porridge, canned fruits, and arrowroot biscuits were all on the standard menu for such men. Lemonade was also highly valued, where lemons could be had. For a variety of reasons, the hospital fund system did not ensure the smooth production of such a special diet. For example, unscrupulous doctors might steal items intended for it or buy delicacies or liquor for their personal use. Then again, the transfer from regimental account to hospital account did not occur until the hospital had been in existence for six weeks, leaving the larder empty for a month and a half.[31] And appropriate foods might not be available in the area markets.

The USSC recognized this dearth and supplied hospitals with foods that found their way into USSC depots by several routes. First, women's aid societies sent canned and fresh foods to the supply depots. There the foods might be pickled; for example, cabbage was shredded and made into sauerkraut; some other items were repackaged into barrels. Or the Sanitary Commission might buy necessary foods on the open market and ship them where needed. Mary Livermore described the USSC depot in Chicago, recalling a noisy, smelly environment. "The odors of the place were villainous and a perpetual torment. Codfish and sauer-kraut, pickles and ale, onions and potatoes, smoked salmon and halibut, ginger and whiskey, salt mackerel and tobacco." This was not all—there was also "kerosene for the lamps, benzine for cleansing purposes, black paint to mark the boxes, flannel and unbleached cotton for clothing,—these all concentrated their exhalations in one pungent aroma, that smote the olfactories when one entered, and clung tenaciously to the folds of one's garments when one departed." Livermore also described useless finds, such as crates with broken jam or honey jars, or ill-packaged chicken that was rotten on arrival. Such putrefaction added to the room's ambience. "We called it 'the perfume of the sanitary,' and at last got used to it, as we did to the noise," she reported proudly.[32]

The many contributions of the Sanitary Commission brought great comfort to the hospitalized soldier. They also supplied other furnishings that made the hospital more homelike. The USSC purchased bedside tables, which could hold a glass of water, or the soldier's glasses, or a portrait from home. They bought writing paper and pencils at a time when demand for such items had driven up their price, along with other sickroom staples such as feeding cups with a spout, bedpans, bed nets, and rubber sheets.

The USSC provided hospitals with food and clothing, and it was able to do so as soon as hospitals were set up following a battle. In the first week after the battle of Antietam, for example, while the government was still mobilizing, the USSC "dispatched successfully, by teams, to the scene of battle, from Washington alone, 28,763 pieces of dry-goods, shirts, towels, bed-ticks, pillows, &c., 30 barrels bandages, old linen, &c., 3,188 pounds farina, &c., 2,620 pounds condensed milk, 5,050 pounds beef-stock and canned meats, 3,000 bottles wine and cordials, and several tons of lemons and other fruit, crackers, tea, sugar, rubber-cloth, tin cups, and hospital conve-niences."[33] All told, the USSC received something like $2.5 million worth of donated goods and spent another $2 million buying supplies, which it in turn provided to the army. In a broadside answering the question "What be-comes of the money raised for the sanitary commission?" the USSC reported buying condensed milk and crackers by the ton; beef stock by the barrel; coffee, tea, and sugar by the hogshead; and potatoes by the carload.[34]

Transporting and Feeding the Wounded

At the start of hostilities, the Union Army was ill prepared to deal with the wounded that were a predictable consequence of battle. After the first battle of Bull Run in July 1861, wounded men lay on the battlefield for days, be-cause the small, poorly organized ambulance system proved inadequate to the task of transporting them. In spring 1862, when McClellan pushed up the Virginia peninsula toward Richmond and Grant was victorious at Fort Donelson and Shiloh, there were few plans in place to ferry the wounded out of enemy territory and into hospitals on northern soil. The hospitals had to be constructed, as well, but that is a story for another chapter. Here the focus is on transport, on moving the wounded from landings on the Tennessee, Cumberland, and James rivers to locations farther north.

The first challenge came after the battle of Fort Donelson, in mid-February 1862. In addition to nearly 2,000 Union wounded, there were more than 1,000 wounded rebels who, along with their nearly 7,000 captured com-rades, needed transport out of Tennessee.

When telegraphs sent news of the battle to St. Louis, Chicago, Cincinnati, and Louisville, the local citizens mobilized to send help. Sanitary Commis-sion workers were among the most active in response. One remembered, "When the news reached Cincinnati, a steamer was promptly transferred to the Commission by General Buell. In two hours, three thousand dollars were spontaneously given to pay her expenses; and she started with nurses and supplies to the relief of the wounded." Such steamers were readily avail-able, since most commercial enterprise on the Ohio and Mississippi rivers was at a halt; the lower valley was blocked by the Confederacy. The cooperation between the USSC and the army was typical; here the vessel was under the command of the military but was transferred to the USSC for the movement of the wounded. The USSC worker continued, "At Louisville, the Associate Secretary for the west, Dr. J. S. Newberry, joined the expedition; and this relief was swelled by more coming down from other branches of the Commission at Cleveland, Chicago, and other points."[35] State governments likewise sent relief ships to provide care and transport for their respective regiments.

The Sanitary Commission's presence transformed the soldiers' existence in a moment. After one USSC agent endured hostility and red tape from the army medical director on the scene, he finally received permission to move the wounded men. They had been lying on the open ground, many without blankets or food, and wearing the dirty, bloody uniforms in which they had been wounded. Few wounds had even been dressed. With the arrival of male and female USSC volunteers, "each was placed in a clean and comfortable bed; their soiled and bloody clothing removed; they were washed with water throughout, including their feet." Once cleaned, they were supplied with "new and clean underclothing, with socks, and, when needed, slippers were furnished to all; food, nourishing and palatable and delicacies to which they had long been strangers, were supplied to them." Even on shipboard, the domestic sickbed was created. "In short, in all things they were nursed and served as though they had been our brothers and sons."[36]

The grim news of Shiloh in 1862 met a similar response. The Sanitary Commission outfitted ships in St. Louis, Chicago, Louisville, and Cincinnati and sent them to Pittsburg Landing to pick up the wounded, this time quadruple the census of the Fort Donelson battle. States also sent ships to pick up their men. The Sanitary Commission agent described the scene, saying most of the wounded they took on board were from Michigan regiments, because the steamers sent from Ohio, Illinois, Indiana, and Kentucky had "taken away the wounded of those States nearly as fast as they gathered there." The reporter of this account lamented, "How sad that is! Think of those poor Michigan boys left languishing and watching . . . ! Had they no sense, as they lay there in their blood shed for their country of the unnatural wickedness and selfishness of section? Thank God for the Sanitary Commission at that moment, if at no other."[37]

The situation in Virginia was worse. In spring 1862 General McClellan's army landed on the tip of the James River peninsula and began to drive toward Richmond. The area, cut through by rivers and creeks, was a festival ground for mosquitoes, malaria, and typhoid fever. Dubbed Chickahominy fever after the largest creek in the area, these febrile diseases were joined by measles and diarrheal illnesses to create an unexpectedly long sick list. Regimental hospitals could not begin to handle this burden, and the closest general hospitals were in Washington, Philadelphia, New York, and Boston. When the first battles of the spring added wounded men to the pile, the Sanitary Commission stepped in to help the inefficient army transport system move men off the peninsula and to northern hospitals.[38]

The USSC chartered its first East Coast ships in April 1862 and, with the reluctant permission of the commanding officers (who were loath to lose control of their troops), began picking up men at a landing near Yorktown and taking them north. Few provisions had been made for the sick and wounded at the landing, and USSC workers found them in pitiable condition. After the battle of Fair Oaks in the first week of June, the problems multiplied. Most of the wounded arrived by train, "packed as closely as they could be stowed in the common freight cars, without beds, without straw, at most with a wisp of hay under heads. Many of the lighter cases came on the roof of the cars." Worse was their physical condition. "They arrived, dead and living together, in the same close box, many with awful wounds festering and swarming with maggots." It was a hot, bright day in Virginia, and the stench was "such as to produce vomiting with some of our strong men, habituated to the duty of attending the sick."[39] There was no receiving hospital at the ships' landing, so the men were unloaded onto the ground until they could be put on board ship.

The army's transport ships offered little comfort. USSC workers crossed into one of them to offer what help they could. "We went on board, and such a scene as we entered and lived in for two days I trust never to see again. Men in every condition of horror, shattered and shrieking, were brought in on stretchers . . . [and the attendants] dumped them anywhere, banged the stretchers against pillars and posts, and walked over the men without compassion." No one appeared to be in charge, to distribute the men to wards or beds, and they had not eaten in two days. There was no food on board the ship to feed them. The Sanitary Commission set to work, acquiring a cow to make beef soup and ingredients to bake bread. Suddenly another 150 men arrived at the landing, with no plan for their care or place to put them on ship or shore. Everywhere the USSC workers found confusion, conflicting commands from officers uncertain of their duty, lack of planning, and men in pitiable condition.[40]

Bringing order to this chaos, the USSC set up tents on shore to shelter wounded men awaiting transport. Their own ships were fitted out with beds, appropriate foods, nurses, and adequately ventilated wards. They fought army red tape, which at times bound the wounded men into a bureaucratic ball of competing jurisdictions. By summer's end the USSC had moved eight thousand men from the Virginia peninsula to northern hospitals and had demonstrated the value of providing a receiving hospital safely at the rear to care for men in transit and also to perhaps cure slightly wounded or ill men and return them to their regiments. The USSC touted this work in fund-raising appeals, emphasizing the need for USSC discipline in bringing comfort to the wounded and sick, especially where the army medicos were clearly incompetent. The reports of the USSC must always be read with an understanding of their function as fund-raising documents. The USSC always had an eye toward maintaining favorable public relations.[41]

In the latter years of the war, the Union Army greatly improved its systems for the management of the wounded, as described in chapter 1. Surgeon General Hammond recommended the creation of a permanent hospital and ambulance corps in 1862 but made little headway toward getting it off the ground. He did persuade General McClellan to appoint Jonathan Letterman as medical director of the Army of the Potomac, and in August 1862 McClellan authorized Letterman to create an ambulance corps for that army, as mentioned earlier. Four months later he had established an efficient system of transport for the wounded, which included a trained and dedicated ambulance corps and a system of field hospitals.[42] With Letterman's system in place, the USSC got out of the transportation business. But the organization could still be helpful in the aftermath of battles, when it was sometimes able to move much faster than the lumbering army medical apparatus.

Depots of Supplies

The key to the USSC's ability to provide swift response lay in its system of medical supply depots and in the fact that it provided its own transportation for those supplies. Army medical men relied on the quartermaster to transport the supplies in urgent need just after a battle, such as bandages, chloroform, opium, and whiskey. But the quartermaster's wagons might be serving other purposes, such as bringing up ammunition or carrying food.

The medical officers had no control over transportation vehicles of their own, and this situation persisted until the end of the war, despite repeated protests. The USSC's ability to move more quickly than official responders was evident following the battles of Antietam, Perryville, and Gettysburg.

Antietam

The battle of Antietam was fought near Sharpsburg, Maryland, on September 17, 1862. The bloodiest encounter of the war up to that time, it pitched the troops of George McClellan against Robert E. Lee's army, and while the battle was not decisive, Lee's retreat to Virginia gave Abraham Lincoln enough of a victory to issue the Emancipation Proclamation.[43] More than two thousand Union soldiers died that day, and nearly ten thousand were wounded. The Union medical supplies were ready and loaded but stuck on the other side of the Monacacy River, where a bridge formed a bottleneck between Washington and the battlefield. The Sanitary Commission agent, W. M. Chamberlain, had loaded his wagons from a depot in Washington, (some seventy-five miles away) and had come by a different route. It was from his wagons that eager Union surgeons acquired the necessities for immediate treatment of their men.[44]

Telegrams and quickly scrawled notes poured into the nearby USSC office in Frederick, Maryland. Louis Steiner was the USSC's medical director for the region, and some of this correspondence survives in his papers. H. A. Dubois, the assistant surgeon in charge of a field hospital at Burkettsville, Maryland, wrote Steiner two days after the battle, "I am in charge of 600 wounded & sick at this place, I have used all my bandages and medicines and have been able to obtain only 350 rations from the US during the 4 or 5 days that I have had charge of this place. If you have medical stores I should be very glad to get bandages as many as possible, lint, old linen, salt, quinine and any other things that you may have. Morphia."[45]

Chamberlain toured the many field hospitals set up in the wake of the battle, distributing whiskey, shirts, drawers, blankets, bandages, old linen, cushions, pads, and farinaceous food. His supplies were quickly exhausted, and he urgently telegraphed to Washington for more. USSC inspector C. R. Agnew, arriving with additional supplies on the next day, saw appalling sights. "The wounded were mainly clustered about barns, occupying the barnyards, and floors, and stables," he reported to USSC headquarters. "I saw fifteen hundred wounded men lying upon the straw about two barns, within sight of each other! Indeed, there is not a barn, or farm-house, or store, or church, or schoolhouse . . . that is not gorged with wounded—rebel and union. Even the corn-cribs, and in many instances the cow-stable, and in one place the mangers were filled." The army surgeons were working tirelessly but lacked essential supplies. The USSC handed out thirty pounds of chloroform within three days of the battle. *"Our chloroform saved at least fifty lives, and saved several hundred from the pain of severe operations,"* Agnew claimed proudly in a letter published by the USSC.[46]

Chloroform was a basic need for the amputations essential to the wounded men's survival, but other items were likewise desperately needed. "Everything in the way of medical supplies was deficient," wrote inspector Agnew. "Poor fellows, with lacerated and broken thighs, had to be carried out of barns into the open fields to answer a call of nature; men, suffering the agony of terrible wounds, were without opiates; tourniquets were wanting in many instances; stimulants very deficient; concentrated food also scanty. . . . In fact, everything was wanting that wounded men need, except a place to lie down, and the attentions of personally devoted surgeons." The USSC did its best to fill these needs until the army's medical wagons finally arrived. One can only hope (since it is not specifically recorded) that Dr. Dubois's hospital was among those that received the blessed relief of chloroform and morphine.[47]

Perryville

The situation was even more dire a month later, when General Don Carlos Buell's Union troops engaged Confederate general Braxton Bragg's army as it was driving north toward Louisville, Kentucky. The conflict began by accident, as units desperate to find water in a drought-ridden landscape clashed near the Kentucky village of Perryville on October 7, 1862.[48] Buell had ordered that medical supplies be left behind, because he wanted to move quickly to cut off Bragg. The medical officers, supplied only with what could be slung over a horse's back, faced nearly three thousand wounded men lying without shelter on the parched battlefield. The Sanitary Commission had a depot stocked with medical stores in Louisville, eighty-eight miles to the north and west of Perryville. When news of the battle reached Louisville, USSC agent A. N. Read set out at once with three wagons and twenty-one ambulances stocked with supplies from the commission. More wagons followed, along with volunteer physicians from Louisville and Cincinnati who sought to help the overburdened army physicians.[49]

When Read arrived on October 13, there were "some 1,800 wounded in and about Perryville. They were all very dirty, few had straw or other bedding, some were without blankets, others had no shirts," Read reported. Five days after the battle, "some were being brought in from temporary places of shelter whose wounds had not yet been dressed." He learned worse news when talking to the Union surgeons: the wounded had very little to eat. "I found several of the regimental surgeons with no medicines whatever, and they informed me that they had received strict orders not to take any. Some of them told me they had a few medicines which they carried on their persons." The conditions grew worse the next morning: "On looking from my window, I saw that the ground was covered with snow to the depth of six inches, and the whole landscape had the aspect of mid-winter. My first thought was the sad one, that my son and 100,000 sons of other parents had passed the previous night within a few miles of the place where I was, with no covering but their blankets and this white mantle."[50]

After Read had supplied the surgeons immediately around Perryville, he moved on to Danville, a few miles away, where he had heard more wounded men could be found. Danville was the county seat, with a large courthouse that had been turned into a makeshift hospital housing hundreds of wounded men. The men had nothing to eat, and Read inquired if soup could be made for them. "The surgeons thought not, but kindly gave me authority to get it if I could." Some local Union men offered to help out. "There was no *beef* in the city, but the butcher agreed to bring in an animal, kill it, and have it ready in two hours." The drought had been so severe that town wells were dry, but the same butcher agreed to have barrels of water hauled in from out of town. Then there were no kettles to be bought, as the rebels had taken them all, but finally one was found two miles out of town. The owner agreed to loan it, because he would not need it until hog-killing time. "No *pails* were to be had for love or money, but I bought some covered firkins with handles, a wash tub and a spade, then dug trenches and laid stones with my own hands, and thus set both" containers. After rounding up some firewood, he scrounged some pepper and salt, and by ten-thirty he had "32-gallon kettles of nutritious and palatable soup ready for distribution." By October 24 another ten tons of food, clothing, bedding, and other necessities were on their way from Louisville to the Perryville area. And slowly the wounded men were transferred by ambulance or train to the hospitals in the North.[51]

Summer 1863

The Sanitary Commission continued its work of emergency relief with the Union Army into summer 1863. When Robert E. Lee began to move his army toward Maryland and Pennsylvania in early June, the Union Army broke camp in northern Virginia to follow him. Ten thousand sick and wounded, heretofore cared for in regimental hospitals near their units, were transferred to general hospitals so their regiments would be unencumbered. They were moved by rail to a port on Aquia Creek, a tributary of the Potomac, and from there by steamboats to Washington. The army had apparently still not learned that men need to eat while on their way from one destination to the other. "The agents of the Commission, with a considerable volunteer force engaged for the occasion, labored night and day to provide for their wants," reported Frederick Law Olmsted, the USSC agent on the spot. "A kitchen having previously been established on the wharf, cauldrons of hot beef soup and coffee, with bread by the wagon-load, were kept constantly ready, and served to all as soon as they arrived, and as often as needed while they remained." In two days they fed more than eight thousand men, and "most of the patients thus received the only nourishment they obtained from the time they left the camps on the Rappahannock till they reached their destination in the fixed hospitals."[52]

In retrospect, these sick men were the lucky ones, because their comrades' march ended at a little town in Pennsylvania and the greatest battle of the war. At Gettysburg on July 4, the armies of Robert E. Lee and the current Union commander, General George Meade, met in a bloody encounter that killed three thousand Union soldiers and wounded more than fifteen thousand others. The Confederates suffered similar casualties and left many of them for the Union surgeons to deal with as Lee hastily retreated into Maryland and Virginia. Agents of the Sanitary Commission responded quickly, moving supplies from their depot in Frederick, Maryland, forward even as the battle raged. Four USSC workers drew so close that they were taken prisoner by the Confederates, confusing the USSC effort on the battlefield.[53]

In spite of losing these men, the USSC was able to provide major assistance during and after the battle. As one of the USSC wagons "came to a point where several hundred sufferers had been taken from the ambulances and laid upon the ground behind a barn and in an orchard, less than five hundreds yards in the rear of our line of battle, . . . a surgeon was seen to throw up his arms, exclaiming 'Thank God! here comes the Sanitary Commission. Now we shall be able to do something," reported Olmstead. The surgeon had "exhausted nearly all of his supplies; and the brandy, beef soup, sponges, chloroform, lint, and bandages, which were at once furnished him, were undoubtedly the means of saving many lives."[54]

Once the battle was over and the railroad lines had been restored into the town of Gettysburg, the USSC set up a major aid station. USSC agents moved supplies in and helped transport the wounded out, to hospitals elsewhere. As the wounded "arrived much faster than they could be taken away, they were laid on the ground exposed to the rain, or to the direct rays of the July sun, without food." Volunteer nurse Cornelia Hancock found them there, hungry and thirsty. As wagons of provisions arrived, perhaps sent by the USSC, she took it on herself to distribute jelly sandwiches and punch made from whiskey and condensed milk.[55] By the second day, the "Commission's agents in Baltimore . . . had a complete relief station, on a large scale, in operation, at the temporary terminus of the railroad. It consisted of several tents and awnings, with a kitchen and other conveniences." This USSC station fed between one thousand and two thousand men daily. In the weeks that followed, the USSC used refrigerated cars to move some sixty tons of perishable food into the area. Included were 116 boxes of lemons and 46 boxes of oranges. The USSC also provided thousands of bottles of whiskey, brandy, and wine to give the wounded the sort of stimulants that were considered best suited for the wounded man.[56]

This pattern of relief was repeated after many battles, large and small, in the ensuing two years. As the army became better organized in the distribution of medical supplies and food, the USSC's supplementation was less needed, especially in Virginia, where the federal army was located near supply routes to major northern cities. William Tecumseh Sherman's army, however, was frequently poorly supplied and poorly prepared to deal with the aftermath of battle. But Sherman was intolerant of the Sanitary Commission, which he lumped together with the Christian Commission and other "fussy do-gooder" civilian organizations that he believed got in the way of an efficiently functioning army. In 1864 the Sanitary Commission did not follow the march through Georgia, occupy Atlanta, or celebrate Christmas in Savannah. Sherman yielded only to the pleas of Mother Bickerdyke, a nurse who organized hospitals and tended men with such efficiency that she earned his grudging respect. When she requested that USSC supplies, at least, be allowed into Chattanooga in the winter of 1863–64, he authorized two train cars a day to carry USSC goods for the hospital.[57]

In 1865 the leaders of the Sanitary Commission turned their efforts to compiling a medical history of the war that would demonstrate the power of commission interventions in saving lives and strengthening a fighting force. But one episode demonstrated that army leaders had not yet learned the clear lessons taught by the war with regard, at least, to minimal dietary requirements. In summer 1865, twenty-five thousand black troops were sent to the southern tip of Texas, to stabilize the border with Mexico and confirm the subjugation of Texas by the Union Army. These men were stationed in desert conditions and were poorly supplied. Their white officers were markedly inept, if not corrupt, and as a result scurvy erupted in unprecedented numbers, afflicting 50 percent to 75 percent of the men in some black regiments. Physicians of the Sanitary Commission became aware of the problem when they saw wards full of men with scurvy in New Orleans, men who had been shipped from the overburdened hospitals in Texas. The USSC began shipping fruits and vegetables to Texas immediately, and the situation improved. USSC commentators lamented (and gloated, perhaps) that the army could still not be counted on to feed its men properly without the assistance of a civilian relief agency.[58]

Education and Inspection

The Sanitary Commission supplied general and field hospitals—and it also printed and distributed hundreds of thousands of pamphlets to educate officers and surgeons on the basics of disease prevention and treatment. One prepared for the individual soldier told him how to stay healthy. Historian Wilson Smillie described these pamphlets as "a historical landmark, for they initiated a special technique of health education which, in later years, developed into enormous proportions and became one of the standard procedures of health education in America."[59] Other pamphlets advised surgeons on practical matters, such as how to treat a fracture, how to control bleeding, the value of vaccination for smallpox and quinine as a prophylactic in malaria, and how to treat and control venereal diseases. There was *Advice as to Camping*, a brochure first issued by the British Sanitary Commission, which described the best construction of campsites in regard to water supply, drainage, the location of latrines, and the disposition of kitchen offal. Other pamphlets focused on specific diseases, such as dysentery, malaria, pneumonia, and yellow fever.[60] By the end of 1861, the USSC had distributed some 150,000 of these pamphlets; by the end of the war, thousands more had been put into the hands of army surgeons. Newspapers published extracts from these brochures.[61]

The USSC was especially proud of the booklet for soldiers, which it tried to make available to every man fighting in the Union Army, or at least his commanding officers. This booklet advised the soldiers to get vaccinated before they entered the army and not to hide illness so they could enlist; informed them that their commanding officers could order issues of fresh vegetables, sauerkraut, rice, and vinegar, which would help prevent scurvy; told them they should eat at regular hours when possible and to cook their food thoroughly; and recommended cooking soup for five hours, to render the vegetables fully digestible, because "those great scourges of a camp life, the scurvy and diarrhoea, more frequently result from want of skill in cooking than from the badness of the ration, or from any other cause whatever." The men were directed to always dig a trench for "calls of nature" and scolded, "Men should never be allowed to void their excrement elsewhere than in the regularly established sinks. In a well regulated camp the sinks are visited daily by a police party, and a layer of earth thrown in, and lime and other disinfecting agents employed to prevent them from becoming offensive and unhealthy." Tents should be aired frequently, and one should seek high, dry ground for sleeping. The men should wash their feet daily and their whole bodies (and underwear) weekly.[62] They were urged, in a phrase, to act like civilized gentlemen.

The USSC reinforced these lessons in its camp inspections. As initially conceived, these inspections were to be the centerpiece of USSC function, with the hope that the inspectors could both discover defects and have the enforcement powers to correct them. But the government did not yield executive power to the commission, so its inspectors acted instead by admonition and instruction.[63] The USSC hired physicians, many of them prominent members of the medical communities of Philadelphia, New York, Baltimore, and Boston, to travel to regimental camps and inquire into their health. A form was devised, with a set of some forty-five questions about the men's tents, food, latrines, and medical care; the list later grew to 180 queries.[64] How far were the horses kept from the tents, and how was their waste dealt with? Was there group cooking, or did each man fend for himself? Did the officers reinforce rules about using the latrines? Was the medical officer attentive to the men's needs? Had the men been vaccinated? And on and on.[65] One colonel complained, "One of your Inspectors came about my camp, and put me through the closest set of questions I ever had to answer. I do believe there were hundreds of them. Well, I did not like it, and I believe I told him I hoped I might never see him in my camp again."[66]

The USSC leaders were aware that they walked a fine line here. Their inspectors could be seen as prying busybodies, who could get a commanding officer or his medical staff in trouble with the surgeon general if deficits were found. But they did not want to make enemies through the process of inspection. Instead, they hoped to educate officers and promote better conditions for the troops. Inspectors were instructed to carefully respect the order of command, speaking first to the commanding officer and only then to the highest-ranking medical officer. Upon arriving at a regiment, the inspector was to present his credentials to the colonel. "Be careful that he understands that your office is not to interfere with his, but to aid him in preserving his men from demoralization and loss." The inspector was then supposed to walk around the camp with the captains, asking questions and pointing out deficiencies. The questions themselves, and the fact the inspector made a note on his form about them, would serve to educate the captain, the surgeon, and other officers as to their duties, the manual for camp inspectors advised. The goal was to educate without arousing resentment.[67]

"Special Relief"

The USSC performed a variety of other services for the needy during the war. Soldiers who were sent home on furlough at times ended up stranded in the major cities of the Northeast. They might become too ill to travel, or not know how to get home, or run out of money. There were men who needed "suitable food, lodging, care, and assistance" when they had been "discharged from the general hospitals" in Washington but were "delayed for a number of days in the city before they obtained their papers and pay." The USSC also provided assistance, information, and transportation for men who had become separated from their regiments. The USSC set up soldiers' aid stations, which consisted of simple boarding houses that provided bed and meals, and a doctor's care if necessary. The commission also helped soldiers find their way in the army's bureaucracy. Sometimes the USSC agent might contact a family member to come when a soldier was too weak to travel home alone. And at times the soldiers were badly in need of clean underwear, a coat, or other clothing, and the USSC provided it. In general the USSC helped soldiers who were "out of pocket" and needed some form of assistance to find their way home or back to their regiments.[68]

The USSC was generous to families in need as well. In Washington housing was at a premium because new jobs had been created by the war, and many troops were stationed in or near the capital. The army erected multiple general hospitals in the city, attracting family members who came to visit their sons, husbands, and brothers. Many of these visitors had limited funds and could not begin to afford the hotels or rooming houses of the city. The USSC set up hostels for such travelers and was particularly solicitous of women traveling alone. It also created hospital directories, so that an arriving visitor could find where his or her boy was hospitalized (or if, sad to say, he had died). The USSC established registers of prisoner-of-war camps, as well, to help families track their soldier relatives. And at the end of the war, the agency set up temporary lodging for disabled soldiers who had nowhere else to go.[69]

Conclusion

The USSC had an undeniably positive impact on the Union war effort. One official history of the USSC claimed that "its direct and tangible results are many thousands of lives saved, an incalculable amount of suffering relieved or mitigated, smallpox and scurvy checked in camps and hospitals by cargoes of vegetables, and by timely supplies of vaccine—and succor, comfort and relief freely given to hundreds of thousands when they could be obtained from no other source." As a consequence the USSC "materially strengthened the National forces, contributed to the National cause, and added a certain number of thousand bayonets to the available strength of the Army" each month for a period of two years.[70] The USSC author compared the value of the men saved with the amount of money spent by the commission. "Leaving out of view all its other work, the Commission certainly saved not less than one thousand lives within forty-eight hours after Antietam. If each of these was worth as much to the country as the average South Carolina field hand to his owner, then the Commission, by its work at this one point, returned to the country more than an equivalent, in money value, for the nine hundred thousand dollars the country has given its Central Treasury during the last two years."[71]

An 1864 history of the Sanitary Commission compared the Mexican-American War to the Civil War in terms of the number of lives lost to disease and took some credit for the reduction. While acknowledging the sacrifices borne by every city and town in the United States, this account stated that the

Civil War "death-rate from disease [was] less than was ever before known in the annals of great campaigns, and actually less than one-third the percentage of mortality from sickness in our volunteer forces in the Mexican war." While giving credit to "the unprecedented success of medical and surgical treatment in our military hospitals," the author likewise acknowledged the power of Providence and the "sanitary service" in bringing about "health and manly vigor as never before was enjoyed by volunteer soldiers in such long campaigns."[72] Although it is impossible to say what might have occurred if the USSC had not taken action, particularly early in the war when the Medical Department was so badly organized, it seems indisputable that the efforts of its agents saved many lives and made many of those lives more comfortable in illness and in need. As the channel of the female benevolence and knowledge of sickbed needs, the USSC succeeded in improving the lot of the Union's sick and wounded.

CHAPTER FIVE

The Sanitary Commission and Its Critics

If the U.S. Sanitary Commission was so great, why did so many contemporaries attack it? The record of the USSC during the Civil War was one of sustained benevolence and significant reform. Why, then, did the agency spend so much effort during the four years of its existence defending its record and actions? How could such a beneficent organization earn opprobrium from contemporaries and, later, historians? The USSC was both patriotic and, in a peculiar sense, antiwar. It served as a reminder that although war might be glorious, war also maimed and sickened men. The soldier might begin as a brave, strong fellow, but the war machine would spit him out dead or as helpless as a child. This hard fact of war was not popular then—or now. It was in trumpeting this unwelcome fact that the USSC raised money to relieve the suffering. It took the woman's perspective on war, and the womanly task was not to shoot and fight but to clean up all the mess left afterward. The dissonance between masculine and feminine approaches to war created ambivalence toward the USSC, which was in turn forced to defend itself.

By its very existence, the commission chided the army's Medical Department and proved to be an irritant throughout the war, in spite of continued rhetoric on both sides about cooperation. The commission also challenged prevailing attitudes toward philanthropy and charity, raising concerns that unreimbursed aid would breed dependency. The USSC pushed a doctrine of nationalism in consonance with the major theme of the war, but states'-rights attitudes did not cease in Union states just because the South had seceded. The states of the Midwest particularly resisted the dominance of the commissioners, whose worldview radiated from New York City. Finally, there was the fussy self-righteousness that infused so much of the USSC's advice and action; it was an attitude that could not help but raise hackles. The USSC did a world of good—and generated clouds of annoyance.

Like Mother Was Here

In representing the women's role in the war, the Sanitary Commission struck at the deep ambivalence of men toward their dependence upon women. War—fighting, glory, and conquest—was the preeminent male domain. Yet much of the day-to-day life in the army demanded that men take on duties that women had done at home, such as cooking, cleaning, sewing, and washing. In the aftermath of battle, traditional female roles of nursing, feeding, and comforting came to the forefront. As the channel that brought women's influence to the army, the Sanitary Commission both basked in gratitude for that kindly presence and endured resentment that such an insertion of female work was needed at all in the manly world of the army. Sometimes the USSC and similar state relief agencies brought women to the wounded or ill men; at other times men took on these female roles in their guise as USSC workers.

When they showed up to inspect camps, Sanitary Commission agents must have come across to the annoyed officers as scolding women. They expected the men to take baths, keep their clothing in order, and clean up their living quarters. They chided the men to use the latrines and not pee behind the nearest bush. The Sanitary Commission was appalled that so many men lacked toothbrushes, and it instructed the soldiers to eat their vegetables. All together, the USSC brought a mother's discipline to a disorderly bachelor campground, seeking to train not just the enlisted men but also the officers in the importance of domestic order to the preservation of health.

Their initial inspections of the army camps dismayed the Sanitary Commission agents: "In many regiments the discipline is so lax that the men avoid the use of the sinks, and the whole neighborhood is rendered filthy and pestilential. From the ammoniacal odor frequently perceptible in some camps, it is obvious that the men are allowed to void their urine . . . wherever convenient." The men were dirty and often literally lousy, with filthy clothing. The raw food provided by the army was abundant but "atrociously cooked and wickedly wasted."[1] A later report concluded, "Slovenliness is our most characteristic national vice."[2] The Union Army grew tenfold in 1861, and the growth spurt brought both enlisted men and officers into the field with little prior training in military discipline. Many of these volunteers had not left home before; they had never lived away from their families. As one (probably female) USSC author noted after working with the sick and wounded on the peninsula in summer 1862, "Of this hundred thousand men, I suppose not ten thousand were ever entirely without a mother's, a sister's, or a wife's domestic care before. They are wonderfully like schoolboys."[3] The men lacked the will and the knowledge to maintain sanitary order in camp and in the hospital.

Women who contributed sewn and baked goods to the USSC for needy soldiers reinforced this image of the Sanitary Commission as supplying female domestic care to the troops. Mary Livermore described the notes tucked into the boxes of goods sent by the local aid societies. "MY DEAR FRIEND," one began. "You are not *my* husband or son; but you are the husband of some woman who undoubtedly loves you as I love mine."[4] Since it was often not practical for women to send care packages directly to their family members in the army, the USSC displaced and directed such beneficence to anonymous targets. Thus the sender became all mothers, sisters, wives, and the recipients all the beloved sons, brothers, husbands far away in the army.

Sanitary agent A. N. Read told of an encounter he had in fall 1862 near Elizabethtown, Kentucky. A regiment on the move had left its wounded with only army rations and nothing else. Read supplied the men with USSC stores from Louisville. "Among the number sick was a young man sick with fever and delirious. He was taken from the floor, and put upon a clean bed with sheets and pillows." The next day the boy recovered consciousness and spoke to the army surgeon on his rounds. Asked how he felt, the young man said, "I feel better, it seems as if Mother had been here." Read reported that "remembrance of this strengthened me in many a weary hour. It was enough for me to receive such a compliment—and his mother had been here, as I was only her representative."[5] The USSC heard similar responses from men on the hospital transports taking them north after the peninsula campaigns. "O, that's good! It's just as if mother was here."[6] The USSC was more likely to send men than women to the dangerous areas near battlefields, but these agents carried "mother's touch" all the same.

But even if they were referred to as "our brave boys" or "our fighting lads," manly men fighting wars were not supposed to need their mommies. And armies that were properly organized should have camp discipline and efficient hospitals and not need a scolding nanny to tell them how to behave. The Sanitary Commission might be able to stir northern women to contribute their time, energies, and money with rhetoric that emphasized the importance of women's role to the war effort, but they were unlikely to find favor with the army's leaders in the process. The women of the North probably knew well enough what kind of messes men would make left on their

own, but the men were tired of hearing about it. Army sanitation, the ambulance service, and conditions in hospitals all improved over the course of the
war, but one suspects the army's leaders' reluctance to credit the USSC with
the improvement was like a sulky child's unwillingness to thank his mother
after she has forced him to clean up his room (even if he is secretly pleased
with the outcome).

Dis-ease with Charity

In that it distributed food, clothing, and bedding to men who did not pay for
them, the USSC was a charity, and it aroused all of the ambivalent feelings
that nineteenth-century Americans felt about such practices. From the colo-
nial era, American welfare policy had advocated local control of care for the
needy and had decried giving alms to the shiftless, the stranger, or the criminal. The "worthy" poor should receive help, and a means test that included
not just need but moral stature was part of the equation of welfare benefits.
Most often the targets of public beneficence were the widowed, the aged, the
very young, and the sick or disabled. Common belief held that the sturdy
able-bodied adult male could work, or starve, but did not deserve access to
the public dole or private philanthropy. In fact giving him alms would create dependency and should be avoided at all costs. Most hospitals intended
to serve the worthy poor; only those with a letter of recommendation from
a citizen attesting the potential patient's moral worth were admitted to the
Massachusetts General Hospital in antebellum Boston. Yet the soldiers in
the Union Army were by definition able-bodied, at least on enrollment, and
they often ended up in the hospital even though their families might not be
impoverished. The Sanitary Commission ignored local demarcations, even
distribution by state, and called instead for a national response to help a
national army. The army situation turned usual assumptions about charity
topsy-turvy.[7]

Leaders of the USSC were explicitly anxious on this score. They told
inspectors to be careful not to weaken the men by raising expectations of
comfort. For example, when discussing cooking stoves, the instructions to
inspectors admonished, "But do not encourage the opinion that any of [these
styles of cooking stoves], or anything not provided for by the Government, is
necessary to the comfort or efficiency of the soldier. Foster a spirit of simplicity, frugality, and hardihood in this as in all things."[8] In a circular letter to
women's aid societies, Bellows advised against sending care packets to the

individual soldier and urged the group to instead forward them to the USSC depots, where their contents could be distributed to the neediest. "Let the homes of the land abandon the preparation of comforts and packages for *individual* soldiers," he ordered. "They only load down and embarrass him. If they contain eatables, they commonly spoil; if they do not spoil, they enervate the soldier." On the contrary, the sick soldier needed such attentions and would profit from them without danger. He was a worthy recipient of such charitable packages.[9]

The USSC struggled to fight rumors that the goods donated by the women's aid groups were going not to the noble sick but to the greedy well, especially to doctors, nurses, and stewards in the hospitals and to officers in camp. Mary Livermore described notes tucked into contributions threatening vengeance on those who misdirected the supplies. "These cookies are expressly for the sick soldiers, and if anybody else eats them, *I hope they will choke him*," wrote one woman. Another said, "He that would steal from a sick or wounded man would rob hen-roosts, or filch pennies from the eyes of a corpse!" A third ordered, "Surgeons and nurses, hands off! These things are not for you, but for your patients—our sick and wounded boys!"[10] The Sanitary Commission denied that such goods were misappropriated, although it admitted that a surgeon here and there might borrow a hospital shirt while his was being washed, or eat hospital stores when nothing else was at hand. They even sent detectives to seek out malfeasance.[11] With the donation of so much labor and goodwill went a determination on the part of women that only the worthy should receive the gifts.

The man who had lost a limb from battle, had been blinded by an explosion, or was weakened beyond recovery by disease was an ambiguous figure for a philanthropic worldview that saw most indigence as a sign of moral failing and personal sloth. At the same time, the disabled soldier was a hero, a dissonant figure who challenged this antebellum stereotype. The leaders of the USSC took up the question in August 1862, already seeing that the war would create a generation of disabled men who deserved help from the people who had sent them off to war. Henry Bellows pointed out to a colleague charged with gathering information on European military pension schemes that within a year there would be "not less than a hundred thousand men, of impaired vigor, maimed, or broken in body and spirit, [who would] be thrown on the country." When it became evident that these men were going to be joined at war's end by "another hundred thousand men, demoralized for civil life by military habits," Bellows added, "It is easy to see what a trial to the order, industry, and security of society, and what a burden to its already strained resources, there is in store for us." Bellows warned that that the country must be careful to oppose sentimentality and not "favor a policy which will undermine self-respect, self-support, and the true American pride of personal independence." The Sanitary Commission sought to discover the disability plan that would best promote self-reliance and "the healthy absorption of the invalid class into the homes, and into the ordinary industry of the country."[12]

The USSC was specifically opposed to the creation of homes for disabled veterans, for this ran counter to the goals of supporting reliance and mainstreaming in the community. Its leaders saw the benefit of temporary shelters for the convalescing soldier but believed that such institutions might breed dependency and wanted stays to be brief and focused on rehabilitation. They promoted a national pension scheme offering soldiers direct payments from the national government that matched the degree of disability generated by the war. The first pension law was enacted in 1863, allowing disabled men who had demonstrated their deficits to a physician to be paid a monthly stipend reflecting the degree of their disability. While throughout the North the poor were mostly allowed relief only in almshouses (to discourage dependence on the government), the disabled soldier's worthiness earned him a government payment and the comfort of his own home. The war created a new category of needy persons who had earned their deserving status from service in the war. The Revolutionary War soldiers had likewise been awarded pensions, but the numbers of men who served in the Civil War created a whole new level of national government relief assistance.[13]

By mixing the efforts of volunteers and paid agents, the Sanitary Commission at times aroused animosity from those who felt that philanthropic work should not be the means of a regular paycheck for some when others gave of their time freely. Historians of women's role in the Sanitary Commission have been particularly keen to point out that the work of tens of thousands of women who sewed and canned and gathered farm goods for the USSC was unpaid, while (male) doctors were hired as inspectors and agents and other men worked for pay as clerks or did the heavy work at depots. By 1864 around 450 people were employed by the USSC, and the agency's monthly payroll came to $28,000.[14] Some of these employees were female: Mary Livermore described women hired as seamstresses to work on an upper floor of the Chicago USSC office, where three dozen sewing machines were used to make bedding and clothing.[15] But by and large women were expected to donate their domestic labor, while men received paychecks, a state of affairs that irked not a few female volunteers.[16]

Official writings of the USSC defended the use of paid agents as the only way to run such a complex national organization. "Without a corps of agents who understand their work, give their whole time to it, and are bound to perform definite service during a definite period, loss, waste, and misapplication of supplies are inevitable," one report argued in 1863. "This branch of the Commission's work may fairly be compared with that of our largest railroads and express companies, and is at least as worthy of being well and economically done. But how long would any railroad corporation keep out of the hands of a Receiver, if it confided its freight business to volunteers?"[17] Managing the USSC's $7 million in cash donations and an estimated $25 million in donated goods required full-time, skilled employees. The first large-scale philanthropy, the USSC set the pattern for philanthropic organizations of subsequent years, ones that would be managed like businesses, on a multistate scale. This transition to the modern style of organization created discomfort over the fact that what had begun as a charity was now an entity that looked perhaps too much like a profit-making business.[18]

Fighting States' Rights

The USSC fought a war against states' rights within the broader Civil War struggle and generated not a little animosity in the process. Not surprisingly, given its base in Washington and New York City, its hold on the eastern tier of states was stronger than on those west of the Appalachians. The difficulties of the USSC reflected the overall structure of the Union Army itself and its tension between federal and state control. Except for black regiments established in 1863 or later, the volunteer regiments were all organized by state. Each state appointed a surgeon general, and initially at least, those appointees responded to battles by organizing relief efforts targeting their states' own men. So during the spring campaign of 1862, for example, Wisconsin boats went for Wisconsin's wounded, and Illinois sent help to its own regiments.

The USSC deplored this selectivity, for it emphasized states' rights over the Union, was an inefficient form of relief, and was cruel to those deserving help but shunned for belonging to the wrong regiment. USSC agent A. N. Read told the story more than once of a wounded Wisconsin soldier who embodied the proper spirit of nationalism. When a Wisconsin group had come through a Kentucky hospital giving "good things" to Wisconsin soldiers only, this man reported, "It made me feel so bad" when they "passed by that Illinois boy on the next bed there, who needed it just as much as I did; but I made it square for I divided what I got with him." The same story was told by an Ohio commander in a letter to Read several months later, who drew the conclusion, "Brave noble fellow—his was the true spirit of a soldier of the United States. We have a common country, language, religion, interests, & destiny & we should more closely weave the net of our national unity, so that the Genius of Liberty may, like Him, who went about doing good, wear a seamless garment." Even though this tale in the retelling took on a mythic quality, it made the useful point (for the USSC) that only a truly national agency was appropriate for a national army.[19]

In 1862 and 1863 the USSC continued to compete with the local agencies that offered to funnel the bounty of each state or locality directly to local men. There were state groups, such as the Indiana State Sanitary Commission, founded in March 1862 as the battles of Donelson and Shiloh brought midwestern troops into the thick of war, which cared only for Indiana regiments. Even some city-based units of the USSC threatened to split off and go their own way, as did the offices in Cincinnati and Chicago in 1862.[20] Such organizations drew upon the long tradition of local charity and gave at least the illusion of greater control to donors who distrusted the national body. The USSC in its 1863 summary report railed against this tendency to splinter relief. "The Commission studiously ignores sections and State lines, and knows soldiers from Missouri or from Massachusetts only as in the National Service," the document proclaimed. "The Maryland or Illinois volunteer who has been rescued from misery and the prospect of death, by clothing, food, stimulants and chloroform, that came to him on the field or in some ill-provided hospital, through the Commission, from some remote corner of New England or Pennsylvania, is likely for the rest of his days to think of himself less as a Marylander or as a Western man, and more as a citizen of the United States; and though he will not value his State less, he will love his country more." Even rebel prisoners would acquire a nascent Unionism from such aid. In contrast, "That of local or State agencies tends to foster, in contributor, agent and beneficiary alike, the very spirit of sectionalism and '*State-ish-ness*' to which we owe all our troubles."[21]

The strongest challenge to the USSC's hegemony over battlefield relief came from the Western Sanitary Commission (WSC), established in St. Louis in 1861 by William Greenleaf Eliot under a charter issued by Major General John C. Fremont, then commander of the West. Historian Robert Bender found enough material in this conflict to fill a doctoral dissertation.[22] The WSC had the gall to advertise for funds in the eastern newspapers, and its agents traveled through New England soliciting support, much to the USSC's ire.[23] Such economic competition was no small matter, as the USSC continually scraped for funds to continue its operations. Bender sees a larger philosophical distinction between the organizations, however. What the USSC considered nationalistic, efficient philanthropy, others saw as elitist and overly centralized. Local citizens could better direct the stream of money and goods than some far-off body, westerners believed. Bender summarized, "Whereas the United States Sanitary Commission seems more influenced by the missionary idea of directing the activities of supposedly benighted regions, the Western Sanitary Commission emphasized the concept of self-reliance and Christian duty to the community."[24]

Nevertheless, the Western Sanitary Commission acted much like its national counterpart in providing relief to sick and wounded soldiers. Its main enterprise was the construction and supply of hospitals in St. Louis, which quickly grew to become one of the principal hospital cities of the West. Safely under Union control from late 1861, the city was a natural destination for wounded brought in by boat from the western theaters of war, as well as a city of refuge for civilians fleeing conflict. The WSC outfitted hospital ships and railroad cars to transport the wounded from Donelson, Shiloh, and later battles. It sent hospital supplies to Union troops in Arkansas and Missouri, including bedding, clothing, and delicacies. The WSC established a "soldier's home" in St. Louis for sick Union soldiers en route from the army to home, and it set up others in Vicksburg, Little Rock, and Memphis as the war went on. While its reach did not extend much east of the Mississippi, the WSC remedied the same medical deficiencies that the USSC dealt with elsewhere.[25]

The leaders of the USSC reacted with immediate hostility when they learned about the creation of the WSC. They appealed first to President Lincoln and then Secretary of War Cameron to cancel the WSC's government commission, using arguments about efficiency and the need to serve the national army, not regional segments of it. But the WSC lobbied successfully in its own behalf. Eliot met with USSC's leaders in fall 1861 in an attempt at reconciliation, but this encounter only resulted in greater animosity. The

USSC's approach was one of uncompromising self-righteousness, and the WSC saw no reason to yield local autonomy, so the rivalry between the two large philanthropic agencies persisted.[26]

In 1864 women leaders of multiple local sanitary agencies staged sanitary fairs to fill the coffers and allow the work to continue. Chicago led the way with a huge fair in fall 1863; the next year Philadelphia, New York, and Boston followed suit. In St. Louis the WSC hosted the Mississippi Valley Sanitary Fair. By the war's end, at least thirty such fairs had been held in northern cities. These events were significant entertainment venues, with music, display of industrial products (the newest harvesters! the latest in sewing machines!), and the sale of food, artwork, and sewn goods created by volunteers. People donated animals, both for farm use and as pets. The fairs took in enormous revenues, reaching nearly $100,000 in Chicago, $150,000 in Philadelphia, and more than $1 million in New York City. All together, the fairs raised $4.4 million, with a little over half of that going to the national offices of the USSC and the WSC garnering the money from the St. Louis fairs. The various USSC branches put their profits to work in purchasing the food, bedding, and clothing that local branches sent to the USSC depots.[27] While appreciating the money raised by the fairs, the national office was frustrated at how much of the donated money remained out of their hands. The USSC never successfully countered the impulse toward local control and never fully won the confidence and endorsement of the nation's people for the national enterprise it attempted to fashion.

Taking on the Old Man

Not surprisingly, the USSC created more enemies in its crusade to reform the army's Medical Department. Its very existence was an affront to the department. If the army had been able to adequately police the health of camps and provide succor to the wounded, there would have been no reason for the USSC to exist. But the founders of the USSC saw glaring deficiencies in the Medical Department and vowed that the great loss of life seen in the Crimean war, when the British army displayed similar deficits, should not happen to the brave volunteers for the Union cause. To improve medical care for the soldiers, the USSC first had to harp on how bad the current situation was. The army's medical hierarchy felt they were competent to handle the situation and did not take kindly to this assault. In fall 1861 and spring 1862, the USSC launched the campaign to place one of their own in the position of surgeon general, as described in chapter 4. The agency began by meeting with various influential people, including the president, in September 1861, with the plan of nudging Surgeon General Clement Finley to resign. These initial, quiet efforts failed.

So the USSC took to the newspapers. In early November 1861, Cornelius Agnew, a member of the USSC board, contacted his friend Manton Marble, editor of the *New York World,* and enlisted his help in ousting the surgeon general. Marble published anonymous articles praising the Sanitary Commission and condemning the Medical Department.[28] The first, which appeared November 8, lauded the general works of the Sanitary Commission, claiming it "combine[d] the ripest results of sanitary science with the broadest common sense" and was bent on applying the lessons of military sanitation learned in the Crimean, Indian, and Italian campaigns. The article twice emphasized the monetary value of the soldier and declared that a life lost to disease was just as heroic, and just as ended, as a life lost in battle. The article acknowledged the heroism and goodwill of the army's volunteer officers, while gently pointing out that they had little training for their jobs and almost none in the science of sanitary prevention.[29]

A week later this encomium was followed by a blistering attack on the surgeon general. The second anonymous article accused him of being haughty, distant, and protected by the tangled vines of red tape, leaving the newly militarized volunteer surgeons without guidance as to their duties. He had also been slow to ensure that the surgeons had the proper qualifications. And he was equally derelict in regard to the construction of hospitals. "Science says, build small and isolated hospitals of wood for temporary purposes, and allow at least a thousand cubic feet of fresh air for each bed." Alas, however: "Science and experience [had] been ignored," and old hotels, seminaries, and warehouses had been retrofitted as hospitals. The surgeon general had pled economy as justification of these moves, but if the loss of soldiers' lives was properly valued, the money saved would be seen instead as the foolish expense of the wastrel. "The loss of life in the general hospitals of our army due to neglect of sanitary precautions would more than pay for the erection of barrack hospitals upon approved models. But how humiliating to deal with the subject thus." The article concluded, "If we had at the head of the Medical Bureau the man that the emergency demands all these evils would vanish. Mr. Secretary, do your duty in placing such a man there."[30]

This was too much for Henry J. Raymond, editor of the *New York Times*. He had published articles generally laudatory of the Sanitary Commission's work before, but now he felt the *New York World* had gone too far.[31] He blasted the USSC "gentlemen who have the direction of its affairs" for "neglecting their proper sphere of duty and meddling mischievously with matters which are already in competent and experienced hands." The USSC's assigned task was the sanitary condition of the camps, and its leaders should see to it and leave the hospitals in the competent hands of the Medical Department. Raymond averred that the charges against the surgeon general were without foundation and that the surgeon general had far more experience in these matters than the men of the Sanitary Commission. In fact, the problems in army medical care were happening not in the hospitals but in the camps, and the Sanitary Commission would do well to increase its vigilance in that regard. The commission should no more take on the surgeon general with justice than a coterie of officers should seek to overthrow their commanding general. Raymond admonished that the Sanitary Commission would "make a fatal mistake if it [aimed] to become the rival instead of the ally of the Medical Bureau."[32]

The *New York World* fired back the next day, claiming, "The case is as bad as we represented it to be, and the half has not been told." It reiterated the charges of the previous column and provided detailed examples of the Medical Department's inactivity. The surgeon general had failed to order that all troops be vaccinated, in spite of the outbreak of smallpox in Washington hospitals. The Sanitary Commission, however, had procured five thousand doses of vaccine and distributed them for application. The army had only five hundred bedticks in stock to furnish hospital beds in the Washington area, even though a single battle might leave thousands of wounded men in need of hospital care. This was the sort of inactivity and poor planning the current Medical Department exhibited. The previous surgeon general had at least worked with the Sanitary Commission to try to remedy deficits in medical supply, but the current occupant of that position not only was incompetent but took on a "dog-in-the-manger's prevention of action by others."[33]

On December 4 Raymond escalated the hostilities. He regretted "the ill-advised and eccentric conduct" of the Sanitary Commission, which had attempted "to misuse power, obstruct discipline, [and] seek patronage, by stabbing the reputation of so able, energetic, experienced and well-qualified an officer as our Surgeon-General." He called upon the USSC leaders to give an account of themselves, to say exactly what they had been doing with all the money and goods the public had lavished upon them, and to stop interfering with the medical officers of the army.[34] In that same issue, a letter appeared that assailed the Sanitary Commission with even more vigor. Signed TRUTH, the letter charged that the *New York World* articles displayed "malice and presumption and narrow judgment of affairs here," that is, in the hospital where TRUTH served as a nurse. The hospitals around Washington were "commodious and comfortable buildings" that were "the admiration of visitors to them for their comfort and cleanliness." TRUTH snarled at the "one clergyman and two or three citizen physicians who *manage*[d] the affairs of the Sanitary Commission" for daring to challenge "a distinguished medical officer who [had] been in more battles than one" and knew his business better than they could. The Sanitary Commission was ignorant of the hospital fund system, TRUTH claimed, and unfamiliar with the boards that already examined physicians for the regular army.[35]

TRUTH was equally snide in describing the work of the commission. Instead of sending sheets and pillowcases to the hospitals where they were needed, the USSC sent the bedding to camps, where it could not be washed and was not necessary. "They have a very slipshod way of doing their little bit of work of distributing jellies, &c.," TRUTH further charged, "though the secretaries, operatives, agents, and all that are engaged, get rather good salaries for the times, and all out of the pockets of the 'generous public.'" TRUTH thought it a rather good idea that a "Board of Inquiry" be appointed "as to whether the labors of the Commission really pay for supporting it." TRUTH's letter took the controversy beyond discussion of the USSC's campaign against the surgeon general to accusations of poor management and perhaps even dishonesty on the part of the USSC itself. Two days later TRUTH published a second letter, which decried the construction of "model" hospitals near Washington and defended various actions of the surgeon general.[36]

Frederick Law Olmsted responded to this barrage by a brief communication to the *Times* on December 7. He suggested that TRUTH visit the Sanitary Commission headquarters to view the receipts, which proved that USSC had distributed thousands of dollars of supplies to the hospitals. He also pointed out that the USSC was diligently at work in the field, as Raymond had urged them to be.[37] After the *Times* published yet another editorial castigating the USSC for failing to supply enough smallpox vaccine and showing no results from their many camp inspections, Olmsted met with Raymond to try to quell his rancor.[38] Raymond responded that the surgeon general was his friend, and if the USSC had a beef against him they should proceed against him legitimately. Olmsted pointed out that this was exactly the USSC plan. He further revealed that the Sanitary Commission was about to release a report on the scope of its activities that would answer the false charges of TRUTH in detail.[39] That report appeared in early January and was excerpted in the *Times* on January 9, 1863. A short article accompanying it admitted, "The report leaves no doubt of the zeal and industry, if not the uniform discretion of the Commission."[40]

Only one further attack against the Sanitary Commission's attempt to overthrow the surgeon general appears in the *Times*. On February 2 an article sarcastically compared the inexperience of Henry Bellows to the acumen of the Medical Department. But when the law reorganizing the Medical Department succeeded and Lincoln appointed William A. Hammond surgeon general in April, only a brief paragraph with no editorial comment appeared to note the transition.[41] Who had tamed the crusading editor? His irate wife, that's who.

By January Olmsted had learned that TRUTH was Elizabeth Powell, a nurse working in one of the Washington hospitals. He told Bellows, "Miss Powell is doing much mischief propagating falsehood and slander very industriously. It is very desirable to squelch her."[42] What was Powell's relation to Raymond, who was a married man? Although in retrospect Powell called it a "silly platonic relationship," she also admitted that he was "passionately in love" with her and had written her some 150 letters over the five years of their romance. In any event, he cared enough about her to publish the TRUTH letters and attack the Sanitary Commission in her honor. Apparently Finley had visited her hospital and made a good impression, displaying appropriate solicitude for the sick soldiers and particularly taking bold action in one case and saving the man's life. But some Sanitary Commission functionary had not accorded her equal courtesy. She was also certain, even forty years later, that Frederick Law Olmsted of the USSC sought only to *"get his hands on the United S. Treasury"* and that this financial motive was the main reason for ousting Finley.[43] Powell's conviction that Raymond was her gallant knight and that together they fought injustice and greed persisted until her death in 1902.

The Sanitary Commission was not helpless in this matter. Powell claimed that Olmsted "had it committed to the ears of Mrs. Raymond thro' some lady officers of clubs organized to send supplies to the San. Com. That there was a girl in Washington who had 'an undue influence' on the Editor of the Times and was using it to hurt the San. Com." Although Raymond had pledged his newspaper's support in the campaign against the USSC, he reneged when his wife discovered the affair. He appealed to Powell "to release him from his promise for the sake of 'escaping hell' within his own doors."[44] Later that spring, George Templeton Strong heard gossip about quite a row at the Raymond home. Powell attended a party there, and Mrs. Raymond confronted her. "I hear that Miss P. is found out, and that Raymond has had a grand scene with that lady which culminated in intervention by the Police, who coerced the lady to quit Mr. Raymond's premises," Strong chortled.[45] One wonders if the *Times*'s avoidance of USSC stories thereafter was a term of Raymond's peace agreement with his wife. He even sent a check to the USSC accompanied by a letter acknowledging the value of their work.[46] However important the USSC seemed to him in December 1861, he made no mention of it in his history of the Lincoln administration, published in 1864.[47] Powell married in 1867, and Raymond died of a heart attack or stroke in 1869, after an evening spent in the company of an actress.[48]

It was not the last time that the USSC would face the charges raised in this contretemps, however. It was bold, if not downright arrogant, for a voluntary organization to take on the army's Medical Department—and win. Suspicions that its agents were too well paid, that the delicacies and constructed items were misdirected, and that its efforts were governed by only amateur pretensions, continued to dog the commission. And while it had triumphed over the old surgeon general, it was less successful in later encounters with the army's medical administration.

The Surgeon General

Taking over as surgeon general in April 1862, William A. Hammond had an enormous job and limited tools for it. He also lost no time in making enemies within the Lincoln administration. Hammond's personality clashed with Secretary of War Edwin Stanton's early in his tenure, and Stanton made his animosity toward Sanitary Commission meddling obvious in an interview with Bellows in May 1862.[49] To complete the triangle, the Sanitary Commission leaders at times became annoyed with Hammond, finding him too weak in his fight for reform and too slow to resolve persistent issues within the Medical Department.

Hammond oversaw the vast expansion of the federal hospital system and, with it, the appropriate supply of general hospitals. Having been a hospital inspector before taking over the surgeon general's office, Hammond was aware of hospital needs and key aspects of hospital design. He did what he could with the aged buildings serving as hospitals in major northern cities and had new pavilion-style hospitals built according to the more modern plan that emphasized ventilation and light. Washington's hospitals grew from two thousand to twenty thousand beds by the end of September 1862; at the peak in December 1864, the federal government was running 190 hospitals with 120,561 beds.[50]

Hammond's very success put the Sanitary Commission in an odd position. It had spent the first month of its existence criticizing deficiencies in the Medical Department and often filling in those deficits with their work and supplies. Now the USSC man was in power, and still the medical needs of soldiers often went unmet, as was evident after the battles of Antietam and Perryville. Who could the agency blame now? Well, just about anyone except the surgeon general and the physicians of the medical corps. After Perryville, J. S. Newberry asked in his report, "Who then, is responsible for the facts, that at the battles of Fort Donelson, Shiloh, and Perryville, no adequate provision was made before hand for the care of the wounded; that proper supplies of medicines and hospital stores and an abundance of appropriate food were not on hand or within easy reach?"[51] Newberry blamed not the Medical Department but the failure of the system that entrusted too much power to the quartermaster and commissary departments and not enough to the surgeons. Secretary of War Stanton, now a widely acknowledged enemy of the USSC, likewise became a target of sanitary venom. In fall 1862 the administration came under increasing fire for the management of the war, so the USSC backed off, feeling that support for Lincoln was more important than taking on Stanton, even if it had a chance of unseating the secretary of war, which was unlikely.[52]

The dissonance was resolved in 1863, albeit not in a way that pleased the Sanitary Commission. Even though by all historical assessments Hammond was a competent and honest surgeon general, by the end of 1863 he had been forced from office, and in 1864 he was successfully impeached on a trumped-up charge of chicanery. Hammond had made too many enemies. Stanton, who apparently was at least initially neutral in regard to Hammond in April 1862, had so closely linked him to the pernicious USSC by 1863 that by then he recalled only opposition to the doctor. Stanton appointed a committee in July to find problems in the surgeon general's execution of his duties and bring charges of negligence. Hammond was all the more vulnerable because he had angered army physicians with his infamous circular number 6, which removed calomel from the army's standard supply tables. Insulting his medical staff was not the best strategy for winning loyalty.[53]

In August Stanton sent Hammond on a trip to New Orleans and the western theater, presumably to inspect conditions there, and appointed Joseph K. Barnes, a longtime army surgeon and his own personal physician, as acting surgeon general. In January Stanton had Hammond arrested, and the court-martial that followed played out Stanton's antipathy and dragged Hammond's name through the slime of revenge. Hammond was ultimately convicted of improperly acquiring bedding for the army, of exceeding his authority in these purchases, and of being rude to an officer under his command. Stripped of his rank, discharged from the army, and barred from further government service, Hammond left the capital in disgrace and went home to open a specialty practice in New York. The proceeding was so fraught with petty malice and hairsplitting that Congress revoked the action in 1878, and President Rutherford B. Hayes signed the pardon.[54]

Joseph K. Barnes, who formally succeeded Hammond, served as surgeon general until 1882. He was Stanton's darling and continued to oppose the USSC while implementing many of their recommendations for improving the Medical Department.

In 1865 the feud between the Sanitary Commission and Barnes reached a new level of pettiness in the competition over how the medical history of the war was to be written. Barnes reported in 1864 that the collection of data for a medical and surgical history of the war was ongoing; ultimately the Medical Department would publish twelve volumes of information that included morbidity and mortality statistics, essays on individual diseases, and reams of case reports on surgical cases.[55]

In spring 1865, the Sanitary Commission began collecting information for its medical history of the war. It sought information from army surgeons via circular, with the plan of preserving the new and extensive experiences these men would have had with disease, wounds, and therapeutics during the war. The USSC also enlisted its many agents and inspectors in the search for data. "The leading design of your investigations will be to ascertain by observation, research, & faithful inquiry, what are & have been the most effectual means of preventing disease & needless mortality, & of mitigating the suffering & perils from sickness & wounds," they were told.[56] The commission ultimately published three volumes based on this research, one on diseases, one on surgical issues, and a final volume containing anthropometric work done by B. A. Gould.[57]

Barnes did all he could to block access by the Sanitary Commission to the surgeons in the Union Army. In June 1865 he sent a telegram to all army surgeons warning them against the USSC circular and forbidding them to answer it. Elisha Harris wrote Cornelius Agnew that the sympathies of the surgeons were with the USSC and against Barnes in this matter. "I mentioned to you the ukase that we heard promulgated from the Med. Bureau by telegraph," he wrote cheerfully, using a term applied to the Russian czar's decrees. He said they would proceed anyway, as if Barnes "were [their] best friend & the Bureau a promoter of science & humane improvement."[58] The resulting history did not carp on the deficiencies of the Medical Department but instead focused on scientific investigations and conclusions.

The "war of the histories" fizzled out over the summer. When Elisha Harris investigated the outbreak of scurvy among black troops in Texas during summer 1865, he could not help concluding, "The *crime* of causing all this suffering rests solely upon the head of the Hon. Secy. of War and on his alter ego in the responsibilities of the Medical Bureau." He urged the USSC not only to send vegetables to ward off scurvy but also to record how and why it was done. He wanted to get one more blow in against the surgeon general, although the resulting publications about the Texas disaster were much more muted in their assignment of blame.[59] As for the vast *Medical and Surgical History of the War of the Rebellion*—if it mentions the Sanitary Commission at all, the indexer did not find it. Many Union surgeons contributed to the USSC volumes, although most were "formerly" in service by the time the books appeared, and hence beyond the dominion of the surgeon general.

Conclusion

If contemporaries were ambivalent about the work of the USSC, historians have been equally divided. Walter Trattner, whose history of social welfare in America is in its sixth edition, likened it to the Red Cross and accorded the commission "universal acclaim."[60] Robert Bremner, well-known historian of American philanthropy, credited the USSC with saving many lives and generally provided a positive description of its efforts.[61] Allan Nevins glorified the USSC as the "most powerful organization for lessening the horrors and reducing the losses of war which mankind has so far produced."[62] The one modern book-length treatment of the USSC, published by William Quentin Maxwell in 1956, is generally sympathetic but dwells so heavily on the political bickering and infighting of the commission that the positive assessment is blunted.[63] The two historians who have focused on women in the USSC, Jeanie Attie and Judith Ann Giesberg, judge the USSC largely on the basis of its impact on women. Since women's aid societies formed the vast base that created $25 million worth of food, clothing, bedding, and supplies, the experience of women as part of the USSC is a relevant question. But whether the USSC experience was positive or negative for those women is a side issue to any judgment of the commission's role in the war.[64]

The most influential, and harshest, critic of the USSC among modern historians has been George Fredrickson. In his book *The Inner Civil War,* Frederickson sets out to counter the view of the USSC as "a kindly beacon, standing out against the dark background of war." First of all, the executive committee bore the taint of both elitism and conservatism, fatal flaws from the perspective of a liberal historian writing in 1965. They were callous enough to put a monetary value on the loss of a soldier, and Bellows was so insensitive as to carry a femur scavenged from a battlefield around with him when lecturing. (Fredrickson misses the context of the money value of a soldier; the point was that it was cheaper to donate to the USSC and prevent the death than to keep throwing money away training and equipping soldiers only to have them die of neglect.) The USSC emphasized discipline and education over sentiment and spontaneous benevolence. But I'll agree, the femur as a stage prop is peculiar. Fredrickson exposed the arrogance of USSC leaders and described them as aspiring at times to take over the government, removing it from the incompetent country bumpkin who occupied the White House. Yet even Fredrickson admitted, "With all their ulterior motives, however, the commissioners undeniably performed a valuable work." And like others, he concluded, "The highly organized and 'scientific' benevolence of the commission set an important precedent for the operation of postwar philanthropy."[65]

There is an obvious disconnect between the goal of the executive committee of the USSC to impose order and the actual and numerous chaotic microcosms that made up the Civil War. As J. Matthew Gallman has pointed out in regard to Philadelphia, the USSC was never able to impose the order and lack of sentimentality on the local branches that Fredrickson's rhetoric described; this lack of control was most evident in that city's sanitary fair, which was ripe with female sentimentality and idiosyncratic diversity.[66] USSC agents in Maryland were condemned for rudeness, drunkenness, and visiting prostitutes; in his answering letter, one agent admitted to visiting the houses of well-known Washington madams but denied partaking of the goods on offer.[67] J. S. Newberry in Cincinnati was always late with his reports (his summary document on the USSC in the Midwest did not appear in print until 1871). USSC men were even captured by the enemy at Gettysburg. Their ideal of order and discipline was worth preaching, however, and while it is easy to see the men of the USSC as prissy elitists, it is worth remembering the values of cleanliness, proper food, and attentive care that their message brought to the soldiers' hospitals.

Throughout its many printed documents and published volumes, the USSC defended its actions against a fairly constant barrage of criticism. Mary Livermore at the Chicago USSC office said a typical week almost always included a visit from a women's aid society delegation sent to inspect the USSC and probe for impropriety.[68] But even with all the suspicion, the Union states voted their confidence in the USSC with a persistent stream of money and material support. The commission may have lost the battle with the surgeon general's office, but it was popular with surgeons in the field, who knew how crucial USSC supplies could be. In the face of more than a half million Union deaths during the war, the USSC offered the civilian North a way to fight back, to help, to support the war effort. This organization of the northern impulse of benevolence was key to maintaining morale, and it prevailed in spite of the many doubters who questioned whether disinterested benevolence, pursued on such a grand scale, could really be possible.

The USSC had multiple facets, and the presence of one tendency did not rule out another. Historian Robert Wiebe's classic phrase "the search for order" as a characteristic of American elites faced with a bristling society of industrialization, urbanization, and immigration in the nineteenth century certainly defines one impulse guiding the leaders of the USSC.[69] Just as historian Lorien Foote highlights the struggle of Union officers to bring order to rough enlisted men,[70] the USSC sought to bring order at multiple levels—from training those rough men to pee in the sinks, to training their officers to create a hygienic camp, to training the Union's physicians in the latest aspects of medical therapeutics, to bringing order to medical transport, to reorganizing the Medical Department at the highest levels. In all these efforts, the leadership countered disorder and was likely to come off as bossy, elitist, and prissy. In these elitist acts the USSC crossed into gendered realms as well. If "mother" were here, she would feed the sick and comfort the wounded—but she would also likely enforce cleanliness and proper behavior. In retrospect, it seems that the USSC generally did know what was best and was a force that improved the lives of northern soldiers.

The general hospital has often been mentioned in this and previous chapters. In the best-run and best-equipped general hospitals, wounded and sick soldiers found the closest approximation to the home sickroom possible in wartime. The USSC did much to create that environment, with its endless lists of bedticks, nightshirts, and feeding cups. Here the delicacies gathered for the sick were spooned into feverish patients, often by women filling the role of maternal caregiver. My argument in chapters 4 and 5 has been that the USSC filled a gendered role of female caregiver, even when it might be men who carried the messages of nutrition and hygiene to camp and field hospitals. Now we will turn to the northern general hospital, where women had a more direct role, and where the best of mid-nineteenth-century medical care was provided.

CHAPTER SIX

The Union's General Hospital

In 1876 Philadelphia hosted the Centennial International Exhibition, ostensibly to celebrate the one-hundredth birthday of the Declaration of Independence but actually to showcase the industrial and agricultural prowess of the United States. Attracting some 10 million visitors (at a time when the U.S. population was 46 million), this first American "World's Fair" illustrated the marvels of the age.[1] Among the scientific exhibits was a room devoted to medicine, but as historian Bert Hansen has pointed out, it did not highlight the progress that latter-day historians might anticipate. There was no mention of ether or chloroform anesthesia, no discussion of the work of Louis Pasteur or Joseph Lister on the relationship of microorganisms to disease, no presentation of new techniques in surgery. Instead, the U.S. Army erected a model post hospital, with a wing dedicated to the hospital designs and medical-transport vehicles created during the American Civil War. The greatest advance in nineteenth-century medicine, as envisioned by the creators of the exhibit, was the clean, orderly, well-ventilated hospital ward.[2]

Although the 1876 exhibition portrayed a post hospital—the relatively small institution associated with an army base on the frontier—it drew upon the Union Army's medical experience with the general hospitals, care centers located far from the field of battle, safely behind the lines. These hospitals were set in large cities such as Washington, Philadelphia, Louisville, Annapolis, and St. Louis, cities in contact with battlefields by rail or water. The distinction between field, post, and general hospitals was not absolute, as we saw in chapter 1. At St. Louis, for example, there were post and general hospitals side by side. Officially, the *general* hospital of the North was unassigned to particular regiments, corps, or sections of the army, while the post, field, and brigade hospitals were reserved to certain military units. Early in the war, both sides commandeered buildings that had previously been schools, almshouses, armories, or industrial sites; but as the war went on, the army increasingly, especially in the North, built brand-new hospitals, most of them as variations on the pavilion scheme. These massive institutions housed thousands of patients and often included grounds that allowed for tents to hold overflow or contagious patients.

At their best, these hospitals were a remarkable achievement. The 1876 exhibition was rightly proud of them, for well-run hospitals approximated the home sickroom and offered wounded and ill men the best available healing environment. Although some drugs were effective, it was those aspects associated with the female attentions of home—cleanliness, nutrition, appropriate clothing and bedding—that made the difference between the effective hospital and the hospital that men sought desperately to escape. That the North was better able than the South to create such healing institutions is evident from a comparison between a typical northern institution, Satterlee Hospital, and its southern counterparts (discussed in chapters 7 and 8).

My use of Satterlee Hospital as a case study has revealed to me how much information can be found on these large general hospitals, while my broader reading has convinced me that Satterlee was representative of many of the northern general hospitals, especially after 1862. Although it is easy to find the dreadful in Civil War medical care, this chapter emphasizes the positive aspects of the northern general hospital. Thousands of men died of wounds and disease in the war, but most of them did not die in the general hospital. It is true, of course, that this fact reflects the winnowing process of the transfer from battlefield or regimental hospitals; furthermore, many men died after they were discharged from hospitals as permanently disabled. Nevertheless, most soldier correspondents writing from northern general hospitals did not say they hated the place, except when contrasting it with home, and their complacency reveals their general respect for the institution and its healing powers.

Satterlee Hospital

Upon assuming his office, Surgeon General William Hammond immediately set to work building pavilion hospitals. In Philadelphia the first one was on a rise of land in the western part of the city. Called in its inaugural year simply the West Philadelphia Hospital, it was renamed Satterlee in June 1863 in honor of Surgeon Richard S. Satterlee, medical purveyor at New York.[3] A second large pavilion hospital named Mower, after former army surgeon Thomas Mower, opened a few months later in Chestnut Hill, Pennsylvania.[4] Others followed in the District of Columbia, Indiana, and Maryland. The pavilion hospital had a variety of designs. Satterlee had a central stripe of grass lawn with a few small structures and then two long corridors on either side of it, with wards sticking off the corridors—like two combs lined up with their backs to a ribbon. Mower's center was a rectangle with rounded corners, and forty-nine single-story wards radiated off to all sides. Lincoln Hospital in Washington had a connecting hallway shaped like a boomerang, with pavilion spikes radiating out, making the whole look like the Statue of Liberty's crown. Hammond Hospital, at Point Lookout, Maryland, was a wheel with its spokes radiating outward. All shared in common the form of single wards only loosely connected to one another by some sort of covered walkway. These hospitals used tents on the grounds for overflow, since a hospital could swell suddenly in population after the infusion of wounded men from a battle, only to shrink again when these men recovered, died, or otherwise moved on. Hospitals also used tents or separate wards to segregate black troops; integrated wards were unthinkable even when black troops became common and abolition became a familiar war aim. Officers could be kept separate from enlisted men as well. When epidemics of hospital gangrene or other diseases recognized as contagious broke out, the hospitals could use these separate structures to isolate patients.

Satterlee serves particularly well as a representative of the northern general hospital for several reasons. First, it was located in the hub of the nineteenth-century U.S. medical universe, Philadelphia. The country's premier medical school, the University of Pennsylvania, stood a few blocks away, and leading academic physicians consulted on Satterlee's wards. Built from scratch, Satterlee epitomized the ideal pavilion hospital. There are two brief histories of the hospital, written by its chaplain during its wartime existence. The director of the hospital was a famous arctic explorer, whose life has been the subject of a detailed biography. And the hospital supported its own newspaper, which offers a window into day-to-day hospital life that is rare for such institutions. These various sources allow the historian to paint an unusually rich and detailed picture of life and culture in this Union general hospital.[5]

Satterlee stood on grounds that had recently been farmland but were later encompassed by Forty-Second, Forty-Fifth, Baltimore, and Pine streets. Work began on May 1, 1862; forty days later patients filled the first six wards, although the building was not dedicated until October 8. At that point the hospital boasted "28 wards 167 feet long by 24 wide, each of which will accom-

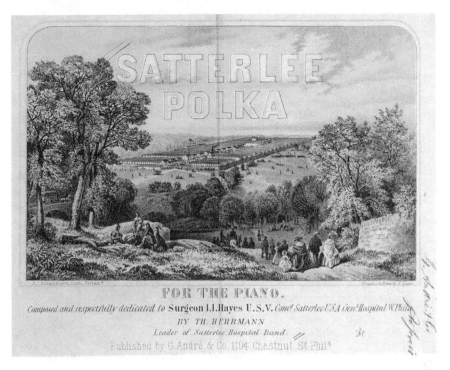

Satterlee Polka. Lithograph by P. S. Duval, drawn on stone by J. Queen. Created ca. March 1864. Bird's-eye view of army hospital and grounds, with patients and families. Note the long sheds of pavilion-style wards around the central buildings. The sprightly music of the polka, written by the bandmaster of the hospital, conveys hope, energy and happiness. Library of Congress, LC-DIG cph.3b04515. The polka has been recorded on a CD entitled *Drexel University Presents The Civil War as It Came to Philadelphia* (2011), a project coordinated by Kathryn Steen, Department of History and Politics, Drexel University, Philadelphia.

modate 70 patients."[6] A communicating corridor 775 feet long connected the clusters of fourteen wards each and served as a dining hall. There were separate buildings for kitchens, laundry, quarters for staff and guards, the post office, and other functions. One building contained three thousand cubicles for storing backpacks and other personal possessions. Covering fifteen acres, the grounds could also accommodate hundreds of tents. The hospital was later enlarged so that at its peak there was space for forty-five hundred beds. Amenities included a spacious chapel that was also used for meetings and performances, a library, a reading room with a piano, a smoking room, and even a billiard table. It was in effect a small town; the guard barracks alone housed 132 men.[7] In August 1864, when the population of the hospital was near its peak after Grant's summer campaigns, the number of letters and packages flowing through the post office reached twenty thousand.[8]

Hammond chose Dr. Isaac Hayes to direct the construction and operations of the new hospital. Hayes was a famous man, but not for his medical accomplishments. Born to a Pennsylvania Quaker farming family in 1832, he had a classical education at a Quaker boarding school, briefly served there as a mathematics teacher, and then attended medical school at the University of Pennsylvania. Upon graduation, he practiced medicine in Philadelphia for only a few weeks before joining Elisha Kane's expedition to explore the Arctic that departed on May 30, 1853. Kane sought the elusive Northwest Passage, an open sea route across the top of North America that would connect the Atlantic to the Pacific; his voyage also, like many others, searched for the lost explorer John Franklin and his team. After many adventures and hardships, Hayes and most of his companions returned to the United States in fall 1855. Hayes then took to the lecture circuit, not only describing the perilous events of the journey but offering scientific observations on flora, fauna, and the indigenous peoples found en route.

Despite losing several toes to frostbite, Hayes resolved to go again, to seek points further north and perhaps discover the elusive open Arctic sea. He was the leader this time, and he met many influential men in the process of raising money for this second trip. William Hammond saw Hayes as a fellow naturalist and invited him to lecture at the Academy of Natural Sciences in 1859 on scientific lessons gleaned from his adventures, including the importance of fresh meat in preventing scurvy.[9] Finally successful in his fundraising, Hayes led his own Arctic expedition in 1860 and 1861 and learned that the Civil War had broken out only on his return voyage, when an "old Dane" at a Greenland port told him that "Oh! De Sout' States dey go agin' the Nort' States, and dere's plenty fight." The Hayes expedition had also failed to find the Northwest Passage, and Hayes arrived home deeply in debt.[10]

He responded to his penury with several strategies. One was to hit the lecture circuit again, where Captain Thomas Low heard him in Washington. "I have just returned from listening to a lecture at the Smithsonian, delivered by Dr Hayes, of Arctic fame," he told his diary on February 19, 1862. "The lecture was replete with valuable information."[11] A second possibility, not immediately followed, would have been to write an account of his travels (as he had after the first expedition) and hope to profit from book sales. But there was no time for that, because the third was to lobby his various influential friends for a high-ranking medical appointment in the Union Army.

Lincoln's secretary of state William Seward had been one of the supporters of Kane's expedition (and later earned ridicule for the purchase of Alaska); Lincoln himself wrote Secretary of War Edwin Stanton urging that Hayes be given an appointment as a brigade surgeon. Senators and scientists chimed in about Hayes's qualifications. He was commissioned a brigade surgeon on April 4, 1862, and on May 5 Hammond ordered Hayes to Philadelphia to oversee the building of a brand-new hospital for soldiers in West Philadelphia. Although Hayes's biographer argues that "few candidates could have offered credentials as suitable for the assignment as Dr. Isaac Israel Hayes," it is hard to justify this conclusion. Hayes had practiced general office medicine for only a few weeks, and although his service in the Arctic made him an expert in scurvy and frostbite, perhaps, his overall medical experience was limited. He had organized a major expedition, however, and may have learned important lessons about morale and human management during those dark Arctic winters. And like Hammond, he was one of a small coterie of men in Boston, New York, Philadelphia, and Washington who thought of themselves as "scientists."[12]

Hayes had the hospital up and running within an impressive forty days. By 1863 construction costs had already exceeded two hundred thousand dollars, exclusive of furniture, an expense that reflected the quality design features that characterized its layout.[13] Each ward had an iron stove for heating and ingenious roof vents to facilitate ventilation. At one end was a water closet, with a twelve-foot trough one foot deep and nineteen inches wide to receive waste. It was flushed several times a day (although perhaps too infrequently to prevent its odors from permeating the ward), directly into a sewer. Hot water flowed to each ward, supplying bathtubs and basins. Hayes oversaw the construction of a kitchen that could turn out meals for thousands, which meant hiring cooks and buying vast soup kettles. He hired twelve pharmacists, who set up a dedicated pharmacy space and established a system for the order and delivery of medications to the wards on physicians' instructions. The hospital needed a laundry, furnished with hot water, and workers to keep the bed linen and clothing of thousands of patients clean. They scrubbed the linens and clothes in large cauldrons and washing machines and had wringers to assist in drying them.[14] There were thousands of iron bedsteads to be found, complete with mattresses, pillows, and bedside tables and chairs.

Hayes eventually had forty physicians under his command at the hospital, who were assisted by another forty medical "cadets," medical students who did grunt work in exchange for the opportunity to learn on the wards. The University of Pennsylvania Medical School had 410 students when Hayes entered it in 1851; even with the departure of southern students and the many men who went to war, there was still a plentiful supply of medical students eager to learn in Civil War Philadelphia. Jefferson Medical College likewise filled this pool of unpaid labor.[15] His connections to the Penn medical school and the medical elite of the city enabled Hayes to bring consultants into the hospital on contract who also continued their civilian practices. Alfred Stillé, John Shaw Billings, S. Weir Mitchell, D. Hayes Agnew, Joseph Leidy, and Jacob da Costa all walked the halls of Satterlee Hospital.[16] (And Leidy did the autopsies, as later reported in the *Medical and Surgical History of the War*.)[17] In total, Hayes oversaw 537 personnel by 1863, a staff that included doctors, medical students, cooks, nurses, chaplains, guards, mechanics, and pharmacists.[18] Satterlee Hospital and Mower Hospital, in nearby Chestnut Hill, were the largest such institutions created by the Union Army during the war.

Therapeutics

Hayes has not left us any general summary of his therapeutic philosophy, but the practices followed at the hospital can be harvested from the many Satterlee case reports in the *Medical and Surgical History of the War of the Rebellion*. Given his training in Philadelphia in the 1850s, it is unlikely that he wholly relied on the "healing power of nature," but in any case, since extant letters and documents rarely mention bleeding, it seems probable that the heavy dependence on venesection common a few decades earlier had subsided to almost nothing. The sort of chronically ill patients received at Satterlee were commonly suffering from, as one surgeon put it, "hypaemia, cachexia, and adynamia, conditions almost universally present in disease as observed at military hospitals." Such labels meant the men were anemic, wasted, and weak. It was clear that such cases required "a supporting and tonic treatment," which would have included nourishing food, alcohol, iron compounds, and tonics such as quinine.[19] Even for pneumonia, a disease felt to be highly inflammatory and one formerly treated with venesection, Satterlee doctors followed "expectant" therapy, while supporting the patient with morphia.[20]

One New York private, hospitalized at Satterlee for jaundice, kidney trouble, splenic enlargement, and a hacking cough said to be caused by remittent fever, received an impressive list of medications including iodide of potassium, bicarbonate of potash, fluid extract of taraxacum, tincture of iodine, a codeine cough mixture, tincture of cannabis indica, extract of uva ursi taken in mint water, and cod-liver oil. After two months of slow recovery, the soldier was well enough to go back to his regiment, where he subsequently lost a finger in a Tennessee battle.[21] It is likely that he received quinine and alcohol as well for his "low" condition, since they were commonly given to other patients.[22] Calomel was most likely little used. Hammond and Hayes were friends, sharing as they did the "scientific viewpoint," and Hayes likely followed his friend in condemning at least the overuse of the drug. The newspaper verses ridiculing calomel (see chapter 1) are another sign that the drug was held in low esteem by hospital physicians. A poem that laughed at homeopathy (copied from *Harper's Weekly*) similarly signals that such remedies found little favor within the hospital.[23]

The Nurses

Hayes made an interesting choice for his nursing staff. Himself a Quaker, he asked permission from Hammond to bring twenty-five Roman Catholic nuns to the hospital to serve as the core nursing staff. The brother of his college roommate was apparently the source of the idea, although by summer 1862 Catholic nuns had served as nurses in multiple Civil War locations. There was a tradition of nursing sisters extending back to the Revolutionary War, and by 1860 nuns had established twenty-eight hospitals in the United States. The nuns who ultimately served at Satterlee came from Emmitsburg, Maryland, where Elizabeth Seton had founded the first American order, the Sisters of Charity of St. Vincent de Paul, in 1809. These Sisters of Charity ran schools and orphanages, and they opened the first Catholic Hospital in St. Louis in 1828.[24] So it is not surprising that after the first battle of Bull Run (which occurred eighty miles from the mother house in Emmitsburg), the sisters volunteered their services to care for the injured. Confederates took wounded Union prisoners from the battle to Richmond, where they were housed under guard on the top floor of the newly built almshouse. The sisters volunteered to care for these unwelcome guests, and as a result these unfortunates received better care than the Confederate patients on the floors below. The sisters did not take sides in the conflict; they nursed Confederates as well as Union patients. Their fame spread among Civil War surgeons as nurses who could be relied upon for obedience, hard work, efficiency, and tender care of hospitalized soldiers. They were, as historian

Mary Maher has noted, the most educated and experienced body of nurses available in the United States as the war began. Other women might aspire to be Florence Nightingale, but these women knew how to fill her role with competence.[25]

During the war 617 Catholic sisters served as nurses for both sides; of these, 232 were Sisters of Charity (also called Daughters of Charity); of this group, 91 worked at Satterlee under the overall supervision of Sister Mary Gonzaga.[26] Hayes later characterized their work as "unceasing and valuable."[27] Each ward was supervised by a nursing sister with three assistant nurses, all of whom were male. Convalescing men worked as nurses and also did other tasks, such as delivering food to the bedridden and distributing meals to patients on special diets. There were also "contract men," who were hired and paid wages to serve in the hospital under the direction of the nursing sisters.[28] Laywomen volunteered in the hospital but put their efforts into the reading room, arranging concerts and other entertainments, decorating the hospital and grounds for special events like the Fourth of July celebration and Christmas, and creating feasts for Thanksgiving, Christmas, and other holidays. When these laywomen asked Hayes why they could not take on nursing, too, because there were "'ever so many' ladies who would be happy to do that service," he responded that he "considered the Sisters of Charity the only women in the world capable of nursing the sick properly."[29]

It is likely that this attitude caused considerable friction between local Philadelphia women volunteers and the nuns. At one time, "a committee of ladies from an association waited upon him [Hayes] with letters of recommendation, accompanied by a large donation of clothing, fruits, preserves, etc." This committee eagerly offered "their services as an organized body to attend to the affairs of the Hospital, and take turns in nursing the sick." Hayes declined, saying that the hospital "already had an excellent organized body working under his own eyes so well that he was unwilling to change it for another." Hayes then asked an orderly to deliver the welcome delicacies "to the Sisters in the Donation Room." Not surprisingly, "the ladies left, rather embarrassed by their cool reception."[30] Hayes did not apparently entertain the notion of having nuns in charge of some wards and laywomen in charge of others. Sister Gonzaga would still have been head nurse, and it is unlikely such an arrangement could have persisted long without difficulties. These reports come from the nuns' recollections of life in Satterlee, which may be biased, but there is no evidence to contradict their version of events.

There were some laywomen on the wards, however: the female relatives of the patients themselves. Family members commonly traveled to see their hospitalized sons, brothers, and husbands. The Sanitary Commission offered housing for those too poor to pay Philadelphia boarding rates and kept rosters of hospitalized patients in the city to help family members locate their soldiers. From one anecdote we know that family members occasionally were allowed to spend the night in the hospital, when the patient was near death. Each ward had a small nurse's office, which included a cot that could be offered for respite to such grieving kin. In this story the nun gloated that while the Protestant wife slept, she (the nun) converted the husband to Catholicism and baptized him before death. The sisters' memoirs are full of such conversion stories, stories that also reveal interesting details about ward culture.[31]

It is clear from these memoirs that the nuns' charge went beyond the provision of good nursing care. They were there to save souls and carefully recorded each patient brought to conversion and baptism. Particularly memorable was the man whose heart had turned to God but was undecided whether the Protestant or the Catholic faith was the best choice. The Protestant Reverend Nathaniel West visited him one day, and Father Peter McGrane the next. When the man chose to be baptized by Father McGrane, Reverend West was "very much nettled." The nuns distributed a variety of so-called sacramentals, holy objects to be carried into battle or kept near the heart to secure the blessings of sacred protection. These included holy medals (small medallions with images of the Virgin Mary or other saints, to be worn as necklaces or pins), scapulars (tiny laminated pictures of sacred images, the size of extra-large postage stamps, worn on strings under the shirt), and the Agnus Dei (a locket with a wax disc impressed with the image of Christ as the "lamb of God"; the wax supposedly came from candle stubs of St. Peter's in Rome and had been blessed by the Pope). Even Protestants sought these talismans, touted as powerful good-luck charms, and "promised to treat them with respect."[32]

While a goodly number of northern troops were Catholics and likely familiar with the sight of nuns, many Union and Confederate men were surprised by their appearance and put off by their religion. Only three decades before, a mob had burned a convent in Charleston, Massachusetts, upon hearing rumors that women were being held there against their will.[33] When five Daughters of Charity arrived in Marietta, Georgia, to nurse Confederate wounded in 1863, they were surrounded by a group of locals who cried "What or who are they?" The people were unsure: "Are they men or women?" And when one sister attempted to gain information, the people cried, "She spoke! She spoke!"[34] The nuns often wore peculiar headdresses, adding to their alien appearance. One Michigan private described them to his wife after arriving at Satterlee. "There are lots of Sisters of Charity here in the Hospitals. [T]hey seam to be nice women and very good to the sick. They all wear black dresses and white collars and white aprons and the dogondest looking bonnets you ever saw." In a later letter he admitted "They are very good to the sick but I don't approve of such nurces."[35]

Historian Mary Maher has argued that exposure to nursing sisters during the American Civil War helped ease the suspicion and disdain they had previously garnered from Protestants. This seems likely, and the impression of one New York man (a Methodist, if his burial site in 1871 is any indication) was probably common. Of the nursing sisters at a hospital in Washington, D.C., he wrote, "Some may say that they are superstitious, it may be, but I have seen [them] sit and bathe the fevered brow of many a poor fellow, that they never had seen till he was brought into the hospital. I have seen them wash and cleanse sores[, and] do for the comfort of us, all what money could not buy." He approved of their proselytizing efforts as well. "I have heard them talk to the poor wretch of a savior of heaven or the joys of a better world all this I have seen them do irrespective of the faith of the patient. [T]hey never tire or feel but keep on doing good & kindness. [I]t is a superstition that would not hurt the world if there was more of it."[36]

Most famous of the Satterlee nurses was their director, Sister Mary Gonzaga. Age 50 when she assumed charge of the nursing sisters who arrived before the hospital's first patients in June 1862, Gonzaga had learned how to run a large institution as manager of the St. Joseph orphan asylum in Philadelphia, as well as from other postings. Although only positive depictions of Sister Gonzaga survive in the records, one must make allowances for the mists of sentiment that adorn such memoirs. One sister remembered an occasion when "Doctor Hayes accompanied by Sister Gonzaga made the round of the Hospital. . . . You may be sure more than a thousand times was it said the superior looks like a real mother to us all. It was a strange contrast. The proud Military Officer in full uniform and the humble daughter of St. Vincent to whom he showed every respect and attention." Another testimonial declared, "Mother Gonzaga was a mother of sixty thousand soldiers, as patients under treatment in Satterlee United States Army Hospital." This soldier, who wrote to a Philadelphia newspaper after reading of Gonzaga's serious illness in 1897, had been a patient there. "Those who were under her care . . . will recognize her familiar countenance surrounded by that white-winged hood or cowl, just bending her form to hear the faint breath of some fever patient. . . . Those who recall these scenes I say think of her truly as an angel of peace and sweetness."[37] Never mind that this "real mother" was celibate and at first unfamiliar to many of the men; she and her nursing sisters were just as able as laywomen to fill the roles of surrogate family, if not more so because of the insulating garb that legitimated the intimacies of their actions.

The Patients

Who ended up in the large general hospitals of the northern cities? In general there were two different streams that fed the vast lake of Satterlee. In the first were men who had traveled a long way, from camp to regimental hospital to regional hospital and finally to Philadelphia. The army kept hoping that such men would recover and return to their ranks, although many of the men ultimately left Satterlee with a discharge, the army's admission that they would never be able to resume the rigors of field life. The second stream was a sudden torrent, draining the battlefields nearby of the recently wounded; this one brought a flood of wounded men for a few weeks, then dried up until the next major encounter. So Satterlee had many men who were chronically ill, many men who were slowly recovering from wounds (and, perhaps, repeated surgeries), and occasionally massive numbers of men just arrived from the battlefield.

While it would be helpful to have summary statistics of patients, describing their diseases or wounds, their time in the hospital, and their outcomes, such records are not available.[38] We do have one accounting from midcourse in the hospital's history, however. The chaplain accumulated some statistics in his brief history of the hospital that was published in late 1863. He noted that 5,847 patients had been admitted from October 8, 1862 (when the hospital was first considered complete), to October 8, 1863. This time period excluded the 1,500 patients admitted after the second battle of Bull Run in late August 1862 but included the 4,602 who arrived after the Battle of Gettysburg (July 1863). The hospital's most fatal month during that year followed in August 1863, when 25 men died. This is extraordinary. The total death rate for this year was less than 2 percent (110/5,847). If all the deaths in August were from Gettysburg wounds, fewer than 1 percent of those men died after reaching the hospital.[39]

The initial cadre of nursing sisters who opened the hospital remembered their first patients in June 1862 as suffering from "typhoid, severe camp fever, chronic dysentery, etc." These were men who had failed to recover nearer their regiments and had been sent for prolonged rehabilitation at Satterlee.[40] The most common diseases of the Union Army were wasting diseases such as chronic diarrhea and dysentery. Surgeon Richard Penrose, who served at Satterlee, reported that the hospital had from its onset been "almost entirely filled with chronic cases of disease." It received, "as it were, the siftings of other hospitals." Most cases would have recovered at hospitals closer to their camps. Penrose explained, "Those which are severe, or occur in men of weak or bad constitution, resist treatment, and after the lapse of weeks or months are crowded out and sent off to try the effects of change of air." He believed chronic dysentery was among "the most numerous as well as the most indomitable and intractable of the various diseases coming under . . . observation" at Satterlee. Penrose treated the men with bland diets, opium, external heat, and "the mineral acids." Of forty-seven cases treated in his ward over the first year, only two died.[41] Typhoid fever was common at Satterlee and was probably a frequent cause of death. Joseph Leidy autopsied six men who died of typhoid in August 1862 alone.[42] According to one set of summary statistics, 6 percent of the men admitted to the hospital with diarrhea or dysentery died and another 16 percent were discharged from the service for disability.[43]

Patients were very aware of their digestive health and often described in letters home how much weight they had lost, as a gauge of their condition. And when the loose stools decreased, their appetite recovered, and they could reliably keep food down, they told their families that they were on the mend.[44] E. T. Stetson, who wrote "Mother McKim" from a Baltimore hospital, was representative of many other Satterlee patients. "I am in the hospital and have been ever since last Oct. & expect to be fore some time to come," he told her in May 1863. "I suppose you will want to know what the matter is with me it is the Dyspepsia I have eat bread and milk three times a day almost, ever since I can remember." He judged the treatment to be appropriate. "I guess I shall get fat on it so far I weigh 130 lbs now I used to weigh 160." Three months later he reported "I am in the hospital yet but am better than I was in the spring, I do not throw up my victals now. But am taking medicine now which consists of three drinks of whiskey every day."[45] He does not tell her whether this regimen had put on any pounds yet.

Men in the hospitals were often well enough to go for walks around town, although they lacked the strength to return to their regiments. Army life demanded endurance that many modern, generally healthy Americans lack—the ability to hike many miles a day, carrying a heavy pack, and subsist on rations perhaps adequate in calories but lacking in vitamins. Army surgeons, at least in the North, were conservative about sending fragile men back to camp. No doubt there were also many patients who exaggerated the severity of their illnesses so they could stay in the hospital. Known as "hospital bummers" or "hospital loafers," they earned scorn from the seriously ill.[46] But many men probably fell between the categories of severely ill and completely well, and they could spend months in the hospital. Soldiers worried about being misjudged as malingerers. William Fullerton told his sister Lydia in February 1863, "My helth is good now, that is I feel well my appetite is good and I am fleshy." After thus reassuring her, he went on to qualify the statement. "But some how I have no Stranth, that is not a quarter what I ought to have for one that looks as well as I do. I am affrade the Doct thinks I look altogether too well to be in the Hospital." A month later he was still in the hospital, but due to go to the convalescent camp the following week, a sort of halfway station for men on their way back to active duty. "I know one thing[:] that every one that comes to a Hospital don't want to leave it, and they have good reason for wishing to stay." Come July, he was still in the convalescent camp, where he endured "as dull a 4th of July as I ever saw." His regiment, the Twelfth New Hampshire Infantry, had a more interesting day near Gettysburg, Pennsylvania. Fullerton spent most of the war in and out of hospitals, although the only ailment specifically mentioned in his letters is erysipelas of the face. It is clear that he worried about being thought a slacker, but doctors reviewing his case must have concluded, repeatedly, that he was not healthy enough for camp life.[47]

After battles, men arrived who had endured hours or days since their wounding, perhaps had had surgery, and most likely were suffering from blood loss, exposure, exhaustion, dehydration, and not enough food. Surgery in the field hospital was hurried and dirty; wounds commonly became infected and drained. Some wounds required further surgery as they festered, reopened, or bled. One man wounded at Gettysburg and brought to Satterlee had experienced field amputation of his left arm at midhumerus. As the wound suppurated, the artery's ligature loosened, and the surgeons took the man back to the operating room to control bleeding. This time they removed the rest of his upper arm, trying to close the wound in an area that was not yet infected so that it would heal. In the middle of the night the closure failed, and the axillary artery at the shoulder began spurting blood. The surgeons could not cauterize or tie it effectively, so medical cadets were enlisted to put pressure on the wound in the hopes that an occluding clot would form. Fourteen cadets worked in thirty-minute shifts, but the hemorrhage continued, and the man died of blood loss during the night, a month after his initial injury.[48]

Other wounds were less immediately deadly but slow to heal, keeping a man in the hospital for months. Cyrus Beamenderfer of the Eighty-Fourth Pennsylvania Infantry was wounded in the face and neck at the Battle of the Wilderness in early May 1864. He was well enough to write a friend two weeks later with the news of his wound, saying, "I am getting along as good as can be expected." By June 9 he reported, "Writing you is middling hard with me I cant sit up long enough to write a letter." From a hospital in Alexandria, Virginia, he described the course of his care. "I'm getting along very good[.] [Y]esterday they [per]formed the third operation on me[.] [T]hey nearly killed me with chloroform and ether[.] [T]hey took a piece of bone out of my jaw again." He wanted to come home but knew he could not stand the shaking of the railroad cars. "I must lay on my back the whole time with my head very high." By July 7 Beamenderfer was in Satterlee and reported, "I had to have my neck cut open again. . . . I can't get out of bed hardly it was two months yesterday since I was wounded and am still in a Bad condition." Beamenderfer was still in Satterlee on Halloween and had been well enough to go home on furlough. It is likely that he left the hospital, and the service, shortly after that.[49]

The hospital's patient population had grown sicker by 1864. One sister's diary entry from May 23, 1864, recorded, "We have now nearly three thousand four hundred patients in our Hospital. Indeed, it is as if we were in the midst of a little city. Everywhere we turn we meet crowds of the maimed, the lame, and the blind, going through the corridors and yards as best they can. The wards are quickly filling up with rows of beds in the centre, as new arrivals are coming in every day from recent battles." Not only was the hospital crowded, but, she thought, the mortality rate was rising. "Many more are dying this summer than last summer. The cause is supposed to be that the poor soldiers are worn out with privations before being wounded or falling ill."[50]

One disease that plagued the hospital that summer and fall was smallpox. The disease was flourishing in general throughout the hospitals of both North and South; vaccination was not always effective in preventing it.[51] Initially the surgeons in charge immediately transferred cases of sporadic smallpox to a specialized smallpox hospital that was several miles outside the city. As the number of cases grew, Dr. Hayes designated one corner of the Satterlee grounds as a smallpox camp, and the sisters helped set up the tents there. "The tents were made very comfortable, with good large stoves to heat them, and 'flys' (double covers) over the tops," their memoir recorded. A nun who had previously had smallpox was assigned to nurse the men. The sisters counted ninety cases, of whom nine or ten died. Their account indicated, "We had, I may say, entire charge of the poor sufferers, as the physician who attended them seldom paid them a visit." The nuns brought them treats of cakes, fruit, and jellies, and the men believed (according to the nuns) that "it was the sisters that cured them and not the doctors."[52] While it may be that the sisters presented their service in overly glowing terms, the mortality rate from smallpox recorded there (10%) certainly bested the official figures, showing that a man's chance of dying of smallpox at Satterlee was one-quarter of the likelihood at other Union hospitals.[53] If true, this may well reflect the value of dedicated nursing. It may also reflect the attenuating effect of prior vaccination, which was perhaps more routine and effective among the northeastern troops who were common at Satterlee than among soldiers farther west.

A second dread disease that made its appearance in 1864 was hospital gangrene. Gangrene was an imprecise term, but even in the 1860s it was understood as something different from the common wound infection. Gangrene turned tissue black with necrosis. As the blackness crept up the limb, the peripheral tissues sloughed off. It was nasty, painful, and often fatal. While it is hard in retrospect to assign a specific organism to these infections, it is likely that Clostridium species were commonly involved, along with mixtures of other anaerobic and aerobic bacteria.[54] Even though only 2,642 cases of gangrene among Union troops were officially reported for all years of the war, it attracted major concern from surgeons. As at Satterlee, the largest number of gangrenous wounds in the Union Army occurred in 1864 (1,611). Most of these began in wounds of the extremities, and nearly half were fatal.[55]

The common opinion was that wounds became gangrenous when exposed to the fetid atmosphere of a badly ventilated hospital. So an outbreak of hospital gangrene was a marker of such poor ventilation and a sign that the hospital management was derelict in securing a healthful environment. This seems to have happened twice at Satterlee, once in summer 1863 and again in summer 1864. Both seasons produced a dense crop of wounded men, who were crowded together in the wards as they arrived in the hundreds from the battlefields of Gettysburg (1863) and Virginia (1864). Such cases generated a particularly foul atmosphere. "It would be difficult to exaggerate the offensiveness of the discharge from wounds in this condition of gangrene," wrote one surgeon who had worked at Satterlee. "There is a peculiarly nauseous smell over and above that of mere putridity."[56] The common assumption was that one case infected another and that the best remedy for the epidemic was to move the men out to tents and away from the ward population. Surgeons treated the cases with debridement, both mechanical and chemical, using nitric acid or the bromine treatment made popular by Dr. William Goldsmith in the western military hospitals. Surgeon William Keen remembered taking strong measures to prevent infection; he ordered increased ventilation and directed the nurses to allot "a sponge for each man," "to dress the gangrenous and unhealthy sores last of all, and to wash their hands carefully in dilute chlorinated soda afterward; [and to use] no dressing or bandage a second time."[57]

Thus, the wards at Satterlee had patients of all sorts. Some were struggling to recover from chronic infections that left them drained and weak. Others had wounds that were in various stages of healing, draining, hemorrhaging, becoming newly infected, and generally causing a diverse range of trouble and pain. Some men were confined to bed, others were weakly ambulatory, and still others seem to have been entirely capable of walking (and carousing) for blocks. But what of the cultural aspects of "this little city" whose "crowds of the maimed, the lame, and the blind" needed both to be tamed and to have their morale maintained?

Hospital Life

Discipline was important at Satterlee Hospital. The patients by definition were soldiers, still in the army and bound by their enlistment agreements to stay under army rule until discharged. They were not to leave the hospital grounds without a pass, a rule enforced by those 132 guards. Most patients were ambulatory, at least to the point of being able to walk down the ward to the mess tables for meals. Many were healthy enough to go into town, when they did get a pass. And not a few deserted from the hospital. While cumulative figures are not available for Satterlee, at the general hospital nearby in York, Pennsylvania, 156 men (out of 4,704 admitted) had deserted in the first twenty-one months of the hospital's existence.[58] Occasionally the Satterlee newspaper, the *Hospital Register,* would publish weekly statistics of admissions, discharges, furloughs, deserters, and deaths, and the numbers of deserters ranged from the twenties to the sixties per week. At least once the *Philadelphia Inquirer* printed the names of deserters from Satterlee, perhaps in an attempt to shame those considering the same course.[59] A letter to the *Hospital Register* suggested that some of these statistics might be inaccurate. E. S. tells of a man who went home on furlough from the hospital (a common occurrence for the chronically ill), died there, and was unfairly labeled by the army as a deserter because he did not return when the furlough expired.[60] Many "deserters" may have been merely "lost to follow-up." Or a furloughed man counted as a deserter one week could show up late the next; this must have happened commonly in an era of irregular transportation.

The guards did more than just control the gates in and out; they imprisoned men who became drunk or otherwise disorderly. Serving as guards was one possible assignment for the so-called Invalid Corps, men who had recovered from wounds but for various reasons (such as the loss of a limb) were unable to return to regular duty.[61] New Hampshire private J. H. Moody ended up in that capacity in summer 1864 at Satterlee. He told a friend, "Putting men in the Guard House is the business I follow for a living now, in plain words I am a Turnkey, a regular jailor, over a prison that has its cells. . . . It is a nice job." He describes a man recently brought in. "You ought to see the bottle of whiskey I took from him. It was a large quart flask and was full, and I guess that he had as much more in himself." It was not uncommon to see "some of the men . . . brought in dead drunk, some crazy[,] in fact the[y] [came] one and all, and in all sorts of conditions, but they soon [got] sober on bread and water." Moody liked the job, for he worked only every other day and frequently received passes to visit ladies in the city. "I have got a widow bewitched that works in a tailors shop here. Her husband is in the western army so he is not in *My way.*"[62] Presumably Moody used the term *widow* to refer to the temporary loss of her husband to the (far away) Union Army.

Controlling excessive alcohol consumption among the troops was a constant issue for the military's leaders, and ambulatory patients were no exception. While the men might be prescribed whiskey as a stimulant on the wards, they were discouraged from acquiring it outside. (According to one tabulation, the army dispensed 1.9 million bottles of medicinal whiskey, in thirty-two-ounce bottles.)[63] In Annapolis, where General Hospital Division no. 1 hosted the publication of a newspaper named the *Crutch,* the commanding officer of the post attempted to forbid the local sale of liquor. "The evils resulting from the unrestricted sale of Spiritous Liquors to Soldiers, as now practiced, are such as to destroy, to a great extent, military discipline . . . and to seriously impair the health and efficiency of the men," he wrote in May 1863. He ordered, "All persons residing in the District of Annapolis, are hereby most positively forbidden to sell, or give, to any enlisted man in the military service of the United States spirituous liquors or intoxicating drinks of any kind whatever, without the written permission of a medical officer."[64] The fact that advertisements in the *Crutch,* such as one for Mullavell's Bowling Saloon, continued to include pitches such as "The best of Wines, Liquors and Cigars always on hand" implies that this order was not enforced.[65]

The behavior of the Satterlee patients while out on the town became something of a scandal. The *Hospital Register* contained numerous advertisements for eating and drinking establishments, and apparently some men overindulged. A local newspaper reported in August 1864, "On a certain street in the Twenty-fourth Ward, along which the men pass from and to the Hospital, they are continually moving at all hours of the day and night, and it is notorious in that neighborhood that the language they use is so lewd and blasphemous as to shock the ears of all decent residents." The newspaper charged that the hospital patients were fit for active duty and should not be loafing about the city, causing a nuisance. "These soldiers spend their time in the city in dissipation, that rapidly disqualifies them for the business for which they enlisted. They go to taverns, get drunk, become boisterous and profane, engage in quarrels and fight, and seriously disturb the public peace." The indignant reply in the Satterlee *Hospital Register* (which quoted the *Philadelphia Evening Standard* article) called these words a "base slander" that could emerge only from the pen of a "disloyal citizen." Of course there were "some cases of drunkenness and bad conduct, but these are exceptional." Few men in the hospital were ready to return to field service, and a hospitalized soldier deserved the chance to "get a change of air and space, or to see his friends, or to have a little pleasure, that the monotony of his weary life of suffering may be broken."[66] It is hard to know whether the author, like Hamlet's mother, protested too much.[67]

In contrast, the hospital newspaper's editor (known only as "the canvas back chair," who occasionally had colloquies with "the camp stool," an unnamed military officer) wrote a year earlier that drinking "was the chief pleasure of the heroes of Valhalla, and appears to constitute the greatest happiness of the majority of the members of that most august body of men, the

United States Army." He even claimed that the preeminence of man over other animals was that man is "an animal that gets drunk."[68] Another author wrote a satirical poem about receiving a pass, going to a bar, and becoming intoxicated. He and his friend were soon

> stretched side by side on the floor;
> Our passes were lost and our money went too,
> Oh! never before were we in such a stew.

The barkeeper called the military police, who

> locked us up in a cell
> Because we were patients and did not look well.
>
> .   .   .   .
>
> The next day I was marched to the Hospital gate,
> And from there to the Office—but was not too late,
> To be placed on the list with my friend and colleague,
> To serve for a week on the "drunken fatigue."[69]

Fatigue duty usually meant manual labor, building or digging or hauling, sometimes ordered as a punishment. Although this author may have been a sadder but wiser man, his story was humorous to him, not tragic. Some soldiers formed a temperance society in spring 1864, but only thirty-four men signed the pledge.[70]

The men had other sorts of fun while outside the hospital. They went swimming at the newly opened natatorium, an indoor pool recommended by a local physician for those suffering from "nervous, debilitated, and dyspeptic conditions." They bowled, played billiards, ate at restaurants, and visited the many merchants who advertised photo albums, clothing, and the best cigars. By 1865 there was a hospital baseball team, which played at least one game against the "Mercantile Club." (They lost, in part because it was hard to keep a team going when their best players kept being sent back to their regiments.) Many patients, no doubt, spent their time in "dissipation," as the local newspaper charged; they visited prostitutes. One Michigan private who had been wounded above the eye in early June 1864 ended up in Satterlee in February 1865. He was disgusted with the behavior of his wardmates (and fearful that he would soon become like them). He told his wife Nan, "I do hate this place. . . . The grub is good enough but it is a mean low lifed place. There is nothing talked about or thought of by two thirds of the sick and wounded here only horeing. Nan it is honestly a place that deserves the curse

Soldiers in Camp Visiting the Sutler's Store. Note the prominence of alcoholic beverages and cigars among the sutler's wares. The sign says "Liquor sold." Modern touches include the man using the counter as a "drive through" window and the early version of the modern sign that says "In God we trust. All others pay cash." The Turkish dress reflects the unusual uniforms of regiments called Zouaves. Paul F. Mottelay and T. Campbell-Copeland, eds., *Frank Leslie's "The Soldier in Our Civil War": A Pictorial History of the Conflict, 1861–1865, Illustrating the Valor of the Soldier as Displayed on the Battle-field*, Columbian Memorial edition, 2 vols. (New York : S. Bradley, 1893) 1:365.

of God if any place deserves it." He went on to explain the system in place for local entertainment. "They give passes to all the boys that want them to go down to the city and stay 24 hours. some of them gets drunk and raises all old ned. The hores are aloud to come in the hospital any time they are amind to. Some of the boys in the ward has got the C[la]p. Since They come here it is the weekedest place I ever was in for a hospital." Then he complained that he had not yet received a pass, and closed with "I would still ask your praers to help me in time of temptation. Your true and loveing husband."[71]

It seems unlikely that a hospital overseen by Sister Mary Gonzaga allowed "hores" to "come into the hospital anytime they [were] amind to," but there is no doubt that houses of prostitution flourished around military bases and hospitals in the Civil War. Official reports recorded that eighty-two men per thousand contracted some form of venereal disease during their time in service and noted that men stationed in cities were more vulnerable than those farther afield. A well-known prostitution quarter flourished near City Point Hospital in Virginia, for example. The transfer of venereal microbes went both ways, of course, and men did not have to pay for sex to acquire infection. Cyrus Beamenderfer went home on furlough from Satterlee in fall 1864. On his return to the hospital, he wrote his friend Daniel Musser, "Give my best respects to all the girls around town but tell Sally Hurburt I got the clap from her like all hell[.] But am getting better of it[.] [D]on't let your sister see this letter. Burn it for I don't want any body to know it around." Beamenderfer apparently fancied himself quite the ladies' man, in spite of the face and neck wound that refused to heal. He closed the letter by saying, "Tell Rach[el] Trump to write to me if you see her and I will send my picture to her and also to Molly Fox."[72]

There can be no doubt that there were rough men and rough behavior in the Satterlee Hospital. One sergeant from Massachusetts who had been "wounded by a bursting shell, fracturing the skull," was annoyed by his neighbors in the hospital. "In the short intervals [between] his paroxysms of pain," one friend reported, "he often spoke of a number of the occupants of the same ward, who seemed to be making great ado about very slight wounds . . . when their surgeon was present; but who seemed to forget them all upon his leaving." The sergeant was not just aggravated by their "shamming" but also was "very much annoyed by their noise over their cards—their rough talking and laughing." It is likely that elite attitudes favoring social control, a mind-set classically described by Robert Wiebe in his volume *The Search for Order,* figured largely in the response by officers and others who oversaw these men within the hospital.[73] The Protestant chaplain certainly tried his best to tame their spirits, offering admonitory sermons on Sunday, frequently visiting the sick, and publishing articles in the "Chaplain's Corner" of the newspaper. The Catholic priest Father McGrane likewise offered Sunday services that would have urged good behavior. And of course there was the guard house for recalcitrants. But a third path was through the introduction of refined, cultural entertainments into the hospital.

The avowed purpose of these entertainments was to stave off boredom, identified as the principal evil, after pain, for the hospitalized patient. Imagine a modern hospital without televisions, and then add the circumstance of stays lasting weeks or months. Most feared was the soldier who lapsed into "nostalgia," a condition described in chapter 1, that is, homesickness or despondency that was dangerous to the patient. One article that appeared in the Annapolis Division 1 Hospital paper and was then reprinted in the Satterlee *Hospital Register* described this complaint and particularly called for letters from home as a sure palliative. "There is no cure for it in the *materia medica*. Many a gallant fellow in the ranks of the Union army dies of it." The soldier was pining for home, and especially for news from friends and family, claimed the article. "It pains me to think that more than one man has let his life slip out of a grasp too weak to hold it because his dearest friends did not send him a prescription once a week, price three cents—a letter from home."[74]

Even soldiers with steady correspondents suffered otherwise from ennui. One humorist from Ward H summarized his plight as follows:

> Ho hum! Dear me, what shall I do
> To pass the weary hours through
> This dull and stupid day?
> I have no pass to go to town—
> I've walked the corridors up and down—
> All games have tried to play.
> For chess I haven't the head;
> My interest in cards has fled;
> With checkers I am through.
> I cannot read—I cannot sing—
> Nor work or play at anything—
> So pray what shall I do?
> Ah! Now I think I have it quite;
> I'll take my pen and ink, and write
> A poem for our paper.[75]

The author continues in this vein, describing all the topics about which he can't think of anything to say. Officers recognized that the men needed diversion, especially those who were ambulatory but not well enough to go back to their regiments, and most general hospitals took steps to provide a variety of distractions.

First were the in-house newspapers themselves. The Satterlee *Hospital Register* printed stories, poems, news items, and editorials on a variety of topics in addition to serving as a record of admissions and discharges. It was likely read not just in Satterlee but in the hospitals all over Philadelphia. It is probably not a coincidence that the hospital's director, Dr. Isaac Hayes, had served as editor of his boarding school's newspaper and even created a handwritten newssheet during the months he spent trapped in the winter ice of his Arctic expedition. He knew the value of creating and controlling the media. Other hospitals began newspapers, too, using the printing presses that were installed within the hospitals to print the innumerable blank forms that kept the organizations running.[76] These newspapers begged their readers to submit material but often fell back on reprinting text from other sources, usually not attributed (as was the newspaper custom of the day). They printed jokes, riddles, sentimental or silly poems, short stories, and other nonmilitary matter. The Satterlee paper officially eschewed politics; others, like the Crutch in Annapolis and the Cartridge Box in York, Pennsylvania, were overtly pro-Republican in the 1864 election. The Satterlee paper rarely reported war news, perhaps feeling that the patients needed more cheerful fare, but other hospital newspapers offered more battle detail. All were supported by advertisements from local proprietors.

Leaders within the general hospital, assisted by local female volunteer groups, organized libraries and reading rooms soon after the hospitals were established. They asked locals for donations of books or money to buy them. Newspaper editors from all over the North sent copies of their papers so that men from Maine and Illinois and Minnesota could read hometown news. Women staffed these libraries and reading rooms, offering conversation and book suggestions and playing the piano for the men's diversion. They probably also fed the goldfish who lived in a "large vase" near the piano.[77] The Satterlee paper documented the civilizing influence of these efforts: "I am assured by the ladies that on no occasion has the slightest rudeness or disrespect been manifested by any one of the many soldiers who have frequented these rooms during the past year."[78] By 1864 the library at Satterlee had more than three thousand volumes.

The hospital also had a stage in its chapel that was the site of entertainments. The hospital band played music daily, and other performers gave concerts featuring voice or string instruments. In April 1864 a group of patients put on a play entitled The Ingrate. There was a debating society that considered such questions as "Resolved, that temperance has caused more misery than war," or "Resolved, that capital punishment should be abolished."[79] At the end of May 1864, "Professor Alfred A. Starr of New York gave a highly instructive and interesting entertainment to the soldiers of this Hospital, in the Chapel. . . . The exhibition consisted of microscopic views of animalcule,

Header for the Satterlee *Hospital Register*. On the left a woman staffs the library; the central oval displays the auditorium with a recital in progress; and on the right men are playing a game of billiards. West Philadephia *Hospital Register* 1 (May 16, 1863): 55.

given through a microscope capable of magnifying an object twenty-three millions of times." Such lectures on microscopy were common fodder for elite soirees in the 1850s; other lecturers used "magic lantern" slides to display pictures of the Holy Land and other travelogue wonders.[80] Perhaps the microscopy talk was too highbrow for some of the audience. The newspaper report continued, "It is to be regretted that those who visited to listen to the lecture, to derive the knowledge that it contained, were prevented from doing so by the uproar of some who thought they were in a 'free show,' and cared nothing for the lecturer or his subject, and had not even proper respect for the place they were in."[81]

With so many men missing arms and legs, it was not lost on the hospital's leadership that these men would need rehabilitation and perhaps reeducation to succeed again in civilian life. Many of them had pursued agricultural or manual trades before the war, work that would be much more difficult or impossible given their wartime injuries. The government did provide prostheses, and men were fitted for them at Satterlee. Soldiers with enough money could patronize local merchants who sold fancier models than the ones offered by the government, and such services were advertised in the hospital paper. One article mentioned that the men were petitioning the government to give them cash to buy limbs on their own rather than limiting them to the designated government purveyor of the items.[82] Hayes and others saw, though, that even with such replacements the men would be well served by education that would fit them for desk jobs. Hayes allowed "the ladies" to set up a school for the patients, which offered the "fundamental branches of an English education—Book-keeping included."[83] Elsewhere in Philadelphia the government opened a new "home for maimed soldiers" in early 1865, "to give men who [had] been disabled by loss of limbs, or other severe injuries, an opportunity for education for some practical pursuits suited to their limited physical capabilities, thus enabling them" to support themselves upon discharge and not become "objects of public charity." The subjects included "telegraph operating, book-keeping, and penmanship."[84]

The funds to buy these various luxuries—books, band instruments, even writing slates for the school—came from a variety of sources. The Philadelphia women associated with the hospital hosted charity fund-raisers and also urged their friends to supply items the hospital needed. A local library society gave a large number of books, which Hayes supplemented from his own library. Local merchants donated furniture for the reading room, and others provided games, stereoscopic viewers, and the billiard tables.[85] The Sanitary Commission and the Christian Commission supplied special foods, reading material, and writing paper, among other supplies. The hospital also had a "slush fund" created by selling unlikely items to raise money. Surgeon William Keen remembered in 1918 that "at the Satterlee Hospital in West Philadelphia . . . the egg-shells were saved and sold. The revenue, if I remember rightly, was $3000 a year! They were used in the manufacture of face powder, as they were pure calcium carbonate."[86]

This "slush fund" was separate from the "hospital fund." The hospital fund "consisted of the credit on the books of the Subsistence Department for those parts of the ration which the sick men were unable to consume." The concept was that the surgeon should use this money to buy "delicacies" that the sick men could eat—milk and eggs, instead of beef and bread, for example. "The 'slush fund' was derived from the sale of bones, fat, stale bread, slops, flour barrels, straw, manure, waste paper, old newspapers, etc." It was this fund (supplemented by the sale of egg shells) that bought the luxuries of books, musical instruments, and other items of entertainment not already supplied by local philanthropy.[87] These two funds could also be used to buy local fruits in season, and the patients in Satterlee and other general hospitals of the Northeast enjoyed strawberries, blackberries, and even the occasional orange.[88] The men in the northern general hospital were, all told, well fed and well treated.

Patients' Opinion of the Hospital

It would be handy to have data from a customer satisfaction survey of the men in general hospitals, but of course that concept awaited another century in hospital administration. So we are left with anecdotal evidence, which largely reveals positive feelings toward the place. Most men would probably have preferred to be at home, but if they had to stay in the army, the hospital was not a bad billet. It is likely that more than one echoed the sentiments of the faux illiterate author of a poem that appeared in the *Hospital Register,* entitled "I'm Too Sic to Go Bak to the War." After describing various other complaints, he concluded,

> There's a mizzery, too, in mi hed,
> and mi legs they kin scarce reche the door;
> im shure id be bter in bed—
> im too sick to go bak to the war.
> im a blacksmith bi trade, an if ev
> er I git to go home enny mawr,
> i wil hamer an slave, but I'll nev
> er be sic az I am uv the war.[89]

Another anonymous article in the paper pointed out that after Bull Run, "cold, hunger, sickness, defeat and death, had to some extent enlightened our minds on the subject of campaigning." He recalled that someone had said, "War is hateful to mothers," and while he allowed, "My mother is anxious on my account," he also stated, "I am equally so. At the same time that we acknowledge the propriety of all the fighting, the most of us find it tiresome."[90] They had seen camp life, many had seen battle, most had seen their friends killed or wounded, and some now found themselves mangled. Enough already. As one Indiana soldier wounded at Petersburg in June 1864 wrote from Satterlee, "I am glad that I got shot, I git to live with the Sisters of Charity."[91] Not once in the hospital paper did a writer express enthusiasm for returning to his regiment.

For one thing, the soldiers ate better in the hospital than in camp. The farms of Pennsylvania and New York amply supplied the kitchens of the general hospitals. The newspaper of York Hospital in eastern Pennsylvania provided a sample menu. "Good dinners are served up every day, in our Mess

Christmas Dinner,

Satterlee U.S.A. General Hospital,

WEST PHILADELPHIA, PA.

*Mrs. Dr. M. C. Egbert, of Petroleum
Centre, Venango County, Penn'a, to the
Soldiers in the Philadelphia Hospitals.*

Christmas, 1864.

ROAST TURKEY.

ROAST BEEF.   ROAST CHICKEN.

Pickles.   Cranberry Sauce.
Chow Chow.   Apple Sauce.

White Potatoes. Sweet Potatoes.
Turnips.   Onions.

Plum Pie.   Apple Pie.
Mince Pie.   Sponge Cake.

Apples.   Cider.   Nuts.   Coffee.

Music by Hospital Band.

DINNER AT THREE, P.M.

Christmas Dinner Menu, Satterlee Hospital, 1864. The men ate particularly well when one donor provided the funds for a splendid Christmas feast in 1864. From the collection of F. Terry Hambrecht, MD. Used with permission.

Room under the auspices of Steward Cheney," one issue bragged. "Monday, Mutton Pot Pie; Tuesday, Pork and Beans; Wednesday, fresh Vegetable Soup; Thursday, Mutton Pot Pie; Friday, Beef, Cabbage and Potatoes; Saturday, Irish Stew, and Sunday, Roast Beef and Mashed Potatoes. Who wouldn't be a soldier?"[92] Compared to the hardtack and salt meat that was often the fare in camp, this was high living. And if they did not like the mess hall, patients could go into town once the paymaster came through. Soldiers in camp could buy extras from the sutlers, but ambulatory men in the general hospitals could choose from a wide range of eating establishments. For Christmas Day 1864, there was a special feast, supported by philanthropic contributions. They ate well.

The sentimental reminiscences of Satterlee, published in its paper, also let us glimpse patient attitudes. One poem written a year after the Battle of Gettysburg recalled the course of events that cost the author his hand and landed him in Satterlee. Arrival at the hospital "seemed like coming home," where the nurses

> brought cold water to my lips,
> That scarce had strength to hope,
> and when my sickened heart would faint,
> They kindly whispered "hope."

He recalled the chaplain who supplied books and tender words, and spoke of the joy of letters from his son. Finally, he closed,

> And if God chooses to bless,
> Me again with ruddy health,
> I'll toss in the air my cap,
> Tattered and torn to see,
> And give one shout for the Union,
> And the next for *Satterlee!*[93]

Another soldier, now back with his regiment, waxed nostalgic in a letter to friends still at Satterlee. "I was learning to love the place—to love its kind people, and even its very scenery. . . . Days, months and perhaps years may roll on before I am permitted to see my second home again." Perhaps he had been in the hospital before its name changed, for he closes, "I shall go to my grave happy in the remembrance of the kindness shown to me while in the West Philadelphia Hospital!"[94]

It is hard to know how seriously to take these encomiums, for the editor would not be likely to publish poetry or prose that offered severe critique or lowered morale. The paper did print light-hearted complaints, such as an article in a spring issue entitled "Ventilation of the West Philadelphia Hospital" that counted up the many cracks in the walls and calculated how many cracks were allocated per patient per ward (all in parody of official proclamations about the cubic feet of air to be guaranteed per bed). Given that there were six and one-seventh cracks per board, he found, "There are about 829 cracks per man. Now the quantity of fresh air introduced through said cracks, when added to the daily supply afforded by the ventilators and windows, has been ascertained by a strict computation to be more than sufficient for the hygienic requirements of 107 elephants."[95] It must have been nippy that March.

The other evidence of patients' attitudes toward the care they received were the gifts they gave to their doctors and other caregivers. The men of Ward V gave their surgeon a new case of surgical instruments in July 1864. Not to be outdone, the men of Ward L gave their surgeon "a fine sword" in appreciation, and they gave a gold pen and pencil set to their ward cadet the next month.[96] Such presentations were commonly mentioned in the *Hospital Register*. Since surgeons and hospital stewards were often ordered to other posts, and medical cadets moved on with their studies, it is likely that these gifts marked the ends of those individuals' periods of service at the hospital. Presenting gifts may have become traditional, in the same way as the modern office retirement party, but they offer at least some evidence of the patients' appreciation of the hospital staff.

In spring 1865, it became more and more evident that the war would soon be over. Instead of being sent to convalescent camp or otherwise delayed for return to duty, men were increasingly discharged home, and the hospital's population slowly dwindled. The institution officially closed on August 3, and in the last week of September, the hospital and all its fittings were sold at auction.[97] Nothing now remains of Satterlee Hospital except a memorial marker to the Gettysburg fallen that stands in Philadelphia's Clark Park, a grassy area carved out of one corner of the hospital's grounds.[98]

Conclusion

The leaders of Civil War hospitals in the North invented the modern American hospital. Many key features were there: a well-organized staff performing specialized duties, including nursing, pharmacy, medical practice, cooking, laundry, and so on; medical students working and learning on the wards from great teachers; patients drawn from across the social spectrum; an "auxiliary" of local women dedicated to providing comforts and entertainments; and a clean, well-ventilated, nourishing environment that promoted healing. Yet the hospitals disbanded at the end of the war, disappearing from the American landscape. Only the Freedmen's Bureau hospitals persisted, and those for just a few years.[99]

Why did the hospitals disappear? The general hospital must be understood as embodying a compromise between having the men cared for at home (the optimum plan) and keeping them under military discipline. Women invaded this military space and recreated the home sickroom as much as possible. In the Satterlee case, the nurses were literally called sisters and mother, a usage that contemporaries were quick to equate with "like being at home." In other circumstances laywomen served as nurses, and the men likewise transferred the domestic roles to them, creating a legitimating language that allowed for women to care for strange men as if they were family members. Many of the sick and wounded African Americans at war's end, either military or refugee, lacked a healing home to return to, and thus the need for their hospitals persisted. But it was not until the surgical innovations of the 1880s that the hospital began to offer something of value that was not available in the homes of the affluent classes.

There can be no doubt that wartime experiences with hospital construction and implementation shaped postwar hospital design. As the medical history of the war concluded, "The experience of the war was decidedly in favor of the pavilion system, each pavilion constituting a single ward isolated from adjacent buildings by somewhat more than its own width, and connected by a covered walk with the other buildings of the hospital. In their aggregation this separation was effected, without removing any of the wards to an inconvenient distance from the administrative and executive buildings, by radiating them around some central point in a form to be determined by the configuration of the ground available for building."[100] Hospitals built after the war, such as Harvard's Peter Bent Brigham, followed the pavilion system, with its emphasis on ventilation and lack of connection between wards. Not until the early twentieth century, when electric fans and active systems of controlling infection based on the germ theory again allowed it, did hospitals begin to build stacked floors and other buildings that could emphasize industrial themes over the all-important ventilation requirement.[101]

The experience at Satterlee and other northern general hospitals was often about abundance—ample and appealing food, extravagances like the piano and the billiard table, and volunteers who brought entertainments and delicacies. The hospital was surrounded by a community where restaurants and markets rarely saw shortages. The city of Philadelphia was threatened only once, when Lee's army made it as far as York, Pennsylvania, one hundred miles away. The situation was very different farther south, in the Confederacy's many hospitals. The Confederates too built general hospitals and favored the pavilion system. But they were much less able to create the rich environment typical of Satterlee and its peer institutions.

Chapters 7 and 8 (at last) take up the Confederate side of the story. Chapter 7 considers the official medical response to the call of wounded and ill rebel troops. It reveals a surgeon general who sought to provide appropriate care for these men but also saw in the structures of his new nation the opportunity to "elevate" the southern medical profession. His goals were ambitious—create a new medical society, publish a journal, distribute educational materials to his surgeons, gather research data from those same men, and enrich the southern pharmacopoeia with intensive research on native plants. Chapter 8 reveals the harsh reality of the Confederate medical system, bedeviled by shortages of all sorts and unable to provide adequate care for the southern troops.

CHAPTER SEVEN

Medicine for a New Nation

Confederate medical leaders struggled with the same challenges as their Union counterparts, but they also embraced the opportunities offered by the war to create a new medical system for their nascent country, one that would rival the North's and encompass the best features of contemporary medicine.[1] These leaders had a fairly clear idea of what progressive medical education, research, and care should be and hoped they could construct a model system within their new country, for which they had such high aspirations. Confederate political leaders looked to the past, basing their claims for legitimacy on a return to the country of George Washington, Thomas Jefferson, and James Madison—southerners all—who founded and led the original United States. Feeling that their country had been corrupted by abolitionist politicians who led it astray, Confederate politicians sought restoration and wrote a constitution that echoed the original almost word for word (with a few additions about slavery).[2] In medicine, however, Confederate leaders looked resolutely forward, seeing models in France, England, and Scotland of what the field of medicine could be. The war offered a fertile ground for the reform and improvement of the southern medical profession.

Southern medical leaders were well aware of the debates and reformist measures that invigorated American medicine in the 1840s and 1850s and wanted the physicians in their region to benefit from this progressive approach to the profession. Many elite American physicians, including some from the South, studied abroad in the two decades leading up to the Civil War, and many more attended medical school at the best American institutions in Philadelphia, New York, and Boston. Particularly in the hospitals of Edinburgh and Paris, students found ample "clinical material" in the indigent wards of large hospitals and had the opportunity to witness autopsies to correlate bedside findings with anatomical evidence.[3] Such research was centered in the hospital, and American students went abroad to study in part because there were relatively few hospitals in the United States and correspondingly little opportunity for this sort of research. To be sure, there was a strong emphasis on anatomical study at the leading schools in Boston, New York, and Philadelphia, where many of the dissected corpses were African American, dead slaves whose bodies were preserved in the South and shipped north in barrels. Southern medical schools also made ample use of such bodies for learning anatomy.[4] But hospital opportunities were limited.

Exposure to the best of European medicine, either firsthand or through the lectures and writings of those so influenced, contributed to a strong reform ethos in American medicine in the 1840s and 1850s. The reform agenda had multiple trajectories. One called for the accurate keeping of vital statistics, so that the true state of disease prevalence and the magnitude of early mortality could be recognized and studied. Mostly this was done on the city level in the antebellum period. Southern reformers who compared the slack information-gathering of their cities to the more complete data available for Boston, New York, and Philadelphia were vocal in their calls for greater statistical awareness.[5] Others recognized the need for hospital study and autopsy experience, and the best medical schools promoted such learning. The American Medical Association was founded in 1847 in part as a reaction to sectarian competition, but it also had at least a stated ambition of raising the quality of American medical practice and education.[6] These voices created a new standard for the best of modern medicine, a standard that the elite medical leaders of both sides in the conflict attempted to put into practice.

In some ways the medical reform impulse in the South mirrored the drive toward industrialization once the war began. Even as southern leaders depicted their country as distinctively and idyllically agricultural (in comparison to the threatening and ugly industrialization of the urban North), others called for the region to begin to modernize or else risk the loss of wealth to outside manufacturing. The war forced the issue; no longer could the southern states buy manufactured goods (including military hardware and ordnance, clothing, shoes, and even glassware) from the North, and foreign importation was limited by the blockade. The explosive growth of southern industry during the war was one of the Confederacy's unlikely success stories; medical men no doubt looked on jealously at the resources plowed by state and federal government into these heretofore neglected endeavors. They wanted to grow medicine too, and to do so with the active contribution of government financing.[7]

The Civil War created new opportunities for medical research in America, and southern medical leaders eagerly embraced this prospect. It is just short of amazing that in the midst of the chaos of war, in the midst of creating a hospital system to provide care for thousands of sick and wounded, with means that were never enough to meet demands, southern medical leaders tried simultaneously to begin a medical reform and a research enterprise. They also realized that many southern physicians possessed deficient medical knowledge, and they promoted education through distributed books and mentorship. It must be remembered that when the Confederate surgeon general, Samuel Preston Moore, and like-minded colleagues instituted these changes early in the war, they did not see the disaster that lay ahead. Moore was initiating a new nation's medical system, and he had grand plans for the future of southern medicine.

Moore was one of a handful of Union Army surgeons who resigned their commissions to follow the South into secession. Born and raised in Charleston, he graduated from the Medical College of South Carolina in 1834. Shortly afterward, he joined the U.S. Army and served subsequently at frontier posts in Kansas, Iowa, and Missouri, as well as at Pensacola in Florida. Service in the Mexican-American War (1846–48) brought him to the attention of Jefferson Davis, who remembered him a decade and a half later when they were fighting a different enemy. Promoted to surgeon and major in 1849, Moore worked at a variety of western posts before the onset of war in 1861. He resigned from the Union Army but appears to have been reluctant to serve the Confederacy. Instead of signing up for military duty, he went west to Little Rock, Arkansas (then part of the Union), and began a private practice. With difficulty Davis lured him into the Confederate surgeon general's position after the first two holders of the chair did not work out. Moore was known as an efficient and hard-working manager but as a brusque and nononsense one as well. He modeled the Confederacy's new medical system on the familiar pattern of the U.S. Army's Medical Department.[8]

Moore's conception of a new medical system for the South was no doubt also influenced by the rhetoric of "states' rights medicine" that had flourished in the 1850s. Antebellum physicians shared a common view that medical treatment needed to be crafted to match a patient's age, gender, body habitus, lifestyle, and geographic location. So a patient who lived in the hot,

humid South or on the boisterous frontier would necessarily require different care than the effete office worker of New York City.[9] Those promoting
southern medical schools took this idea further, proclaiming that men who
sought to practice medicine in the South should learn medicine in the South,
because only there would they find professors with experience in the relevant
characteristics of southern patients and diseases. This rhetoric crested with
the mass withdrawal of southern medical students from Philadelphia and
New York in 1859. More than three hundred students went home in protest
of John Brown's raid and the growing hostility to slavery in northern cities.
Southern schools welcomed them with regional rhetoric that soon became
national as the southern states seceded.[10]

It would be a mistake, however, to think that there was ever a separate
set of medical ideas that characterized southern medicine. There might well
be more malaria and yellow fever in the South, for example, and the well-
trained physician would need to know how to treat these "southern" diseases.[11] The leading physicians in the Confederacy had mostly trained in the
same schools, studied with the same professors, and read the same books
as their colleagues in the Union. Those southern physicians who had attended medical school (as opposed to only studying with a preceptor) had,
many of them, either trained in the North or studied under professors who
were themselves the product of northern schools, especially the University of
Pennsylvania. In North Carolina in 1860, for example, 193 of the 233 members of the reform-minded North Carolina Medical Society had obtained
their professional training in Union states.[12] Leaders of southern medicine
did not want to create different institutions in the South, according to some
new scheme, but rather to emulate the best hospitals and medical schools
of the North and of Europe as they sought to build a new medicine for their
new country. They also built on a rich local tradition of exploring the ecology
of their region to identify information about local disease patterns and to
discover indigenous healing plants.[13]

The reform impulse among Confederate medical leaders can be viewed as
a sort of "medical nationalism." Much has been written about the question
of Confederate nationalism overall: Did it ever truly exist? Was it trumped
by the loyalty to home state embodied in the phrase "states' rights"? Did
it perhaps exist in the heady days of origin but fade over the course of the
war, contributing to a loss of will and decline in morale that in turn led to
Confederate defeat?[14] Medical nationalism within the Confederacy can be
approached with similar questions, and it suffered from some of the same

tensions. Moore wanted to create an organized federal system but had to pay lip service to states' rights to appease local sensitivities. And the objectives of his medical nationalism—providing excellent hospital care, pursuing new knowledge through research, elevating the education level of less-well-prepared practitioners, and creating a native pharmacopoeia—all ultimately foundered on the rocks of Confederate defeat. There is little evidence that the bulk of Confederate physicians, struggling to provide even the minimum standard of care with short supplies, ever took Moore's grand ambitions seriously. Southern medical institutions needed at least a decade to begin to recover from the destruction of war, and not until the twentieth century did southern medicine approach the quality envisioned by Confederate medical reformers.

This phrase *medical nationalism* is not just a concept about how the Medical Department might be organized; it also serves as a label to cover the pride in all things Confederate and medical and to reflect the creation of new institutions within the nascent nation to mirror those left behind in the United States. Hence the need for a new national medical association to replace the American Medical Association and a new national medical journal to replace the AMA's *Transactions*. If the Union created a process for gathering interesting specimens and case histories so that information could be garnered from the war's experiences (as it did), then the Confederacy should do so as well. Medical nationalism, used in this way, does not counter a view of medical practice that was essentially local—based on personal relationships and tied into local environment and influences—but rather adds a layer of ambition and perspective on top of the local perspective. It drew on the ideas of states'-rights medicine expressed in the 1840s and 1850s and diminished after the war into the idea of distinctively southern approaches to disease and healing.[15]

Moore and his colleagues recognized that medical research thrived in the hospital and that the war opened vast new opportunities for such research. The antebellum South had few hospitals, and many that did exist served African American slaves who had been leased for urban work (and thus were away from their master's care) or were in transit for sale in western markets. Touro Infirmary in New Orleans, for example, did a large business in sick slaves who were pending sale in the major slave markets in New Orleans.[16] These black hospitals did not provide sufficient clinical opportunities for southern students, and only a few southern cities had the sort of voluntary charity hospitals for whites that were more common in the North, and especially in the cities of Europe.[17] There were exceptions—Charity Hospital in New Orleans, for example—but only with the war did the South suddenly have thousands of men in hospitals whose bodies replicated in part the opportunities offered by the hospitals abroad. Moore wanted to take every advantage of this situation to promote southern medical research and education.

But first he had to face the same challenges as his Union counterparts—organizing field relief, building hospitals, recruiting and examining physicians, and training those physicians for tasks that may have been unfamiliar in civilian life. Moore saw hospitals as a place for research and training, but he answered to a political and military leadership that saw the hospital very differently. Military leaders were uninterested in research; they needed every man available in the army and wanted their sick and wounded men returned to duty quickly. Given a choice between care in a hospital and care at home, most soldiers would have preferred the latter, but their officers saw the former as the best way of keeping them under military discipline and efficiently returning them to active duty. This chapter explores how Moore and his colleagues met these various challenges and attempted to institute their model of medical reform; chapter 8 describes how it all fell apart. Although Moore briefly created at least the beginnings of his new medicine in Richmond, the center could not hold. The story of Confederate medicine was instead one of despair and ultimately dissolution.

Organization of a Medical Corps and a Hospital System

As in the North, most of the men who served as surgeons in the Confederate army began the war in private practice and learned their trade as military surgeons on the run. Only twenty-four surgeons resigned from the Union Army to join the Confederacy; the remaining three thousand or more Confederate army surgeons came from domestic practice. They were unprepared for tasks that awaited them on the battlefield.[18] As Confederate surgeon Joseph Jones noted after the war, "the surgical staff of the army was composed of general practitioners from all parts of the Southern country whose previous professional life, during the period of unbroken peace that preceded the civil war, 1861–'65, gave them but little surgery, and very seldom presented a gunshot wound."[19] To remedy this ignorance, Confederate surgeon John J. Chisolm, who had studied military surgery during the 1850s in Europe, rushed into print in summer 1861 a guide to performing such surgery, and Moore had it distributed to Confederate surgeons.[20] This volume also served to transmit the rules and regulations of the new medical department. One of the first imperatives of the Confederate medical corps was education.

As in the Union, surgeons associated with regiments were originally chosen from among the community's available doctors as the regiment was forming. But like his Union counterpart, surgeon general S. P. Moore soon put into place a system for examining surgeons and determining their suitability for appointment as contract surgeons or as commissioned assistant surgeons and full surgeons. He set up boards to "scrutinize the moral habits, professional arguments and physical qualifications of the candidates requiring from them certificates that they [were] equal to the arduous duties of the position whether in field or hospital."[21] Ultimately there were five boards spread over the geographic regions of the South. One count puts 3,237 doctors on the Confederate roster, although arriving at such a number requires some guesswork because of lost records.[22] While estimates of the number of soldiers enrolled in the Confederate army vary from six hundred thousand to just over a million, it is certain that much illness prevailed in these ranks. One observer looking at the first two years of the war found that on average each man was either wounded or sick six times over that period.[23] There was plenty of work for those surgeons to do.

Moore had to create a medical hierarchy from scratch, and he followed lines similar to those of the Union Army. But he struggled to impose a federal system of central control on an army organized by states, and confusion about the medical order of command persisted into 1863.[24] He created a bureaucratic system to track supplies, personnel, and hospital expenditure, published as the *Regulations for the Medical Department of the C. S. Army* in 1862.[25] At the same time, the medical system was evolving locally, not exactly beyond the control of Richmond but in response to urgent, immediate needs. The official hierarchy often adapted after the fact to events on the front lines, such as when the government took over hospitals formed in response to emergent medical and surgical needs. While Moore sought to build from the top down, he met many local instances of creation from the ground up.

Most Confederate surgeons crafted their own local medical systems, under the leadership of Moore but not always in accordance with his direction or inclinations. Medical care began with the regiment and fed into the big general hospitals when necessary. North Carolina sergeant Duncan Buie explained this system to his friend Kate McGeachy in July 1862. "I have been from the hospital about one week," he wrote her from the North Carolina low country. He hastened to assure his friend, "The hospital is not such a horrid place as you suppose it is. It is decidedly the best place on this side of home a sick soldier can find. We have in our hospital women to attend us. They are too very kind and attentive. Moreso than you might suppose," he teased (he married Kate in 1866). "We could not get along at Fort Caswell without a hospital. Sickness is so common that it is scarcely noticed." Buie's unit had seen the usual round of childhood sicknesses run through camp— measles and mumps and chicken pox—as well as pneumonia and malaria. Several months later typhoid appeared, and the mortality rate rose. They had plenty to eat, the company of friends, and generally good care on this small scale, a scene repeated around many Confederate regiments.[26] Usually when a regiment went on the march, men in the regimental hospitals were sent to general hospitals, so the unit's surgeons could go with the army.

A higher level of organization began to form naturally enough, as regiments merged their regimental hospitals into state-identified institutions. After a battle the surge of wounded overwhelmed regimental hospitals, just as the armies prepared to move on for another day's fighting, taking their regimental surgeons along. In response medical officers set up state-level hospitals for ill and wounded men left behind by their regiments. In Virginia, home to the first battle of Bull Run and other early encounters in the war, surgeons from the Georgia regiments established a hospital for their men, and surgeons from North Carolina regiments likewise rented a school, church, or hotel for the care of their troops, particularly after a major engagement. Some hospitals were true field hospitals, which served the wounded immediately after a battle and then disbanded as the army and the wounded men moved on. In 1862 and 1863 orders came down to the regimental surgeons, instructing them to organize "field hospitals in houses to be rented if such can be conveniently found within their lines, otherwise under canvas, to which all their sick will be sent to be attended by their own medical officers."[27] Further orders dictated the organization of ambulances and field infirmaries in the wake of battle and described the sort of packed knapsacks and supply containers medical men should have on hand to carry forward for field relief.[28] Hospital patients were frequently on the move, either back to their regiments for duty or forward, to better established hospitals, when their regiment moved on.

Moore sought to bring structure and order to this initially chaotic system. But his attempts at centralization and standardization were stymied by the very philosophy that led to the Confederacy's creation in the first place. Whereas by summer 1862 the Union was moving toward a general hospital system unrelated to state of origin, and the U.S. Sanitary Commission was

promoting a nationalistic attitude toward patient care and community donation, the Confederacy remained officially and actually tied to a state-level
organization of medical care. This was most evident in Richmond and the
other major hospital centers in Virginia. Richmond had 60 percent to 70
percent of the South's hospital beds by 1863, with at least fifty individual
hospitals. Many of these hospitals occupied existing structures, such as factories, schools, almshouses, and hotels. Three of the largest were newly built:
Chimborazo (October 1861), Winder (April 1862), and Jackson (June 1863).
Moore ordered them all constructed according to the pavilion plan, with row
after row of long sheds housing patients. Although it is fair to call Chimborazo a hospital, it was considered in fact a loose confederation of hospitals.
With a capacity at times of more than eight thousand beds, it was made up of
five divisions of thirty buildings each, with additional buildings for support
such as baking, brewing, and washing. Just as the Confederacy consisted
of states loosely governed by a central president, Chimborazo was made up
of the Georgia hospital, the Virginia hospital, the North Carolina hospital,
and so on. Winder opened a few months later on the outskirts of Richmond
and covered 125 acres of land, with a bed capacity at times exceeding five
thousand, and with six divisions. Most of its patients came from Georgia
and North Carolina. Jackson Hospital, located in the western suburbs just
beyond Hollywood Cemetery, housed an additional twenty-five hundred patients and was devoted to North Carolina, South Carolina, Georgia, and Louisiana troops.[29]

Hospital centers were also located at Lynchburg, Danville, and other cities
in Virginia.[30] More than 400,000 Confederates were wounded in Virginia,
and most needed hospital care, if only briefly; hospitals in Virginia cared for
412,958 men between September 1862 and August 1864 alone.[31] But during
1862, when the Union was building large general hospitals in the northern
cities closest to the Confederacy, the Confederate government solidified the
state arrangement into law. Even where such distinctions made little sense,
hospitals had to be renamed and designated. If in Virginia the entire roster
of Confederate states was represented, in South Carolina the troops garri-
soning the coast were largely local. Yet there, too, medical officers had to
rename hospitals by state, when the order came down in 1862: "Hospitals
will be known and numbered as hospitals of a particular State. The sick and
wounded, when not injurious to themselves, or greatly inconvenient to the
service, will be sent to the hospitals representing their respective States, and
to private or State hospitals representing the same."[32] So the Wayside Hos-

pital in Charleston became the First Louisiana Hospital, another hospital in Charleston became the First Virginia, and a hospital that eventually settled in Columbia was the Second North Carolina.[33] This style of organization was defended as being true to states' rights, as making it easier for family members to find patients (there was no USSC creating hospital registries), and as simplifying patient assignment when a large contingent of wounded arrived from the battlefront. As historian Jerrold Moore has said about the Confederate commissary general's attempts to create a centralized system to supply the army, "The entire South was permeated by a persistent affection for local arrangements and local ideas."[34] When because of expediency Maryland men were admitted to a hospital in Richmond designated for Virginians, it caused quite a fuss. Some hospitals were identified by disease (smallpox or gangrene, for example) or other markers (rank or race), but on the whole the Confederate medical system simultaneously clung to states' rights as its leaders sought to impose centralizing order and efficiency.[35]

The benevolent impulses of the South likewise remained on a states'-rights basis. While the USSC decried this attitude—condemning the woman who walked past a Wisconsin bed to tend to an Illinois one—and saw it as being in concert with the very roots of the rebellion, the southern woman who tended the wounded had no difficulty in choosing to help only "our boys." Without a national organization such as the USSC to organize and stream benevolence to hospitals where it was needed most, Confederate women's relief work remained resolutely local. If women traveled to Virginia to nurse, they headed to the hospitals for their states; they likewise channeled food, clothing, and other support to the men of their state regiments.[36] Women founded hospitals on the spot when the wounded came pouring in from nearby battles and there was no institution to care for them. These temporary female-run hospitals retained autonomy longer than similar makeshift care centers for Union troops, precisely because the South lacked the strong centralizing powers of the North. The North had no institutions labeled "the ladies hospital"; the South had many. Although the Confederacy took over many of these hospitals later in the war, some remained under female control to the end, such as Robertson Hospital in Richmond, run by Sally Tompkins. Women performed much the same work in the southern hospital as in the North. But while northern women were mostly far from the battlefront, most southern women saw it come to their front yards and as a consequence had a far more direct role in supplying immediate relief and organizing locally to aid the sick and wounded.[37]

Photograph of Chimborazo Hospital. Prints and Photographs Division, Library of Congress, Washington, DC.

Imposing Order and Systematic Care
on the Confederacy's Hospitals

Moore made major efforts to move toward standardizing hospital practices in the face of the persistence of states'-rights attitudes about the care of the sick. His prescriptions for the ordering of a proper hospital would have sounded quite similar to Union surgeon general Hammond's and were founded on the same set of ideas put forward by Florence Nightingale. He commanded that each bed should have eight hundred cubic feet of air surrounding it and be no closer than six inches to the wall, to secure adequate ventilation. The bed sacks, stuffed with straw or horsehair, should be renewed once a month and all bedding frequently aired and washed. Each ward was to have a medications receptacle, and each patient's medications were to be properly labeled with his name and dose. He should be able to read the posted diet explanation, so that he understood the rationale behind his food choices. Ward matrons were to be in charge of medications and to supervise the nurses; hospital matrons would supervise the ward matrons as well as the kitchens and the laundry. Ward masters saw to the maintenance of cleanliness, including the disposal of wastes, daily cleansing of the floors with sand, and whitewashing of the walls two to three times a year. Hospital rules were posted so that all knew their roles and expected behaviors.[38]

These orders are impressively detailed, and in that attention to minutia one can see either an obsessive determination to micromanage or else the recognition that most Confederate surgeons had little experience with how to order a hospital. There is also the fact that the medical department had come under significant criticism from the Confederate Congress, especially in regard to nursing care, and these rules were perhaps in part an answer to such critics.[39] The only surgeons with any hospital experience would have obtained it in either the slave hospitals or the urban indigent hospitals that did exist in the South, so even these seasoned hospitalists needed to learn to take care of more affluent patients properly and with propriety. Although Confederate hospitals never had the means to duplicate the comforts of the large Union hospitals such as Satterlee (no pianos or billiard tables here), there was a recognition that the hospital environment was key to healing and that the men resident therein might well demand higher standards than the indigent or the slave could have expected before the war. The men complained often in letters to home, at times rioted, and most of all just deserted as a response to hospital conditions that did not please them.[40] Moore's orders can be read as an attempt to counter these problems by rigid organization and cleanliness.

There was also clearly a component of discipline in the surgeon general's orders and supervision. As Michel Foucault's work emphasized, the hospital had some features in common with other institutions, such as prisons.[41] The men were not free to leave, and they were under military orders as to their behavior. So Moore ordered ward masters to call the roll twice a day, make sure that "convalescents do not lounge on the beds," see that the nurses were "quiet, orderly and respectful," and allow each patient only one visit to the dining hall per meal.[42] As desertion became more and more of a problem, greater attention had to be paid to preventing it. A surgeon in southwestern Virginia, for example, advised his colleagues putting patients on trains in Lynchburg for transfer to his hospital to lock the doors of the cars after the men were loaded. Too many men were deserting en route.[43] Each hospital had its guards to watch for escapees, although as the war progressed and manpower became ever more precious, hospital guards were pulled away from this duty and put into the regular army. Physicians in charge of hospitals had to maintain order with the aid of the barely convalescent; a man fully fit for guard duty was also deemed ready to return to his regiment.[44]

Doctors were not exempt from this disciplinary eye. Moore sent a corps of medical inspectors around to report on the state of hospitals and military

medical care and appointed medical directors for various regions to supervise the hospital systems in their charge.[45] He was not shy about calling out
the slackers, although it is unclear whether his protests resulted in remedial action. One letter to a South Carolina hospital from the surgeon general
barked, "It has been reported to this office . . . that medical officers in charge
of wards in hospitals, fail to give the necessary attentions to the patients entrusted to their charge. This information is received with mortification and
deep regret. The lives of officers and soldiers in our armies cannot be left to
the care of careless, indifferent, and incompetent medical officers." Although
it clearly should not need specification, the surgeon general spelled it out: "It
is, therefore, the duty of surgeons of division, to make morning & afternoon
visits to their wards, see each patient, and ascertain from him, if all his wants
have been supplied, and report to Surgeon in charge without delay."[46]

Doctors may have been particularly prone to abandon their duties at
night, as they were drawn off to evening parties or other entertainments.
One caustic letter from a supervising surgeon to a subordinate decried this
practice. "Your presence is frequently required in the Hospital and particularly at night when the sick soldier is too often neglected by the attendants
&tc which has been often detected by myself during my late night visits to
the wards." He went on to remind this miscreant of his duty to the soldiers
and his country. "The Confederate Government provides amply for the sick
soldiers and requires that they should be well taken care of and attended
to. Medical officers are charged with this responsible duty. It cannot be performed if absent from their Hospitals until late hours of the night. Officers &
subordinates, attached to this Hospital must rigidly conform to the Regulations of the army."[47] The diary of John Apperson, a hospital steward with the
Army of Northern Virginia, reveals a surprising variety of social gatherings
that encircled the medical corps (while being frustratingly brief about medical information that might be of interest to the historian). Apperson would
have been a physician if the war had not interrupted his medical studies; he
and his medical colleagues typically joined with the other officers in enjoying
the genteel hospitality of the neighborhood, perhaps at times to the detri-
ment of the men under their care.[48]

Union surgeons may have had more experience at hospital work, since
there were more hospitals in the North, but neither side had large numbers of
physicians with the skills to run the complex organization of a four-thousand-
bed hospital. The southern physician who combined medical knowledge
with previous experience running a plantation may have acquired mana-

gerial skills that were useful in hospital administration. He would, at the least, have had some experience with ordering and producing food in bulk, handling problems of supply and transportation, supervising workers who might well be unwilling, unskilled, or otherwise ill-adapted to their tasks, and so on. Much of hospital life was structured under the supervision of the female hospital matron, and the southern plantation owner may have experienced a similar division of labor on the plantation, where his wife supervised certain aspects of the organization. The hospital-as-plantation has more overlap than, say, the hospital-as-Arctic-expedition. If southern surgeons had not themselves owned plantations, the ruling elite of southern medicine would likely have come from families that did.

Historian Nancy Schurr has emphasized that "Confederate general military hospitals were a microcosm of southern society and that, over time, they evolved into complex institutions that often defied the intentions of their creators."[49] Schurr found frequent reference to the hospital-as-family, especially in the maternal relationship between nurses and the sick men. She also stressed that each actor within the hospital saw it from a different perspective, although it is much easier to recapture the white nurse diarist's or the white male physician's perspective than those of the hospitals' black workers, who could make up 75 percent of the labor force of a hospital. Plantation, rather than family, may be a stronger metaphor for the hospital—although in the South these two were often deeply intertwined.

This history of plantation life with its overtones of oppression and force may have contributed to an attitude toward patients. If the emphasis on order and cleanliness reflected an attitude of respect toward the patient, other indications argue against this perspective. While those in the North may have moved toward recognizing the rights of patients, in the South there was apparently little alarm at the issuing of the following proclamation in May 1864. "Medical Directors will instruct Medical Officers," ordered Surgeon General Samuel Preston Moore, "that in all cases requiring surgical interference, in which no serious risk to life is incurred by such action they have the power (in the opinion of this office, sustained by the approval of the Secretary of War) to compel submission to surgical operations." With these words Moore revealed some anxiety about the legality of the order, but nevertheless he hastened to justify it. "This opinion is expressed with a full appreciation of the mutual obligations resting upon the Soldier and the service, and with the intention of securing the mutual advantages equally derived from a just fulfillment of these obligations. The power vested in the Medical Officer should be exercised firmly, yet with due delicacy and caution."[50] The soldier's body belonged to the army for the time of enlistment, and at least in the Confederacy, he lost as much right to its disposition as the black slave who suffered a similar loss of agency. Perhaps to soften the assertion of power, Moore reminded medical officers that enlisted men were entitled to one pound of tobacco per month.

This document emphasizes the persistence of class structure in the Confederate hospital. As was also the case under Union medical custom, officers were housed in separate hospitals or in separate wards within hospitals. They might lie side by side with the enlisted men in the chaos of the field hospital, but once they reached the peace of town, the niceties reasserted themselves. In the South sick and wounded officers were more likely to be billeted in private homes or sent home on furlough.[51] In Richmond officers could pay for care at the private hospital run by the Medical College of Virginia.[52] In both North and South they were likely to eat better food, but in the South, where hospitalized men were more dependent upon external food sources, officers would have had an even greater advantage. Officers in camp or in the hospital had to pay for their own food; rations were not part of their pay. They would have had more expendable income to purchase luxuries such as fruit and meat, and they would have been more likely to be literate and able to write home to prosperous relatives requesting the shipment of food.

Although it is hard to find statistics regarding Confederate mortality by army rank, a comparison to Union numbers is at least suggestive. Among the officers of Union volunteer troops, 7,035 men died; 48% were killed in action, 22% died of wounds, and 30% died of disease. In contrast, for the enlisted volunteers, the figures were 22.7% killed in action, 13.5% dead of wounds, and 64% lost to disease. According to a nineteenth-century medical source, the case mortality rate among diseased enlisted men was four times as high as that of officers.[53] One modern historian working with records of a selection of New York regiments found that enlisted men were three times as likely to die of disease as officers.[54] Provost Marshal General James M. Fry offered this explanation for the discrepancy: "This remarkable disproportion . . . is owing to several causes. Officers are better sheltered than men, and their food is generally better in quality and more varied in kind. . . . They are not so much crowded together." He also included superior morale and the better treatment received as prisoners of war. Southern officers would have shared these advantages, although as food prices rose, their ability to purchase food of better quality and variety would have diminished. It also may well be that in the South officers died so often in battle that they had less chance to succumb to disease. Joseph Glatthaar's account of Lee's Army of Northern Virginia found that "close to nine of every ten officers in Lee's army were either killed or wounded [and that] about 68 percent of all company-grade officers—lieutenants and captains—were killed, wounded or captured."[55] So it is difficult to pull out the influence of greater affluence, access to food, and more comfortable care from statistics that compare battle mortality with disease mortality. Joseph Jones, our best source on Confederate mortality statistics, includes the comparative numbers for the Union but does not break down the Confederate numbers by rank.[56]

Distinctions by race likewise persisted in hospitals, both northern and southern. The Union had black enlisted men to care for and always segregated them from white soldiers in separate wards.[57] Southerners had black workers, slave and free, working in a variety of roles for the government. They also required hospital care when sick and competed for scarce beds and other hospital commodities. William Davis, surgeon in charge of one segment of Chimborazo hospital, wrote to its director, J. B. McCaw, in March 1862 that the boss of a black ambulance crew had approached him about care for some of the workers who were sick. General Winder had ordered that the men be cared for at Chimborazo. Yet McCaw had just ordered Davis to make ready one hundred beds for wounded Confederates. "If I had received the negroes I should have been obliged to give to them these accommodations, to the exclusion of an equal number of soldiers, and to the annoyance of a larger number, as my hospital buildings are in close proximity. I have no isolated building fit for a negro ward."[58] While there could be no doubt that the white soldier took precedence over the black worker, black workers were highly valued and their labor was much needed in Civil War Virginia.

Southerners had routinely provided care for black slaves before the war, either on the plantation or in special slave hospitals, so this practice was not new to them. Labor became more and more precious as the war dragged on, making those black bodies even more valuable. The provision of medical care for them was a matter not of philanthropy, but of necessity. The medical director of the Army of Northern Virginia made this point clear in a letter to a subordinate physician. "The employment of negroes as teamsters & laborers in the place of enlisted men has rendered it necessary to provide special hospital accommodations for them. You are therefore directed to have two or three hospital tents at each Division Infirmary reserved for the use of the negroes who may require medical & surgical treatment and direct the surgeon in charge to render them all necessary attention."[59] Surgeons struggled to maintain proper social hierarchies amid the forced intimacies of war. They also used such proximity to humiliate the enemy. When captured wounded Union soldiers were hospitalized in Charleston, the southern surgeons took pleasure in putting black and white troops together. The white Union prisoners were disgusted, but the Charlestonians jeered that it served the Yankees right. If they wanted equal rights for blacks, then here it was, they could have it.[60]

Such discussions emphasize the tensions between the inherent disorder of war and the drive to bring order to the hospital system. Both northern and southern surgeons general sought to centralize and make uniform the systems under their commands. Like the North, the South had a standard list of medications provided by the government in the "supply table." Surgeons had preprinted forms to use in placing their orders for medications and supplies (such as bottles and bandages). Union surgeon general Hammond sought to influence physician behavior by manipulating this medication list, most notably in the case of calomel. If Moore ever had such a reform in mind, it will never be known. His major action in altering the supply table was to remove a whole list of drugs and equipment (including stethoscopes and lancets) in April 1864 because shortages of drugs and money meant that the cabinet had to be curtailed.[61] He also urged physicians to seek local plants to supplement the diminishing official supply of drugs.

Moore did take steps to curb the performance of unnecessary or inept surgical procedures. He could do nothing to interfere with the emergency surgeries that immediately followed battle, but he set up boards of review in hospitals for surgeries done there. In August 1863 he ordered, "Medical directors of hospitals will organize at each hospital station a board of medical officers to be consulted by the surgeon in charge, in all important surgical cases, or when an operation is deemed necessary: and when delay is practicable. No important operation will be performed without the sanction of the board."[62] Such a step recognized the value of clustering physicians in hospitals, so that the most experienced could teach the novice. Moore's review boards codified the sort of oversight that later became typical of teaching hospitals. This order also acknowledged the persistence of ineptitude and poor judgment among the medical ranks, since otherwise the order would have been unnecessary.

Moore's research and reform agenda had its greatest influence in hospitals within the purview of his office in Richmond. The situation west of the Appalachians might be within his pen's reach, but it remained out of his control. Historian Glenna Schroeder-Lein has described the work there of medical director Samuel Stout and his daunting task of maintaining a hospital system "on the move." Whereas the hospitals in southern Virginia remained intact and near the battlefront for much of the war, in the West hospitals had to follow the armies and were frequently at risk of being overrun. Stout proved to be an impressive administrator who made the best of daunting circumstances. Moore's medical reforms made few inroads on a system of medical provision that was almost constantly in emergency mode.[63]

Medical Research in the Confederacy

Moore's broadest attempt at gathering information from his surgeons was to ask them to send reports of interesting cases to his office in Richmond, in addition to their required statistical reports about sick and wounded soldiers. Similar requests to Union physicians netted the hundreds of pages of case reports that filled the *Medical and Surgical History of the War of the Rebellion*. In the South, however, it is unclear how many surgeons ever returned these records to the central office, since it burned along with much of Richmond in 1865. Moore sent special medical diaries to his surgeons, to record special cases; one that survives in the Museum of the Confederacy contains regular entries for the first few weeks and then afterward is mostly just a roster of patients in the hospital.[64] Another one, held by the Duke University Library, is as blank as the day that it arrived in surgeon James Tracy's hands.[65] Moore continually struggled to get his surgeons to file even routine reports properly. He chastised them with a circular letter in April 1863. (Circular letters served as memos to the entire medical corps, circulated either through printed copies mailed to the doctors or, as paper became scarce, via text printed in newspapers.) This one noted "the frequency with which Reports of the Sick and Wounded reach[ed] [his] office incomplete and defective, and wholly unsuited to establish accurate and reliable statistics concerning the prevalence of disease throughout the army" and expressed Moore's disapproval. He then instructed his medical officers (again) on the proper ways to fill out the forms. If he could not get his men to comply with the basic reporting guidelines, it seems unlikely that he had much success in obtaining the special case reports.[66]

In September 1863 Moore organized the Association of Army and Navy Surgeons, with the purpose of gathering and sharing the wartime medical experience of Confederate surgeons. Moore was not particularly known be-

fore the war as a medical reformer or as a physician active in medical organizations; until 1861 he was, as mentioned, off on the frontier following his
military surgical practice. So it is unclear whether the ideas for promoting
medical research were his own or whether he was following the suggestions
of other elite surgeons around him. In any event, he began by circulating
questions about wound healing and later interrogated surgeons about chloroform use, hemorrhage, and other surgical topics.[67] One related product of
Moore's zeal was a new journal, the *Confederate States Medical and Surgical
Journal (CSMSJ)*, which first appeared in January 1864.[68] In an unsigned editorial either written by or approved by Moore, the author issued a clarion call
for medical nationalism and research. "Amid the din of war's wild alarms,
when the shock of opposing armies is felt around us, while a new-born nation struggles for its breath, even then the calm, peaceful voice of Science is
heard," the editorialist reassured. "Let all who love her heed the call. . . .To
do justice to the herculean task, to vindicate themselves and their art, the
medical staff of our army must fulfil [sic] its duty to Science, and prepare to
lay before the world the results of its labors."[69] Historian H. H. Cunningham
lists J. B. McCaw as editor of the *CSMSJ*, although his name does not appear on the masthead.[70] Moore hoped that the association's meetings and the
publication of the *CSMSJ* would promote these ambitions. It is likely that the
American Medical Association and its journal, *Transactions of the American
Medical Association,* were models for these steps in professional elevation.[71]

Moore's association first convened at the Medical College of Virginia in
August 1863. Dr. J. B. McCaw, who headed Chimborazo Hospital and was
also a professor at the Medical College, hosted the meeting, which elected
Moore as president of the association. Some ninety-six surgeons paid the
ten dollars to join the group, mostly local men, with a few "corresponding members" away from the city. The association originally planned to
meet every other Saturday night, and minutes record twenty-one meetings
through April 1864.[72] The anonymous article in the *CSMSJ* that reported
the founding of the association praised Moore for not losing "sight of the
permanent advantages that might accrue to science out of the progress of
this war." Moore's efforts would glean the accumulated wisdom of the Confederate surgeons, "creating the means for prosecuting research under most
favorable auspices."[73] The founders hoped that members would bring their
collective experiences to the meetings (or send written reports), that issues
could be debated in turn, and that the association would then vote on the
arguments deemed most convincing.

The journal published some of the proceedings of the medical association, which continued to meet in Richmond. At the January 30, 1864, meeting, President Moore was in the chair. After he read some brief letters about notable cases, a lively debate ensued on the nature and symptoms of tetanus. Eleven physician participants were quoted in the *CSMSJ* article, but no count of those present was included. The association's proceedings were recorded two more times that spring, although it met at least six times in that time period, with the last recorded manuscript minutes in April.[74] At the late-March meeting, a number of letters were read on the subject of tetanus, reflecting the likelihood that these distant surgeons had read the prior discussion as published in February and now wished to respond.[75] Moore must have been pleased; this sort of interchange was exactly what he hoped to sponsor as the association mined the collective wisdom and experience of Confederate surgeons. But the young sprout of collegiality and group learning apparently shriveled afterward, as the gruesome series of battles during summer 1864 overwhelmed the Confederacy's medical men.

The journal was not only to be the repository of information gained by physicians during the war; it also aimed to distribute the latest medical news from abroad, since the traditional avenues for such knowledge transmission, the southern medical press, had all but ceased during the war. Few copies of the journal now exist; that fact probably points to its likely limited impact, although a prospectus heralding the publication of volume 2 claimed that it had a broad circulation. The persistence of optimism evidenced by this prospectus, published in December 1864, is an impressive testament to Confederate hope amid looming disaster.[76]

Moore's drive to gather group knowledge from his surgeons bore fruit in the person of Georgia physician and Confederate surgeon Joseph Jones. He wrote the lead article of the *CSMSJ*, an account of traumatic tetanus, which fueled the debate at the medical meetings. Jones was one of the few Confederate surgeons with skills and interest in medical research, and Moore apparently gave him respite from other duties so that he could, for example, collate hospital statistics in 1863 and investigate conditions at Andersonville in late 1864. Jones eagerly explored the causes of camp diseases, employing his background in chemistry and peering through a microscope at patient discharges. His main research tool, though, was interrogating his fellow physicians, seeking to mine their experiences. He wrote reports for the surgeon general, based on his travels to various hospitals and especially to the Andersonville prison.[77]

Jones was the exception, however, and it is curious that Moore allowed him to do so much research when the Confederacy was short of doctors. Perhaps Jones's persistent ill health protected him from regular duty.[78] The opening editorial of volume 2 of the *CSMSJ* bemoaned the disinterest among southern medical men in its research agenda. "If the Journal fails," wrote the anonymous editorialist, "its conductors have determined that the fault shall be laid to the door of the Southern profession, who refuse to bestir themselves in spite of every inducement." All that was needed was that they "in the midst of tumult, with quiet minds to ponder over the vast fields of knowledge which surround them . . . be faithful to the last to [their] noble art."[79] Moore persisted in this attitude—chiding his surgeons for not return-ing proper forms and case reports—even as, for example, W. T. Sherman was approaching the hospital center at Columbia, South Carolina, and the end was weeks away.[80] Although the *CSMSJ* published numerous case re-ports and research papers stemming from state-supported projects, it never achieved the degree of support from the southern medical profession that it sought. The physicians were otherwise occupied, trying to keep the Con-federate soldier on his feet. It is fair to say that the same sort of interest in research existed among northern and southern elites, but in the South there were fewer such men, and the absence of resources made research even less likely to happen.

Moore's research enterprise generated the most activity in the search to find indigenous southern plants to use as medications, especially for the treatment of malaria.[81] South Carolina physician Francis Peyre Porcher, who had a special interest in this topic, published his *Resources of the Southern Fields and Forests* in 1863 to inform southern physicians of the local plants available for Confederate medical use. Many of these plants contained in-gredients even now recognized as active, but there was no doubt at the time that these herbal remedies were substitutes for preferred drugs that had previously been imported. Central to this discussion was quinine, es-sential for the treatment of malaria and available only through the blockade or by capture of enemy stores. Porcher's recommended concoction of barks (dogwood, poplar, willow) in whiskey was widely tried but soon abandoned; its ineffectiveness was evident even to the most hopeful of physicians.[82] A Dr. Fair experimented with one tree, *Pinckneya pubens,* found in swampy areas from the New River in South Carolina to the gulf coast of North Florida, and discovered that it contained cinchonine, a compound similar to quinine in its effectiveness against malaria. The supply of the bark was not sufficient to put a dent in the Confederacy's need for quinine, and the drug received only passing mention in Confederate documents. It is not known whether extracts from this tree's bark were found to be ineffective or whether southern physicians had insufficient access to the tree because its native habitat was under Union threat or control in the later years of the war.[83]

There was also a flurry of interest in the external application of turpentine as a substitute for quinine, perhaps because the liquid felt cool to the fevered skin. Turpentine is made from pine sap, and pine trees were abundant throughout the Confederacy, especially in North Carolina. In September 1863 the surgeon general sent a special report around to surgeons, urging them to try it and report on the results. The form sent to the hospital in Columbia, South Carolina, where there were many malaria patients and not nearly enough quinine, remained blank and was not returned.[84] But apparently other surgeons were more compliant, as an editorial in August 1864 reported the receipt of "seventy returns, involving over four hundred cases." The anonymous author, either Moore or a subordinate writing with his approval, noted, "The remedy is regarded as one of great power, if not positive efficiency." But he added, "Candor compels us to say that our conclusions are not so positive." He feared that the results were confused by administration of this remedy in the company of others, which might deserve the credit for improvement, or that the patient would have recovered even without the intervention, giving the false impression of therapeutic efficacy. He concluded, "[Turpentine is] one of the large class of agents which may be rendered useful in the treatment of periodic fever, as an adjuvant to other remedies, but . . . it does not deserve to be regarded as a specific in the treatment of such affections."[85] In short, it was no quinine.

Confederate surgeons tested other treatments for common diseases. As smallpox sickened more and more soldiers in 1863 and 1864, Moore urged physicians to try tartar emetic ointment to prevent the scarring of facial eruptions. W. A. Carrington, medical director of the Richmond Hospitals, rebuked one surgeon for his nonparticipation in this study. "You have reported . . . 'no case' in which you had used the tartar emetic ointment as a counter elimination to the eruption of small pox in the face," he wrote accusingly to Surgeon H. Barton of Richmond Hospital no. 13 in February 1864. "You will try it in all skin cases as are diagnosed sufficiently early and forward your notification . . . through this office."[86] Similarly, Carrington ordered McCaw, the surgeon in charge of Chimborazo Hospital, to set up a separate ward for skin diseases. Research on "camp itch" was particularly wanted. "All such cases coming to Richmond will for the present be sent to Chimborazo principally because of the advantages offered in their treatment, by scientifically conducted baths. Let any plausible and well recommended mode of treatment be tried, and report made at the end of the month."[87] McCaw was also ordered, directly by Surgeon General Moore this time, to test "Worthington's Cholera Mixture" and "report the result to this office."[88]

Moore had hopes for surgical research as well. A southern dentist, James Baxter Bean, developed a splint for immobilizing fractures of the lower jaw. As late as March 11, 1865, Moore had W. A. Carrington issue a circular ordering men with maxillary fractures sent to a special hospital in Richmond to try the new apparatus. In the same month, Carrington announced that another specialized hospital had been organized in Richmond to treat "unhealed gunshot wounds of long standing especially ones causing deformities." Surgeons at the new orthopedic hospital would employ experimental instruments and a new type of apparatus. Richmond fell to Union troops three weeks later; it is unknown whether any patients were treated in these announced special hospitals. But their creation demonstrates Moore's determination, to the bitter end, to continue research into medical questions of importance to the military and to southern medicine.[89]

Conclusion

Moore's dream of a budding medical nationalism failed as the Confederacy yielded to Union forces. A revival of the research enterprise, the Association of Medical Officers of the Army and the Navy of the Confederacy, formed in 1874 but foundered almost immediately; it was refounded in the late 1890s. By then it had become an organization dedicated to preserving the medical history of the war rather than to finding and testing new knowledge. Southern physicians rejoined the American Medical Association in the decade after the war and did not create the Southern Medical Association until 1906, when the AMA meetings had grown quite large and the regional group offered a podium for younger and novice presenters. Two years later the organization began publishing its proceedings in its new journal, the *Southern Medical Journal*.[90] Although initially not focused on "southern" medicine per se, the association evolved a subgroup of presenters on such tropical diseases as malaria and hookworm that particularly plagued the South. By 1915 the journal had a feature called "Tropical Diseases and Public Health," which eventually spun off as the *Journal of the National Malaria Society* in 1942.

To some extent this sort of research fulfilled Moore's vision of a southern medical research enterprise, although the broader goal of duplicating the northern record of the war's work never succeeded.[91]

What is perhaps most remarkable about Moore's vision for a new southern medicine is that he persisted in implementing it when the situation inside southern hospitals was growing desperate. The disconnect between Moore's goals and the conditions "on the ground" may speak to his not heeding the reports he was receiving, a stubborn dedication to his higher ambitions, or the attempt to change what he could in the face of a chaotic system that was out of his control. Without Moore's correspondence, it is hard to know.[92] But there can be no doubt that, by and large, Confederate medicine did fail, for reasons far beyond the capacity of one man to remedy the myriad problems plaguing the medical care of the southern soldier.

Robert A. Kinloch, a Charleston surgeon who served as medical director and inspector of hospitals for the Confederacy, was president of the South Carolina Medical Association in 1884. His presidential address echoed Moore's goal of improving the profession, in particular by reforming education and re-vitalizing the society to elevate its scientific functions. He attributed the low state of medicine in the 1880s to the poverty brought about by the war and particularly to the abolition of slavery. Kinloch was referring not to the wealth created by slaves but to their accessibility as clinical and anatomic subject matter. "The school amidst slavery was a valuable one. The young physician had an early opportunity for acquiring confidence, tact, wisdom and ripe experience."[93] Such an accessible cadre of patients led to the valuable work of J. Marion Sims, an exemplar of the profession (in Kinloch's view) who made useful clinical advances in gynecological methods while operating on female slaves.[94] "The school [of slavery] that I have referred to," Kinloch lamented, "was richer in advantages and more prolific in results than the one instituted by the bloody war that swept it away. The experience in military surgery was large, but the benefits to the profession of our State have not, in my view, been abundant and lasting." Kinloch stated that the medical culture in camp and hospital, lacking the "calm" of civilian life, offered few opportunities for learning.[95] Kinloch may well have been right about the Confederate medical experience, as will become clearer as we turn to the many deficits of military health care in the southern states.

CHAPTER EIGHT

Confederate Medicine

Disease, Wounds, and Shortages

While Samuel Preston Moore imagined his southern medical Renaissance in Richmond, the day-to-day reality of the Confederate hospital was one of frequent and widespread shortages that affected every aspect of medical care for soldiers as the war went forward. Hospital beds, medical supplies, surgical instruments, hospital workers of all types, and mules and wagons all were in short supply. Surgeons complained of the lack of books, lack of medicines, lack of bottles to put medicines in (and corks), lack of bedding and clothing, lack of wood, and lack, above all, of sufficient and appropriate food for the sick. All of these deficiencies were tied to the inadequacies of the Confederate monetary system, inadequacies of the southern railway network, and the blockade of southern ports, which limited imports and raised their cost.[1] The printed order forms supplied to southern surgeons in charge of hospitals included not just the expected medicines, but bedding, mosquito nets, corkscrews, ink, coffee mills, bedpans, pencils, pins, thermometers, urinals, pots, shovels, candle snuffers, pokers, splints, matches, chairs, and brooms—not to mention surgical equipment and special kits for field use. Very few of these items could be manufactured in the Confederacy. And while their appearance on the printed forms suggested that the government could and would supply them, they were often requisitioned in vain.[2] Adding to these shortages was the chaos generated by frequent moves (especially west of the Appalachians) and floods of wounded men that came in great pulses after battles. The South never had enough surgeons to meet all these challenges.

The Confederate Army always needed more men. After a burst of volunteers filled the ranks in the heady days of 1861, by April 1862 the government resorted to conscription, and still there were not enough.[3] Slaves served the army as servants and were impressed by the government for construction

and other projects; in 1864 free blacks became officially liable to impress-
ment as well—that is, a use of government force already common in practice
was legalized. Insufficient manpower became part of the "Lost Cause" interpretation of the war, as southern sympathizers depicted the valiant little Confederacy holding out against the overwhelming numbers of northern troops.
The burning of Richmond and resulting loss of enlistment records makes
it impossible to know how many men served in the Confederate Army. Historians' estimates range from 600,000 to 1.1 million; Gary Gallagher puts
the number at around 850,000.[4] The shortage in manpower throughout the
Confederacy undoubtedly was real. In any case it is a constant complaint
in the records on hospital staffing that there are not enough people—male,
female, white, black—to perform the medical work. Convalescent soldiers, a
common labor pool for northern hospitals, were moved out as soon as they
were marginally able to return to camp.

The southern hospital's success (or failure) directly impacted the Confederate war effort. If the hospitals succeeded in healing soldiers and returning
them to duty, then the number of men available for the army rose as effectively as if a recruiting agent had found an untapped pool of southerners to
enroll. This chapter chronicles the many deficiencies of southern hospitals
and speculates on their likely effect on a soldier's health and his ability to
return to duty.

All the evidence supports the proposition that southern hospitals were
less effective than northern ones in creating a healing environment. The
shortages they suffered had a very real impact, even by the medical standards
of the time. The first was in the quality and quantity of food; the second, the
availability of personnel; the third, opportunities for rest and shelter; and the
fourth, the availability of medicines and surgical equipment. These aspects
of medical care did make a difference. Men in the field might be expected to
march miles in a day, carrying a heavy pack. They might have to dig fortifications, endure frostbite or sunstroke, and live on vitamin-deficient rations of
salt pork and cornmeal. The sick soldier in the field, particularly one with
diarrhea, needed simple things: rest, hydration, and someone to help him
stay clean and avoid spreading microbes to his fellows. A little opium to slow
down the flux helped too. The adequately staffed, well constructed hospital
offered healing rest, warmth, nutritious food, and hydration, all the factors
of the home sickroom already discussed in earlier chapters. But whereas the
North—assisted by the U.S. Sanitary Commission and backed by the abundance of the Union agricultural and industrial economy—succeeded in cre-

Animals on Parade. *Harper's Weekly,* June 1, 1861. This cartoon flaunts the availability of food in the North, food that the commissary would provide to Union troops. Lea and Perrins steak sauce was a familiar item in the United States by midcentury—though "Lea" is misspelled "Lee" here.

ating such a healing environment, the southern medical department all too often failed.

It would be satisfying, for a historian, to use comparative hospital statistics of the North versus the South to demonstrate decisively how the southern hospital's deficiencies affected the outcome of the war. If one could compare patient survival rates, for example, in Satterlee Hospital (where food was abundant) and in Winder Hospital in Richmond (where the starving patients ate rats and wormy peas), then the hypothesis that better conditions in northern hospitals meant better outcomes could be tested. Or if scurvy was more prevalent in Virginia than in Pennsylvania in 1864, surely the wounded did more poorly in Confederate hands, and the effect would be evident in outcome statistics. The available numbers do not allow us to draw such conclusions with any certainty, however.

The best Confederate statistics available are for Chimborazo Hospital in Richmond, two hospitals in Atlanta that were situated on the fairgrounds there, and the small Robertson Hospital in Richmond. Union authors writing the *Medical and Surgical History of the War of the Rebellion* had access to the Chimborazo record books, and they summarized the hospital records. They did it twice, once in volume 3 and again in volume 5. The historian seeking quantitative accuracy is disappointed that these two accounts do not gibe. They were compiled by different men—in volume 3 (published in 1879) by J. J. Woodward and in volume 5 (published in 1888) by Charles Smart—although there is no explanation for the different numbers they sup-

TABLE 1

Chimborazo Hospital statistics, version 1

	Admitted	Died	Furloughed	Transferred	Unknown	Returned to duty (?)[a]
Diseased	63,357	3,146	8,052	18,049	7,348	26,762
Wounded	14,532	798	5,028	3,809	1,534	3,363
Aggregate	77,889	3,944	13,080	21,858	8,882	30,125

Source: MSHW, 3:28.
[a]Not specified in text but calculated here.

TABLE 2

Chimborazo Hospital statistics, version 2

	Admitted	Died	Furloughed	Transferred	Deserters or discharged	Returned to duty	Not mentioned
Diseased	50,350						
Wounded	14,661						
Unknown	12,057						
Other[a]	821						
Aggregate	77,889	2,717	5,537	14,464	1,675	19,457	26,501

Source: MSHW, 5:29.
Note: The aggregate does not sum to the total number of patients admitted.
[a]"Convalescents and Malingerers."

plied.[5] The number of patients admitted to Chimborazo, from October 1861 to March 1865, is the same in both accounts—77,889 men. Yet in Woodward's account, 3,944 of those men died, and in volume 5 the number was 2,717. Either way, these numbers are impressively low, yielding a mortality rate per man admitted of 3.5 to 5.1 percent. But the number of unknown outcomes renders these numbers as merely the low end of a mortality *estimate*. Thousands of men were furloughed (13,080 in version one, 5,537 in version two); at least some of those men were sent home to die. The outcomes of those 14,464 (version two) to 21,858 men transferred to other hospitals is not recorded. Version one lists the fates of 8,882 men as unknown and does not include a count for "returned to duty," while version two leaves 26,501 patients unmentioned (tables 1 and 2).

These figures yield a mortality rate from wounds of 5.5%, which compares favorably with a white Union mortality from wounds of 9.2%.[6] Yet if all the Chimborazo men furloughed, transferred, and unknown died of their wounds, then the case mortality rate of wounds in Chimborazo would rise to 77.0%. The true case mortality figure for wounds is thus somewhere between 5.5% and 77.0%, but it is impossible to calculate its value from this information. Survival from disease is equally difficult to estimate. Using the

TABLE 3
Fairground hospitals' statistics, Atlanta

	Admitted	Returned to duty	Died	Furloughed	Transferred	Unknown/ other
Diseased	13,067					
Wounded	5,627					
Total	18,694[a]	7,259	896	3,158	5,301	2,291

Source: Welsh, *Two Confederate Hospitals,* 99, 128.
[a]Aggregate given as 18,694 on p. 99, 18,905 on p. 128; rows sum to 18,905.

data in version one, anywhere from 5% to 58% of patients admitted with disease ultimately died of it, in Chimborazo hospital, in another hospital, or at home. The records of Union troops recorded in the *Medical and Surgical History of the War* are drawn from the individual soldier's "carded record," which means the information followed him from location to location when he moved. Equivalent information for Confederate soldiers, if kept, burned in Richmond in 1865.

A statistical study by historian Jack Welsh of two hospitals in Atlanta whose record books survive is similarly frustrating (table 3). He found that more than 18,600 men had been admitted to these two hospitals from August 1862 through January 1865. Of that number, around 39% returned to duty, around 5% died, and the remaining 58% were unaccounted for due to furloughs, transfers, or missing information.[7] At the Robertson Hospital in Richmond, a small establishment run by Sally Tompkins, there were 1,329 admissions between July 1861 and June 1865 (admissions are now listed in an online database).[8] Of these admissions, 89 died, 312 were furloughed, 153 were transferred, 5 deserted, 3 were paroled, 70 were discharged, and a total of 688 returned to duty. Information on outcome is missing for 9 patients. For these admitted soldiers, we know that 52% returned to duty and 7% died; for the remaining 41% we can draw no conclusion on outcome.

So hospital statistics cannot supply comparative case mortality rates for the North versus the South, with any confidence. Too many bodies are unaccounted for. There are suggestions in the historical record that mortality was higher in the Confederacy, however. One specific disease that should have differed in its outcomes was malaria; the Union surgeon had free access to quinine, and his Confederate colleague had to make do with much less.[9] The case mortality rate of malarial fevers at Chimborazo was 125/1,998, or 6.2 percent.[10] In the Atlanta hospitals, it was 33/1,221, or 3 percent.[11] Compared to the overall case mortality rate for malarial fevers among white Union troops, 0.7 percent, these numbers are high.[12] For the Robertson Hospital, there was 1 death out of 54 cases of malaria (2%), although the outcome of 19 patients (35%) is unknown.

Military surgeon Charles Smart, the author of the *Medical and Surgical History of the War of the Rebellion*'s written analysis of overall mortality in the war (which begins in volume 5 and was written in the 1880s), attempted to compare the disease outcomes for the two sides. He drew on the Chimborazo data and published works by Confederate surgeon Joseph Jones in an effort to form a meaningful comparison, while acknowledging in great detail the deficits and approximations in the numbers. Smart's best estimate was that the death rate from disease among Confederate troops in the first two years of the war was 167.3 per thousand, a figure that surprised him. Comparable white Union rates were 49 per thousand in 1862 and 63 per thousand in 1863. "In brief," he concluded, "so far as comparison can be made with the statistics at command, disease was not only more fatal among the Confederate forces, but the number of cases in proportion to the strength present was considerably greater among them than among the United States troops."[13]

Did mortality rates from disease increase or decline after Jones's detailed statistical period ended in June 1863?[14] Historian Daniel Smith has shown that the deadliest time for rural troops in the New York regiments of the Union Army was the first year; if they survived this time of seasoning, then they did just as well as their urban counterparts.[15] Surgeon General Moore saw a similar pattern among southerners, since "a large majority of the Confederate soldiers, being from rural and up-country districts, had never had the different contagious diseases to which residents of the more populous districts are exposed."[16] Since these men suffered from the urban diseases of childhood (measles, mumps, rubella, smallpox, and typhoid) during their first months in camp, one might expect their health to improve subsequently. Joseph Jones found, for example, that the rates of typhoid fever and measles were at their worst in 1861 and early 1862 and afterward dropped precipitously. This demonstrated, he concluded, that "as a general rule, mankind are afflicted with these diseases but once during a lifetime."[17] Furthermore, survivors were survivors; making it through the first couple of years of the war might well predict that they were now immune to many pathogens and able for whatever reason to survive in the microbe-laden environment of the Civil War camp. Jones noted, "The destruction of the feeble . . . also tended to diminish the lists, . . . and as only the stronger and more hardy were left, the sick list necessarily became smaller."[18] Joseph Glatthaar's statistical study of the Army of Northern Virginia found a similar pattern; hardened soldiers were healthier in 1864 than two years earlier.[19]

However, food was more plentiful early in the war, and malnutrition may have played a greater role in causing and contributing to Confederate disease in later years. Glatthaar found that wealthier men were less likely to be sick in earlier years, reflecting, he believed, their ability to supplement their diets with purchased foods; in later years food was so scarce and expensive that this influence faded.[20] Scurvy in particular would have inhibited wound healing, as vitamin C is essential for building collagen and repairing tissue. Jones reported that "secondary haemorrhage, as well as pyaemia and hospital gangrene, progressively increased with the progress of the war, and after the great battles of the expiring Confederacy, as that of the Wilderness, it was of daily, and at times, almost hourly occurrence in the hospitals."[21] Secondary hemorrhage meant renewed bleeding of a wound after it had been surgically closed; instead of healing, the wound broke open again. Such an outcome was at least in part due to the deficiency of vitamins and other essential nutrients that marked the southern hospital in the later years of the war. Pyaemia and hospital gangrene referred to wound infections, which likewise would be encouraged by malnutrition (and a consequent decline in immune function), as well as by unsanitary wound care. Jones certainly believed that such complications were tied to the "imperfect nutrition" of the troops.[22] Drugs like quinine and opium became scarcer after 1862, likely worsening disease outcomes. And most of all, as the Confederate armies became hungrier and hungrier for men, convalescents and the outright ill may have been kept in the ranks out of desperation, making death from disease (not to mention minié balls) more likely.

All told, it seems likely that Charles Smart was correct: the southern soldier was more prone than the northern soldier to die of disease once he contracted it. The southern hospital was deficient in the essential components of a healing environment. The hospital problems were part of the broader issues of supply that plagued the Confederate armies, and the story must be told within the context of major aspects of the Confederate polity, such as monetary policy, railroads, food supply, manufacturing, and the blockade policy. Even as the medical system absorbed larger and larger sums from the central government (it was the most expensive line item in the 1863 budget), the funds were never enough. The inadequacies of the hospitals contributed to the manpower crisis that ultimately crippled the armies of the Confederacy.[23]

Building the Infrastructure for a New Country

In spring 1861 the seceding states of the new Confederacy faced the challenges of both building a new nation and fighting a war. Cooler heads than P. G. T. Beauregard's might have postponed starting hostilities until at least the basics of the new government had been established. As it was, those fatal shots bombarded Fort Sumter only a month after the ink dried on the new constitution, and by August war was well under way. Wars cost money, and the South had to immediately face the issue of establishing a currency to replace the U.S. money then in circulation. Coining money required gold or silver bullion, and the Confederacy was starting from scratch. The first steps were to float a $15 million bond issue and to levy an export tax on cotton. Although patriots were willing enough to issue pledges, they had trouble laying their hands on enough gold coin to pay their commitments.

In March 1861 the Confederate government began to print banknotes, hoping that the sale of cotton abroad in return for gold coin would shore up the currency. The Confederate government might have shipped large quantities of cotton to Europe early in the war and recouped significant quantities of gold in taxes. Instead, it hoped to force England and France to recognize the Confederacy by withholding the cotton so essential to the prosperity of foreign mills. This policy not only failed to gain the Confederacy the recognition it sought, but it also left much unsold cotton in the South. That cotton was sometimes burned by its owners to avoid benefiting the enemy; sometimes it was actively traded to that enemy; and sometimes it was seized by Union troops as they overran plantations. The Confederate currency never rested on a solid basis, and the value of the Confederate paper dollar eroded steadily as the war went on. The resulting inflation in the price of goods needed by the hospitals—food, leather for harnesses, cotton fabric for sheets, drugs, glass containers, medical instruments, even paper—meant that hospitals rarely had enough money to meet basic needs.[24] By September 1864 the Confederate government lacked the funds to pay the soldiers' wages, much less to buy them food or support their hospitals.[25] Desperate medical officials fell back on barter to purchase even the barest necessities.

The situation of the railroads compounded these shortages. As elsewhere in the country, most railroads in the South had begun as local ventures, railroad lines that ran, say, between two cities or that connected a canal to a lake or ocean port. Local businessmen put up the money, the legislature complied, and the tracks were built. At junctions the lines did not always connect, or the track gauge might be different or the system otherwise not well designed for smooth through traffic. Many stretches of track had only one set of rails, so travel could go in only one direction; oncoming trains had to wait their turn. This system operated fairly well in peacetime, but in war national priorities for the movement of troops, supplies, and military gear demanded more efficient transportation. Yet there was no organization in place to tell a railroad manager to run this car of wounded soldiers or ammunition or hospital supplies instead of that car filled with passengers eager to travel to Raleigh to visit friends. The supply of iron and steel to repair tracks and rolling stock was in short supply, demanded as it was simultaneously for the manufacture of artillery and other military hardware. Consequently, food destined for the troops might rot by the rail-side awaiting transport. Clothing in abundance in, say, North Carolina, might never reach freezing men in Virginia. The wounded could take days to reach a hospital destination, as they were repeatedly shunted to the side to allow another train to go through.[26]

One Confederate private described his difficult journey to a hospital in Charlottesville. It was the winter of 1862, and he had spent days marching with inadequate food and clothing and nights without recovering warmth. Pneumonia and pleurisy followed, with a shaking high fever. "The Dr came to me and tried to stop the chill but there was no go until about 10 o'clock Monday by this time he had about a quart of whiskey and enough opium to stupefy an ox in me and I felt tolerable stronger by this time." Then the sick man was sent on to the hospital. It was a rough trip. "I had to be hauled in an ambulance some 75 miles to . . . the Hospital at Culpepper C[ourt]. H[ouse]." "[After a brief stay I] had to ride in open [railroad] cars and the snow showered in on us beautifully. [W]hen we got to Gordansville we were put right on the train for this place where we arrived the same evening not seeing a spark of fire from time we left Culpepper until we reached this place and it snowing all the time."[27] While the U.S. Sanitary Commission put effort into creating model ambulance trains with comfortable hammocks and stoves in every car, the Confederacy never had the money for such luxuries. Patient transport remained harrowing throughout the war, although the government did build wayside hospitals to support men in transit from one station to another. The Union could rely on both riverboat and railroad transport; the Confederacy depended on its increasingly dilapidated trains and on wagons pulled by the ever-diminishing supply of mules and horses. With their ports bottled up and the Mississippi and Ohio rivers closed to them for most of the war, the Confederates had few options for moving their wounded and diseased men.

The final source of shortage came from the effectiveness of the blockade. The ports of New Orleans, Norfolk, and much of the Carolina coast fell into Union hands by summer 1862. Nashville, a railroad depot that had funneled food from the Midwest into the South, was gone, too. Louisville and Cincinnati never left the Union, and the trade between the northern "breadbasket" states of the Midwest and the southern cotton-growing regions ground to a halt with the onset of hostilities, blasting their economies and shutting off the food that had fed the plantation South. These borders remained porous throughout the war, but the blockade did limit flow and raise prices, prices that those entrepreneurs expected to be paid in gold, not in the increasingly depreciated bills that even Confederates began to call "shinplasters."[28]

Studies have shown that many blockade goods continued to get through, even in late 1864. But they did not necessarily reach the soldiers and hospitalized patients of the Confederacy. In November 1864 one Wilmington, North Carolina, shipping firm official told his Nassau (Bahamas) agent not to send any more chloroform because he could not sell the stock on hand. Perfume would be a better bet, and it could still fetch a high price.[29] It was not that Confederate surgeons no longer needed chloroform but that they no longer had any acceptable currency with which to buy it. Similarly, while Lee's army was starving in Virginia, North Carolina smugglers were selling food to Union agents via the waterways of the coastal lowlands.[30] Even when the items necessary to health were available in the Confederacy—especially food and drugs—their high price and a currency of falling value meant that these products did not reach the troops. Those who were hospitalized were the weakest and the least able to fend for themselves through nonofficial channels—and these were the bodies the hospitals were supposed to restore.

Stability and the Confederate Hospital

Like its Union counterpart, the Confederate military establishment was ill prepared to deal with the war's flood of disease and causalities. The first major battle, at Bull Run in late July 1861, sent a steady stream of wounded men south to Fredericksburg and Richmond. The spring and early summer of 1862 saw the battles of Donelson and Shiloh in the West and the Peninsular Campaign near Richmond in the East. Like their Union opponents, wounded Confederates suffered for want of simple necessities such as water, food, shelter, and even basic first aid for their wounds. And these young men, gathered into camps with little previous experience of community hygiene, led by officers who had never given such things a thought, and bearing the naive immune systems of the rural dweller, began to get sick. The litany is so familiar and yet was so deadly—measles, mumps, diarrheal illnesses of all sorts, malaria, and pneumonia. The Confederate leadership belatedly became aware that this war was not going to be over quickly and that hospitals for the care of the sick and wounded were a necessity.

The "ideal" Confederate hospital looked much like the similar "ideal" under the U.S. flag. Chimborazo hospital in Richmond, for example, was structured very similarly to Satterlee. The hospital consisted of thirty one-story pavilions, with the same emphasis on cubic feet of air per patient and the same glorification of ventilation and light. But the Confederacy built far fewer of these "from scratch" pavilion hospitals. More commonly, surgeons found an existing building, rented it, and converted it to hospital use. The Union did this as well, but it was acknowledged that new construction was preferred. The Union surgeon general would close existing hospitals once new pavilion hospitals could be built. In the South, different decisions had to be made. First, many schools, warehouses, and hotels were empty because of the war's effects, a consequence less felt in northern cities. Second, there was a shortage of lumber and nails and labor for new construction, however desirable it was.[31]

A third reason for the use of existing structures was that many southern hospitals were seen as temporary. With the war on its doorstep, the South could not assume that the space occupied by a hospital today would be in Confederate hands tomorrow. Hospitals in the District of Columbia, Maryland, Pennsylvania, St. Louis, Cincinnati, and Louisville spent most of the war secure from enemy attack. There were exceptions, of course—Confederate troops overran the Union hospital in York, Pennsylvania, in summer 1863 and caused panic in the Philadelphia hospitals before Lee met defeat at Gettysburg.[32] The battle of Perryville, Kentucky, fought in fall 1862, was part of a push toward Louisville, a major Union hospital center.[33] But aside from Richmond, Lynchburg, and Petersburg, which held out until the very end, most southern hospital centers were not worth the investment of large sums of money and effort because they could so easily fall into enemy hands. It was better to rent a hotel or a school and have a wagon train ready to take the supplies and patients to safety when the enemy came. Confederate surgeons also set up tent hospitals or built temporary sheds with board roof and floor and canvas sides, suitable for quick deconstruction and transport to the next safe haven if enemy troops approached.[34]

Hospitals in the South were often overrun, and patients had to defend them. In one instance, Union skirmishers threatened Wilson, North Carolina, in July 1863. The hospital patients there formed a troop to repel the invaders. "The 'Hospital Defenders' flew to arms—to rally in defense of their bunks, their rations, their homes, and all they hold dear (and what is not dear now?)," the surgeon in charge of the Wilson hospital reported. "This glorious band, composed of the halt and the maimed and the blind, or those who had divers miseries in the bowels, in the back, and especially in the breast, were stationed where the mighty Toisnot [River] rolls its languid tide, to guard the bridge there placed, which was the key to this situation." The defenders held off the Yankees until the militia arrived, and the Union forces (which had recently burned nearby Rocky Mount) retreated. The military hierarchy was impressed—so impressed that the men were soon reunited with their regiments.[35]

In Liberty (now Bedford), Virginia, Union troops arrived in late June 1864. The surgeon of the hospital there sent 230 men to nearby Lynchburg to serve in the trenches, and the rest of the men who could march at all were sent to Danville. The remainder "went to the woods & have returned to the Hospital." A wayside hospital near the train depot, which handled the initial care of patients coming in by train, was burned by the enemy, along with a nearby tannery and foundry. The Yankees left the main hospital buildings unmolested, and it seems likely the wayside hospital was collateral damage to the destruction of the industrial sites. In addition to the wayside hospital building, some bedding and hospital clothes were burned, because the surgeon "couldn't get trains from Lynchburg to convey hospital supplies" out of the path of Union forces, and he did not want them to fall into enemy hands. All told, the hospital center was mostly intact, though the surgeon does not comment on the effect of this massive disruption on the patients themselves.[36]

This temporary status meant that the southern hospital was rarely able to move beyond the bare necessities. Remember Satterlee, with its grand piano, its three-thousand-book library, its billiard table, its industrial kitchens? Contrast the Confederate hospital with its cookstoves occasionally under tents because there was not enough lumber to build a cookhouse, perhaps a few books and newspapers, inadequate beds and bedding, and so on. While some hospitals in Virginia, North Carolina, and South Carolina were able to benefit from stability, others were constantly on the move. This was particularly the case west of the Appalachians, where the location of the southern troops shifted so dramatically through the war. Paul F. Eve was a well-known physician, educator, and medical author who began the war at the University of Nashville; by summer 1863 he was running a hospital in Atlanta. One hospital formed in August 1863 began in Cleveland, Tennessee, not far from Chattanooga. From there it moved south to Columbus, Georgia, southwest to Mobile, north from there to Corinth, south again to Meridian, and then through Montgomery, Alabama, to end up in Opelika, Alabama, by January 1865.[37] Not infrequently such mobile Confederate hospitals consisted of little more than tents with cotton sacks on the ground for beds.

In April 1865 the Confederate patients were still moving. James Wilkerson wrote his mother from Greensboro, North Carolina, on April 4: "Mother since I wrote to you last [there] has been several hundred sick and wounded brought to Greensboro. They has made Hospitals of every church and court house and schools nearly in Greensboro." He had been put to work as a nurse. "I was . . . [ordered] to wait on a parcel of wounded men heare in tents. I don't get any rest at all in the day time, and not much some nights. I dress the wounds twice a day. It was a new thing to me."[38] While the Union dealt with such chaos frequently in 1862 and occasionally after major battles in later years, Union surgeons mostly had the resources and the organization to provide orderly care for their sick and wounded. The South found some calm and order in the major hospital centers of Virginia, with Richmond almost the last to go, but more typical was this sort of scramble from beginning to end to make do, set up anew, and recruit the inexperienced to perform necessary duties. Layered on top of all the material shortages was the deficiency of calm, of time, of structural stability to allow a hospital and its staff to mature and learn its business.

There was an overall shortage of hospitals and hospital beds. As southerners scrambled to build and outfit hospitals, the battlefields and camps frequently outpaced supply. Surgeons would receive large numbers of patients on very short notice, or they would be told to ship out their convalescents to make room for a new surge of wounded. Thomas Williams wrote S. P. Moore from Danville, Virginia, in summer 1862, "I have the honor to state that the ambulance train, with 556 patients, arrived here this morning, about half an hour after the reception of your telegraphic dispatch announcing that they had been sent. There were but a few over 200 vacant beds in the Hospital at this place." He went on to remind Moore, "If you want to send us the 1000 pts you recently mentioned—please remember that we have great deficiency in hosp. furniture, having made requisition but none coming. We need beds, plates, knives cups etc."[39] Considering that a battle could generate thousands of wounded in a few hours, hospitals frequently had to take in hundreds of patients on short notice and often ship out hundreds of the less ill to other hospitals farther from the action, which in turn had to deal with a flood of patients.

Rapid change and occasional disaster meant that hospitals were often not the peaceful havens that ill and wounded men needed to promote recovery. Their general instability was exacerbated by the many shortages that plagued the southern hospital system. Foremost among these was the ever-present need for food. Provision of nutritious, digestible, and abundant food was one sure remedy for many of the ailments suffered by the Civil War soldier, and in the South it was frequently in short supply.

Food

Soldiers in the Civil War often went hungry, and many became malnourished. As was also true in northern armies, the southern army diet plan assumed that officers would use leftover ration money to buy foods locally to supplement the government ration, which was heavy on cornmeal, salt meat, and hard crackers. This system depended on two assumptions—that there was food available in the community and that the money officers had to spend was valued by the farmers selling produce. Neither assumption always proved true as the Civil War progressed, especially in battle-worn Virginia.

Scurvy has been well documented among Union troops, in spite of the best efforts of the USSC and the wealth of northern agricultural lands untouched by the war. Early in the war southerners had few problems, since the countryside was well stocked with a variety of foods, and eager female supporters sent boxes of jams or baskets of fruit from home. As the war continued, however, southern agriculture grew ever more devastated by the many actions of the war. By April 1863 Surgeon General Moore was instructing medical officers to "make persevering & well regulated efforts to have collected for use of the sick in the field, affected or threatened with scurvy" appropriate wild native plants, such as "wild mustard, water cress, wild garlic or onion, sassafras, lambs quarter sorrel, shoots of the poke weed (bleached preferred), artichoke, plum of the dandelion, (bleached) garden parsley, peppergrass, [and] wild yam."[40] Lafayette Guild, the medical director for the Army of Northern Virginia, in 1863 transmitted this order to his surgeons, noting that "symptoms of scurvy are manifesting themselves among some of the troops of this army. . . . The only certain method of preventing scurvy is by issuing with the soldier rations a suitable proportion of vegetables & vegetable acids." He also pointed out that surgeons could order "vinegar, pickles, sour crout, dried and canned fruits, horse radish, peas, potatoes and beans" from the commissary.[41] Both orders appear to assume that buying produce from local producers was out of the question, which seems likely, given that spring planting was just getting under way.

Before the war the South routinely imported food so that land could be dedicated to the production of staples for cash export—cotton, sugarcane, rice, and wheat. The food came either through the ocean ports or from the midwestern states, although some corn was also raised locally. The major hog and beef production areas of the South were in Tennessee and Kentucky, states that were largely under Union control after midyear 1862. Although much of the South turned from growing cotton to growing corn after 1862, there was still not enough food produced in the Confederacy to feed its people. Much of the labor that might have produced food was in the army or serving the men in the army, and the fields were directly damaged by soldiers' boots. This shortage of food became particularly acute in Virginia and in the Virginia hospitals.[42] But everywhere the armies tramped and fought, agricultural fields were destroyed and livestock eaten.

The Confederate hospital system can be divided, roughly, into two groups: the hospitals in the east (Virginia, North Carolina, and South Carolina) and the ones across the Appalachians in the western theater (Tennessee, Georgia, Alabama, and Mississippi). The hospitals in southern Virginia especially had the advantage of stability; once established, few of them had to be moved in the face of enemy advancement until the very end of the war. Hospitals in the western part were constantly on the move. As historian Glenna Schroeder-Lein has documented, it was the administrative genius of Dr. Samuel H. Stout that kept the western hospitals functioning at all.[43] Stout's mobile hospitals had one advantage over those in Virginia, however. They were moving across a countryside not yet ravaged by war. There was food to be bought, even if at times the locals took only barter and not Confederate bills.[44] In contrast, patients in the hospitals in Virginia were almost always hungry.

The three largest hospitals in Richmond—Chimborazo, Winder, and Jackson—had 15,500 patients, and there were hundreds more at smaller hos-

As General W. T. Sherman's army moved through Georgia, they stripped the countryside in their path of all the animals and crops they could find. This undated sketch, captioned "Federal Troops Foraging in Georgia," most likely stemmed from Sherman's march in the summer of 1864. Paul F. Mottelay and T. Campbell-Copeland, eds., *Frank Leslie's "The Soldier in Our Civil War": A Pictorial History of the Conflict, 1861–1865, Illustrating the Valor of the Soldier as Displayed on the Battle-field,* Columbian Memorial edition, 2 vols. (New York: S. Bradley, 1893), 2:30.

pitals in the area. Other Virginia cities, such as Fredericksburg, Lynchburg, Charlottesville, and Danville, became hospital centers as well. It did not take long for soldiers and patients to begin to exhaust the food supplies of the state, especially since its principal crop had been tobacco. Wheat was a crop in northern Virginia, but much of that area was under federal control or frequent threat after the first year of the war. In 1862 the hospitals in Virginia seem to have been fairly well supplied, however. One North Carolina soldier who was in the hospital at Charlottesville, within "three hundred yards of the University of Virginia," reported, "[The] food is good too. . . . The fare here is better [than] at any Hospt I have ever seen good food and well prepared such as will strengthen a man when he is able to eat it." Georgia soldier W. S. Shockley wrote from Fredericksburg (after describing a horrific transition from camp to hospital) that he found the hospital there "the best I have seen anywhere everything is kept clean and nice. . . . The fare is very good we have beef bacon potatoes (Irish) peas light wheat bread and good warm corn bread and molasses those that are very sick have rice fruit eggs butter milk and such things."[45] Women volunteers sent food to the Virginia hospitals, where it was stockpiled for the sick men; such "Ladies depositories" supplemented the official fare by providing essential foods for the invalid soldiers, likely including the fruit, eggs, and buttermilk that Shockley enjoyed.[46]

This plenitude of hospital food was not universal even in 1862. Shockley was back in the hospital in October, this time at Winder Hospital in Richmond, where he found, "This wants to be a pleasant place to stay at if there was a little more to eat a well man cannot get enough without he buys it." A week later he was more emphatic: "I cannot stay here they don't give me enough to eat so as a last resort I elected to report and go to my regiment." He did not last long at camp and was in the hospital in Petersburg in November. From there he told his wife, Eliza, how much many foods cost, including that "a good sized Opossum sells for $2.00." This was not just information of general interest, since it was increasingly the case that hospitalized patients who wanted enough to eat in Virginia had to use their own funds to purchase food locally.[47] And money was exactly the problem. As the price of all foods rose, the amount of money available to the hungry patient, or to the hospital administrator, did not grow sufficiently with it. The government did increase the ration-equivalent from $1.00 in 1862 to $1.25 in 1863, to $2.50 in 1864.[48] At the same time, inflation had raced on. In January 1863 a Richmond newspaper showed a comparison of food prices over three years, demonstrating a tenfold increase in cost.[49] By 1864 more than one hospital administrator resorted to barter to buy necessities, because they had little money that held any value. One Lynchburg hospital chaplain acquired three barrels of brown sugar and bartered it one cup at a time to buy items for his patients.[50]

James C. Franklin was in Chimborazo Hospital during summer 1862, and he likewise complained of the inadequacies of the food he received. He asked his wife, Sookey, to bring him "light wheat bread and light corn bread and some butter and chickens and eggs and some tobacco and a piece of soap and some salt and some pies." Franklin did not want them just for himself; he wanted to sell the surplus of his family's farm to his fellow patients. "Tell your father to bring a bushel of potatoes with him. I can sell potatoes and make some money. Bring some chickens too—kill and salt them and you can sell them for 8–10 dollars a piece."[51]

Food grew scarcer in Virginia as the war dragged on. By 1864 J. B. Farthing, a North Carolina soldier sick in Richmond's Winder Hospital, complained sarcastically, "I get very good fare here that is corn bread and soup which is good enough for hospital rats the very name of hospital sickens me." T. G. Richardson, a medical inspector in the western army, visited the Richmond hospitals in March 1864, and after comparing them favorably to the hospitals under Stout's supervision, told Stout, "It is almost impossible for any one to get enough to eat, either outside or inside the hospitals, for the simple reason that it is not to be had at any price." John H. Kinyoun, a North Carolina physician who eventually rose to be an assistant surgeon general in the Confederacy, lamented the shortages of fruits and vegetables in Virginia in August 1864 (even in August!) and was especially disturbed by the insufficiency of brandy for hospitalized patients. He blamed speculating liquor and food dealers for the inability to supply the hospitals.[52]

Virginia could not import food from the North, had its main port at Richmond bottled up early in the war, and could not rely on large quantities of food reaching the state from points south. Already in 1862 a North Carolina man noted that bacon was $0.33 a pound and flour $12.50 in Wilmington. "I don't like to see such high prices," he complained. "It looks like starvation almost." And this was in North Carolina, whose fields were as yet mostly untouched by boots on the ground, and Wilmington, where blockade runners successfully brought in goods through much of the war.[53] Kinyoun had noted in fall 1861 that the people of Wilmington begrudged the soldiers' consumption of the local food. "Most of the citizenry are mad because we are here taking produce from the market, saying that we ought to be made to bring the support from our counties, for they cannot [get] enough to feed us and them." It was not likely that large supplies of food were going to arrive from North Carolina. And the limited transportation system of the Confederacy diminished how much could come from farther south. Thomas Scott wrote the director of one Richmond hospital in December 1862 that he had fifty bags of potatoes and a barrel of soap for him, but he could not get any closer than Petersburg "on account of the transportation of troops to N.C."[54]

In August 1864 Surgeon General Moore urged hospital surgeons to avoid buying food in the market towns and instead to go out in the countryside and negotiate directly with the farmers. He warned, ominously, that "the surgeon in charge [would] be held strictly responsible for any extravagance in price." Yet Surgeon Benjamin Blackford, trying to do that very thing, lacked the necessary wagons. From his hospital west of Lynchburg, he could draw on a part of Virginia largely unscathed by war. "I require two additional wagons to keep up the supplies of vegetables milk, butter, eggs &tc and have made frequent requisitions upon Capt Mallory A[ssistant]. Q[uarter]. M[aster]. for them," he complained to Moore in June, 1864. "With these wagons placed under my control I could keep the hospitals abundantly supplied with provisions at much cheaper rates than it is possible to purchase them in the immediate vicinity of this post."[55] Yet wagons, harnesses, mules, and the men to drive them were all in short supply.

In late 1862 and early 1863, the public began to protest food shortages, at times with violence. For example, there was a critical shortage of salt, which was essential for preserving meat in an era without refrigeration. Women in Greenville, Alabama, growing desperate for salt by November 1862, forced a supply train to distribute salt to them, shouting "Salt or Blood!" In April 1863 women rioted in Richmond, crying "Bread or Blood!" They broke storefronts to steal bread and other food. During that spring there were similar food riots in North Carolina, in Atlanta, and in Petersburg, Virginia.[56] Historian William Blair has argued that food became more plentiful in 1864 as the government finally realized that civil insurrection was at hand if there was not more food to buy in the markets at prices the poor could afford.[57] Increasing the supply meant taking steps that were unpopular with the planter classes, such as impressing food and forcing the sale of foods at set government prices. The government needed simultaneously to buy food for the army, to feed patients in the hospital, and to keep the poor from rioting. Its efforts were only partially successful—the first thing Lee asked for when surrendering to Grant was food for his men.[58]

If Blair is right and the food crisis eased up in summer 1864, such relief is not evident in the accounts of hospital life. Hospital administrators were constrained to buy food with government money and not, at least officially, to force food sales by locals or acquire it without paying. But this did not stop the patients. Whereas in the North men got in trouble for going out, getting drunk, and singing bawdy songs in residential neighborhoods, hospital patients in the South stole food when out on pass from the hospital. There were reports of patients stealing fruit from local orchards or livestock from farms.[59] Ambulatory patients began to scavenge food wherever they could find it. By August 1864 Brigadier General Raleigh E. Colston, commanding at Lynchburg, Virginia, responded indignantly to such news, advising his surgeons, "Numerous complaints [have] been made to these Hd Qtrs of continual depredations committed by soldiers of this command on property of citizens in the city & vicinity of Lynchburg." He reminded the men that it was illegal to pillage from civilians and that punishment by court-martial would follow for those who disobeyed. He limited the passes that allowed men to leave the hospital to two hours and forbade the taking of firearms when men went out on pass. The patients were clearly getting desperate.[60]

In summer 1864, the patients at Winder Hospital in Richmond finally rioted over food shortages. They suspected that the hospital steward was stealing the funds that should have bought food, and they condemned the baker for his ineptness. Emily Mason, a Winder Hospital matron, recalled the day. She reported in memoirs published in 1902 that the hospital had sufficient food at first. "But as the war went on, only peas, dried peas, seemed plentiful, and we made them up in every variety of form which dried peas are capable. In soup they appeared one day; the second day we had cold peas; then they were fried (when we had the grease); baked pease came on the fourth day; and then we began again with the soup." The peas often were infested with worms, and Mason tried her best to pick the worms out before the peas were cooked. Eventually even the pea supply gave out, and "toward the last we lived on corn meal and sorghum, a very coarse molasses, with a happy interval when a blockade runner brought us dried vegetables for soup from our sympathetic English friends. A pint of corn meal and a gill of sorghum was the daily ration." A gill is four fluid ounces. This ration amounted to at best about thirteen hundred calories (and we know from prisoner-of-war complaints that Confederate cornmeal contained much indigestible husk as filler) and was sorely lacking in essential vitamins. It was a starvation ration.[61]

One day the hospital steward came running to Mason, "pale with fright." He told her that the convalescent patients had "stormed the bakery, taken out the half-cooked bread and scattered it about the yard, beaten the baker, and threatened to hang the steward." Mason ran into the courtyard and found "two hundred excited men clamoring for the bread they declared the steward withheld from them from meanness, or stole from them for his own benefit." She vouched for the honesty of the steward and baker and calmed the crowd by reminding them of her own trustworthiness. "Do you accuse me, who have nursed your through months of illness . . . me, who gave you a greater variety in peas than was ever known before, and who latterly stewed your rats when the cook refused to touch them? And this is your gratitude!" Mason quelled the riot, and some of the hospital patients later bought her a ring in honor of her "brave conduct."[62]

Phoebe Pember was also a hospital matron in Richmond in summer 1864, and she likewise commented on the shortage of food and the desperation of patients. She attributed the shortages to worthless money and a

disrupted railroad system that failed to bring gathered food from the countryside before it had spoiled. "The inducements for theft were great in this
season of scarcity of food and clothing," she remembered in her memoir,
composed by 1866.[63] Food was desperately short, and the cooks had to take
a string and measure out the cuts of a pan of cornmeal, starting first with
how many needed to be fed, and then cutting the pan of bread into that many
pieces. While she personally never ate rats, she described with glee her success in catching one. The rats robbed food the patients needed, so turning
them into sustenance had a certain irony. "Epicures sometimes managed
to entrap them and secure a nice broil for supper, declaring that their flesh
was superior to squirrel meat," she declared. "Perhaps some curious *gour-
met* may wish a recipe for the best mode of cooking them. The rat must be
skinned, cleaned, his head cut off and his body laid open upon a square
board . . . then baste with bacon fat and roast before a good fire quickly like
canvas-back ducks."[64]

The sick were also considered to be in need of delicacies, such as tea and
toast, with some jam on it, not just food in general. This was just the sort of
food that the "Ladies dispensatories" supplied in 1862. But while southern
women might be willing to make the jam from locally available fruits, they
lacked sugar to sweeten it, wax to seal it, jars to contain it, and train or wagon
transport. And food was scarce on the home front, too, meaning there was
little surplus to ship to the hospitals. By the end of the war, hospitals were
short of foodstuffs of all kinds and lacked money to buy from local markets,
even where food was available. Christmas 1864 was celebrated with a feast
in Satterlee Hospital in Philadelphia, paid for by a local female philanthro-
pist, but there was no equivalent for the Richmond hospitals. Southern
money was almost valueless, and southern farmers would rather sell their
produce to Yankees for viable money than hand it over to the Confederate
government for worthless payment. Patients in the southern hospitals were
starving.

It is hard to quantify the effect of these short rations on healing and the
southern war effort as a whole, but it is likewise hard to underestimate
the pervasive influence of semistarvation on the southern soldier. Confederate generals needed every man possible in the ranks after the horrendous
losses of the summer campaigns of 1864. Food shortages meant that wounds
did not heal well, men grew weaker from malnutrition, and slight injuries or illnesses became incapacitating. Food shortages contributed to desertion from hospitals as well as regiments, further decreasing manpower. Even

the horses were starving.[65] The lack of nutritious food robbed the hospital of one of its principal healing powers.

Personnel

The labor shortage of the South affected the hospitals as it did every other aspect of Confederate life. The labor of the millions of men and women in bondage required the labor of others to maintain discipline and supervision. Yet the army demanded manpower and kept consuming it at an alarming rate. Whereas the North could replenish its ranks with European immigrants, the South had little inward flow of population. Women took up the slack in North and South, but few women lived lives of leisure in either section before the war started, even if their work was more often done in and around the home than out in the fields or for wages. There was not much spare labor to go around; with labor taken for the war, work was left undone on the home front. This included workers skilled at leatherworking, railroad construction and repair, sawmill operation, the construction of buildings, and glassmaking, for example. Just when the South needed every skilled laborer, the army needed every able-bodied man. As the army's demand for soldiers grew ever more insistent, men were pulled from these essential duties, so railroads continued to decay, soldiers had no shoes, and food production declined. By April 1864 the conscription bureau reported that there were no more men to be found for the army. All exemptions had been canceled, convalescents were called to do every duty possible to free up the able-bodied, and still there were not enough.[66]

The 4 million black southerners, slave and free, did much of the South's physical labor. But during the war, southern slaves escaped whenever possible, and as the areas under Union control grew after summer 1862, more and more slaves left the South and took their labor to the other side. The Confederacy impressed slaves from their masters for work on fortifications and to perform manual labor in the army such as cooking and driving horses. When this was not enough, in 1864 the government began impressing free blacks, the only untapped part of the male population not already under government decree. Masters received their slaves' wages; free blacks received wages themselves but had no choice about being impressed.[67]

The hospitals experienced the labor shortage in several ways. First, it was difficult to hire people to do the many jobs the hospital created. Someone had to care for the sick men in hospitals, and their womenfolk were often far away. People were needed to cook, to wash the sheets and garments, to scrub the floors, and to administer the medicines. The surgeon general's organizational scheme, which outlined the duties of surgeons, matrons, nurses, stewards, cooks, guards, and so on, assumed that there were people to do these jobs. Some of the labor was done by volunteer women, but more women sought wages and the food that came with the position, because of their own hunger and that of their families. Convalescent soldiers were put to work as nurses and ward masters. Surgeons fought to keep them, once they were trained, although as soon as they were well (and best able to perform their new duties), they went back to their regiments.

Thomas H. Williams, conveying orders from Surgeon General Moore, instructed surgeons in late 1861 on how to hire workers for their hospitals. The first choice was to hire civilians and negroes as nurses. If there were not enough of these people available, "officers commanding at the posts where said hospitals [were] situated [were] to *press* free negroes into service." While the Confederate government did not legislate such powers until 1864, it is likely that free blacks were forced into service much earlier, as this order suggests. According to Williams, only when no other options were available should surgeons ask for soldiers to be assigned to the hospital—and those to be returned to their regiments as soon as civilian substitutes were found. One surgeon responded to Williams's orders by noting that his new hospital building was almost ready, although it would have been ready sooner if they had had access to enough laborers and mechanics. He also reported trouble finding nurses. "There are no free negroes to be *pressed* in the service, all of them are in the service as officers' servants," he reported. The problems evident in December 1861 only worsened as the war went on.[68]

By August 1863 a circular went around to the hospitals ordering that boards be convened once a month to examine "all enlisted men . . . who have been detailed for duty in hospital, and all white Males subject to conscription, who may be therein employed." These men were to "return for duty to their commands, or report to the enrolling officer for conscription."[69] Men who were recognized as too weak for service in 1862 were conscripted in 1864 in spite of their hernias or weak constitutions. Blackford in particular fought to keep his clerk of two years, as the man was so familiar with all the duties necessary to his position, but to no avail. Even the disabled were called to serve guard roles in the military and were taken from hospital service.[70] Enlisted men who served as hospital workers were also sent back to the provost guards for general guard duty when the hospital census fell low in the spring, before the summer fighting season began.[71] "Many of the hospitals having been closed temporarily during the season of calm in operations, the nurses were turned over for Provost Guards," the Confederate medical director for Virginia noted in late April 1864. "They cannot be returned to me too quickly. There will be several hundred nurses and cooks needed in the various hospitals of the state. . . . Some of the men are now needed immediately. Can you get me any competent free negroes for this purpose?" he asked a Richmond official. "If you can, I would like to have some 200 at once of them."[72] "Competent negroes" were more valued than the transient white soldier—they were more biddable and more likely to stay in hospital work once trained.

Discipline among convalescent men who worked in the hospital could be difficult. The surgeon in charge of a hospital in Lynchburg ordered in October 1862: "Wardmasters and nurses must not be allowed to leave their wards except on duty without the permission of the . . . Asst Surgeon in charge of the ward. The Surgeon in charge is at a loss to understand how it is possible for the patients to be properly nursed when the ward masters and nurses are absent from their wards as much as hitherto. He therefore requests that upon any infringement of this order the man [is] to be reported to this office for punishment." This physician also wanted to hear about "any ward master or nurse guilty of intemperance in the slightest degree."[73] Perhaps discipline improved when it became increasingly difficult for hospital directors to hold on to men whose condition permitted transfer to other service, as assessing their status was a subjective act at least partially under control of the surgeon in charge. In July 1864 the surgeon general ordered that stricter control of medicinal alcohol be instituted and put in the matron's hands, to avoid abuse by staff as well as to make the best use possible of the limited supply.[74]

The use of slaves and free blacks in hospitals somewhat obviated the difficulties in keeping and disciplining white convalescent troops. Blacks made up the bulk of southern hospital workers; around 75 percent of Chimborazo's labor force was black. Slaves might be "rented" from their owners or issued by the government after it had impressed the hands from their owners.[75] These people were a constant source of contention. Owners often wanted them back, to do work at home, and valued the depreciated money paid for their chattel much less than the labor itself. And contending factions within the Confederacy vied for the work of the impressed slaves. Benjamin Blackford, surgeon in charge of a hospital at Liberty (Bedford), Virginia, constantly fought to maintain his minimal hospital staff, from surgeons down to laundresses. In July 1862 he had eight hundred sick and only five doctors.[76] In October the county sheriffs came to take black slaves who had been hired by the hospital; they were being transferred to work on the Richmond fortifications. "These negroes have been in the hospital some as cooks and others as nurses ever since their establishment," Blackford told Surgeon General Moore. They "are perfectly familiar with their duties and as they are already in the employment of the government. I respectfully request that an order be sent me to forbid any interference on the part of the sheriff." Two years later he wrote Moore to inquire who was going to pay the $150 fine levied against him for not releasing these men to the sheriffs.[77] It is unknown how Moore and the relevant bureaucrats sorted out the dispute.

James Lawrence Cabell, surgeon in charge of the hospital at Charlottesville, likewise importuned Moore in 1862. "Having failed after diligent & persistent efforts to procure colored men and women in sufficient numbers to meet the demands of the Hospital, I respectfully suggest the expediency of authorizing the Quartermaster of this Post to impress the hands of Planters engaged in cultivating Tobacco," he wrote the surgeon general. Cabell recognized the importance of food production but thought slaves could be taken from tobacco cultivation without harm. He saw the tobacco planters as "persons in this region of country who have turned a deaf ear to every appeal to their patriotism" and believed that it "would be fit and proper to make these men bear a share of the necessary burdens of the war." Cabell desperately needed more help. "The need, of nurses to perform the manual work of Hospital is exceedingly urgent. I respectfully request that the proper measures be taken to procure a supply in the only way which the demand can be promptly met."[78] We have only one side of this correspondence, and so we cannot know whether Cabell received any of the tobacco hands as nurses.

Dr. Robert Libby in Charleston, South Carolina, also had trouble keeping his hospital staff. In June 1864 he reported that of the four blacks detailed from the engineer corps to work in his hospital, two had been taken away again. He had identified another candidate, described as "William Beckett a free col'd conscript," and hoped that the general commanding in Charleston would allow him to be added to the hospital staff. Libby went on to reassure the general that this man was unfit for field duty. Libby wrote a pass for another "colored conscript" who was assigned to the hospital for the war. The pass included a physical description and gave the man permission to go home and get his clothes. Libby called this man a conscript, the same term applied to white soldiers who were forced to enlist. It is also evident that when such free blacks were impressed, they had for all intents and purposes been enslaved, for they were not allowed free movement. Of course neither were the white conscripts, who likewise required passes to be absent from camp and were at risk of being labeled a deserter. In other letters Libby requested that specific white soldiers be assigned to the hospital as they, too, were unable to bear the rigors of camp life. This search for labor was a constant theme in Libby's letters to his superiors.[79]

The hospitals needed doctors, who were always in short supply. About three thousand men served the Confederacy as surgeons (the Union had nearly four times as many, for an army at least twice the size of the Confederacy).[80] The directors of various hospitals and medical divisions fought over the trained men needed to care for all their patients and clearly indicated that some surgeons were of much higher quality than others. William Carrington, medical director of the Virginia hospitals, told the surgeon general in May 1864, "Many of the most competent [medical] officers are engaged in executive and supervisory duties," whereas he needed them at the bedside. Within the hospital, "61 of those on duty [were] Contract Physicians, who [had] been employed only on account of the necessity of the case, in knowledge of the fact that they were comparatively inexperienced and incompetent to perform the duties of med officers." Carrington pleaded that more experienced surgeons be assigned to the Richmond area hospitals. "I confidently anticipate that we will have at 10,000 more wounded to provide for in a very short time for which we should have 200 more medical officers," Carrington suggested that surgeons be pulled from regiments. "The ravages of battle have thinned out the regiments so that one med officer can easily perform all the duties required, especially when regiments are brigaded and arrangements made by the Reserve Corps and the Ambulance Committee to care for the wounded."[81]

Such appeals to Moore were common and were commonly unmet. In 1863 Moore told his medical directors that the assignment level for hospitals would be one surgeon to every seventy men, an impressive workload.[82] With thousands of men pouring into the Lynchburg hospitals after the battle of the Wilderness, and some surgeons absent because they had gone to the front, the surgeon in charge, who was working "day and night without taking off [his] clothes," called for help "but was told to do without it."[83] Richard Curry, physician at the Salisbury, North Carolina, prisoner-of-war camp, reported to his medical director in October 1864, "Since the acceptance of 10,000 Federal Prisoners—and 1500 guards to this Garrison, the amount of sickness has so increased that I find it impossible to give that attention to them, unaided, which they require. I have now 200 beds and all filled. . . . I am greatly in need of at least 3 assistant surgeons." When this appeal went unanswered, and after the surgeon had written directly to Moore, medical director P. E. Hines replied, "I have the honor to acknowledge the receipt of your letter of Oct. 14, requesting that three or more Assistant Surgeons be sent to assist you in the Prison Hospital &tc., and to inform you that there are no Assistant Surgeons whom I can send to you."[84]

In the months that followed, Curry did receive an additional surgeon. But then he was pulled from his work by an incident that appears to be the height of wastefulness on the part of the Confederate command. Curry was court-martialed, a proceeding that cost him eleven days away from his patients and four hundred dollars. The charge was that he had illegally given a diagnosis of disability to a soldier. It turned out that this soldier had lied about being wounded at the Wilderness and did not deserve the furlough granted by Curry and the board of examining physicians at Salisbury. The court deemed Curry's examination suitably thorough, exonerated him, and returned him to duty. This incident illustrates that not only were doctors in short supply for the care of patients; they were pulled from caregiving to perform examinations of disabled men in order to grant the certificates that allowed these men to remain on furlough. The military leadership's suspicions in Curry's case also suggest that some doctors collaborated in helping men avoid military duty by giving false certificates.[85]

Physicians were on the front line in policing disability and detecting those who chose malingering (faking illness or injury) as a way out of military service. On one hand, the military authorities were well aware that furloughs were frequently abused; one Confederate senator proclaimed that only one in three soldiers returned to duty from furlough as expected.[86] On the other hand, furloughs were a way of transferring the cost of care and feeding a sick patient from the army to the soldier's family. Physicians had to balance these competing interests. They also had to be alert to the possibility that hospital patients under their care were amplifying their disabilities to remain within hospital walls. Military commanders urged vigilance in discovering the truth in such matters.

In the first year of the war, the Confederate troops apparently used sick leave as a time to escape from military discipline, and the hospital system had only weak controls in place to prevent them from doing so. Col. Carnot Posey of the Sixteenth Mississippi Regiment wrote one hospital inspector in fall 1861, "I have been informed through a reliable source that a number of my men at Warrenton are well enough to do duty, and are *wandering* over the town and country drinking and frolicking. The General of the Brigade detailed a Lieutenant to superintend the conduct of the men but I am told that he has gone to the springs for his health."[87] Another captain complained to Dr. J. B. McCaw at Chimborazo in summer 1862, "Between 20 and 30 of [my] men have left my company from time to time to report at Hospitals in Richmond. Some of them instead of so doing, went immediately to their homes in Hanover [County, Virginia]. Others spent a day or so in some Hospital & then gave the surgeons the slip & likewise went to Hanover." This captain clearly felt that McCaw was in a position to take action. "Some few have had regular sick furloughs, but I have good reason to think that all are now fit for duty. My company being at present very weak in numbers it is very necessary that these men should return."[88]

If a surgeon was to meet the military's expectation, he not only had to diagnose malingering properly but had to prevent desertion. Yet the hospitals did not have sufficient guards, since by definition a man well enough to do guard duty was well enough to be back with his unit. Blackford defended his hospital guards against such transfer. They were convalescent soldiers, well enough to do this job but not well enough to go back to their regiments. "I have the honor to state that these men who are detailed to guard duty are not sufficiently convalescent to be returned to duty, and are only required to stand guard one or two hours at each hospital," he reported to Surgeon General Moore in May 1862. "I find it absolutely necessary to establish a hospital guard, not only to keep visitors without proper permits [out], but to prevent any disturbance among the patients." By December 1864 Surgeon General Moore ordered, "Soldiers who have lost their left hand or arm, and otherwise healthy, but who are incompetent to perform clerical duty, can, in the use of a pistol, act as efficient guards for Hospitals and purveying depots. The majority of the guard can be composed of such men."[89]

In 1863 and 1864, the pressures on physicians to serve as gatekeepers for patients' destinations (home, hospital, camp) increased. Surgeons first were ordered to examine convalescents frequently, then three times a week, to see if they were fit for duty.[90] In the summer months when hospitals were crowded, men whose homes were nearby were to be furloughed there; in at least one instance, an order commanding that transfer was followed by a telegram a day later saying that "only in case the hospitals become crowded" should the men be sent home.[91] In the wake of the battle of Chickamauga in October 1863, E. A. Flewellen, medical director of the Army of Tennessee, asked his corps directors to "at once order one of the most energetic and efficient Med. Officers of [the] corps to report temporarily to Med Officer in charge of the Receiving & shipping hospital at Chickamauga." The man's task was "to aid him in detecting and sending back to their command all officers and men who have been ordered to the rear because of sickness." Flewellen was primarily concerned not about providing care, but rather about triaging men appropriately. "In the performance of this duty a close scrutiny and wise discretion will be necessary to avoid the extremes of returning to their commands men who are fit subjects for hospital treatment, and of permitting malingers and men so slightly sick that they could safely [be] treated in camp to go to the rear."[92] The Confederate surgeons' medical practice was dominated by awareness of constant personnel shortages at all levels.

Shelter and Clothing

Shortages of material goods plagued the Confederate hospitals. Oddly enough for the land of cotton, fabric items of all sorts were in short supply. Uniforms were so precious that when a man died, his clothing was turned over to the quartermaster for reissue.[93] Samuel Cooper, adjutant and inspector general of the Confederacy, advised in orders of August 1864 that the clothing was to be washed before return. So what was the honored deceased to wear on burial? William Carrington dictated in August 1864, "[The] surgeon in charge of the Hospital is authorized to purchase the necessary winding sheet with the hospital fund." He warned that if hospital clothing or bedding was used for this, the cost would be "charged against the officer making the misappropriation."[94] As Confederate money was more and more discounted, and manufactured goods became increasingly dear, the honored dead were buried as they came into the world, naked.[95] Meanwhile their doctors were charged with keeping careful count of the laundry, getting a receipt when handing over the dirty linen, and receiving that precise amount upon return of it.[96]

The men lacked shoes as well, a factor that came into play when they were ordered back to their regiments. One surgeon in Lynchburg, Virginia, reported in late 1862 that he had "not less than two-hundred soldiers in the Hospital under his control, who are fit for duty, but cannot be sent to their

Regiment because they are absolutely barefooted." Not only that, but "during the last week some sixty men were received into Hospital, who had nothing the matter with them, except that their feet were sore from want of shoes." This surgeon had requisitioned shoes some time ago but none had arrived.[97] Lee's army was short of shoes for much of the war. They wore out quickly when men were on the march and could less easily be replaced than clothing, which many family members could manufacture. Lee pulled 271 shoemakers from the ranks of his army and put them to work making shoes in November 1862, using the skins of butchered cattle. Legend has it that one reason Lee pushed north into Pennsylvania was to find shoes in the vicinity of Gettysburg, but that story is based on little foundation.[98]

Although surgeons might be reluctant to send convalescent men back to their units barefooted, Confederate leaders had no such qualms. One order issued February 1864, with snow thick on the ground in Virginia, commanded, "no soldiers detailed or in hospital at this post will be retained at the post beyond the time he is properly here by reason simply of the fact of his being without shoes or suitable clothing unless the condition of the soldier be such that to return him under such circumstances would expose him to danger as to his health." At least the order included this caveat: "As a general thing men are more likely to be provided in the field than at Posts & besides this would sell or dispose of their clothing if they believed that being without them would entitle them to be absent from the field." The shortage of clothing and footwear for Lee's army and other Confederate troops was by this time so commonplace that it could not be a barrier to building regimental strength.[99]

Many other everyday items were needed at the hospitals. Hospital construction was delayed due to a lack of nails and glass.[100] The South still had plenty of trees, but men to cut them and make boards were scarce, and the price of lumber for construction was accordingly high.[101] Candles to light the wards were likewise expensive and scarce; in Atlanta the hospitals used lamps that burned lard oil instead.[102] Wood for burning in hospital stoves was consumed at a great rate in winter time, so that at least in one hospital in Richmond, men who wanted to be warm had to go into the woods and cut firewood for themselves.[103] Cooks lacked for stoves, pots, and sufficient plates, cups, and spoons to set the tables. Even the most mundane household items—needles, thread, corks, paper—became unexpectedly valuable because replacement could be so dear.

Medical Supplies

The most important questions about medical-supply shortages in the Confederacy hinge on the blockade. The Union decided that medicines and other medical supplies were contraband items that could not be legally sold to southerners, so all medicines imported to the Confederacy had to arrive by less open routes. Some historians have argued that the blockade was so inefficient that it made little difference for the availability of medical supplies, but whether the blockade was to blame or not, medical shortages were widespread in the Confederacy.[104] The medical leadership repeatedly mentioned deficiencies in their communications to physicians. In April 1863 Lafayette Guild reminded the surgeons of Lee's army, "The supply of Med. & Hospital stores in the Confederacy is very limited." He ordered them to be as economical as possible in the use of remedies, especially "stimulants, chloroform, morphine, dressings, &tc." Ten months later the medical director of hospitals in South Carolina, Georgia, and Florida scolded his surgeons to be more careful. "Attention is again called to the importance of observing the utmost economy in the expenditure of Medical Supplies, owing to the difficulty of procuring such supplies through the blockade, and Medical Officers are directed to substitute as far as practicable, Indigenous Remedies for those of foreign importation."[105] The drive to procure remedies from local plants was not inspired by dissatisfaction with the usual list of medications, but by the insufficient quantity of those medications.

In retrospect, one is tempted to applaud the turn to botanics as at least offering the men less-toxic drugs than the commonly prescribed mercury compounds and other medicaments that are now seen as ineffective and poisonous. But even at the time, surgeons picked out a few drugs as being indispensable, and their value has stood the test of time. Morphine, ether, and chloroform were at the top of that list, given the large number of wounds and accompanying surgical interventions. Moore supported the establishment of pharmaceutical laboratories to make these medicines where possible; the one in Lincolnton, North Carolina, had a large poppy field growing next to it. Some southerners claimed these southern medicines were as good as imports, but others felt the end product did not meet prevailing standards of efficacy.[106]

Quinine was also a precious drug. Used by Civil War physicians to treat not only malarial fevers but also fevers of all other sorts, it could not be manufactured in the Confederacy. Historians have argued about how short the supply really was, but pharmacy historian Michael Flannery contends conclusively that quinine shortage was a critical problem in the Confederacy by 1864.[107] Detailed inventories of a hospital in Charleston, South Carolina, support this conclusion. The Carolina low country was a hotbed of malaria; here if nowhere else in the Confederacy, an ample supply of quinine was necessary. The Union supply table listed 100 ounces of quinine every three months as the minimal supply for a small hospital serving a hundred men, while the standard Confederate pharmacy list called for 80 to 100 ounces.[108] Yet this hospital in Charleston had only 3 ounces on hand on July 1, 1863, and received only 25 ounces for the whole year. The surgeon recorded receipt of only 12 ounces for the first half of 1864.[109] There can be no doubt that the Confederacy did not have enough quinine to adequately treat malaria.

The Confederacy paid ever-increasing prices to blockade runners for imported medicines. When paid with Confederate dollars, calomel began the war at $8 a pound but rose to $102 by 1864; at its cheapest, morphine was $10 per ounce but went for $180 per ounce in 1864. By the end of that year, Confederate money was worthless. Confederates also acquired medicines by other means. Raided Union supply trains frequently yielded a bounty of medical supplies. Confederates traded cotton for drugs, especially on the western frontier. Women became famous for smuggling quinine under their hoop skirts, although it is impossible to estimate whether the volume of such smuggled goods made much difference. The Museum of the Confederacy has a large doll that, according to the donor's family legend, carried contraband quinine in its hollow body. The persistence of such legends likewise supports the importance of such shortages in southern popular memory.[110]

Confederate surgeons also struggled to maintain other necessary items for the execution of their duties. Alcohol was used as a stimulant in the war, and even this was in short supply as hospital workers appropriated it for their own use. In addition, the grain necessary for its production was commandeered to supply food for men and horses. The Confederacy set up its own whiskey and brandy manufacturing centers, but the supply must not have been sufficient, since surgeons were enjoined to watch their stores carefully and prevent overprescribing or illicit use.[111] Surgeons had trouble maintaining their instruments as well. With all that cutting and amputating, blades broke, became dull, or for other reasons needed repair or replacement. Surgeons in South Carolina were told to send their instruments to Charleston for repair, since there were no new ones to be had. The medical purveyor in Richmond, E. W. Johns, failed to acquire any new instruments to deliver to

Confederate surgeons during the whole of 1864.[112] The image of men operating with dull, broken, or substitute equipment is particularly gruesome.

In 1864 and 1865, as the Confederate government neared collapse, the hospitals stopped receiving funds altogether. In May 1864 a surgeon at a South Carolina hospital told his medical director, "Hospitals at this post are very much embarrassed for want of money. The commissary not being able to procure funds necessary for the purchase of supplies, this renders it absolutely necessary that provisions should be purchased on credit." He asked either that he be allowed to buy on credit or that money be sent.[113] The only response in the hospital's record book was a telegram alerting the surgeon that five hundred men would shortly be arriving, transferred from Richmond hospitals (which were overflowing with casualties from the Battle of the Wilderness).[114] A month later another South Carolina surgeon wrote directly to Surgeon General Moore, saying that the hospital had received no funds to pay workers for four months. "I trust and respectfully request that the Surgeon General will relieve his officers of the difficulties & embarrassment now experienced in the support of Hospital in the South," he beseeched Moore. Hospitals in Georgia were in similarly dire straits. Samuel Stout told Moore in October 1864, "Unless some remedy is applied soon the hospitals of this department will have to furlough all their inmates to secure for them appropriate diet. If the Treasury Department does not furnish the currency the sick and wounded cannot be fed."[115] The money was worthless, and now there was none to be had.

In spring 1865 the hospitals gradually gave up. Some, such as Chimborazo and other hospitals in Richmond, were formally surrendered to Union forces.[116] Others, such as the hospitals in North Carolina, kept on the run as long as there was any hope. North Carolina medical director P. E. Hines wrote increasingly desperate letters about his situation. Where was the transportation for patients from the hospitals in Raleigh? What was he to do with hospital food, medicines, and bedding? What about the wayside hospitals in Tarboro and Weldon, or the general hospitals at Kittrell's Springs or Wake Forest, the latter two housing 340 patients? He could not get access to trains or ambulances—and then finally the letters stop coming. How those patients made out, how they got home, how many died because of lack of medical care, all are questions left as unanswered as Hines's pleas for direction amid dissolution.[117] And what happened to the hospital fittings? The sheets and coverlets and pots and pans that had previously been carefully counted? Novelist Margaret Mitchell has Frank Kennedy appropriating such items in the

Advantage of "Famine Prices." A quinine cartoon, joking about the high price of the drug in the Confederacy: "Sick Boy. 'I know one thing—I wish I was in Dixie.' / Nurse. 'And why do you wish you was in Dixie, you wicked boy?' / Sick Boy. 'Because I read that quinine is worth one hundred and fifty dollars an ounce there; and if it was that here you wouldn't pitch it into me so!'" Cartoons are often indicative of what "everyone knows," since without such common knowledge the humor would be lost. Historians may debate the availability of quinine in the South, but the northern reading public was clearly expected to know how scarce and expensive it was there. This cartoon by an unknown artist originally appeared in *Harper's Weekly*, November 14, 1863. Library of Congress Prints and Photographs Division, LC-DIG-cph.3a48489.

power vacuum of postconquest Atlanta and using them as the basic stock to build his store.[118] Perhaps other enterprising southerners likewise benefited from the war's medical detritus.

Conclusion

Medical care in the southern hospital was inferior to what was offered in the North, especially after 1862. At its best, the hospital environment could provide nourishing food, warmth, rest, hydration, and cleanliness, reproducing the care of the home sickroom. And certain Civil War medicines, such as quinine, narcotics, and anesthetics, made a difference in patient care. But the southern patient often lacked some or all of these aspects of ideal hospital life. Although southern women did what they could, their labor was likely to be needed at home, and hospitals were persistently short-staffed. These hospitals had little to offer compared to the luxuries at Satterlee. This was not due to the inadequacies of the southern physician or any evil intention. There were surely inattentive doctors, insensitive doctors, drunk doctors, and inept doctors aplenty on both sides; I make no claims here about specific behaviors within the hospital. My reading of hospital letter books and diaries shows a cadre of physicians struggling to provide optimal care for their patients and frequently frustrated in achieving even minimal goals for patients' comfort. The shortages within the Confederacy overwhelmed these good intentions and contributed to the South's defeat by depriving it of critical manpower. In summer 1864 more than 102,000 southern soldiers flooded the rebel hospitals in Virginia. Given the poor conditions prevailing in those institutions, few returned to fight in the closing battles against the Union forces.[119]

CHAPTER NINE

Mitigating the Horrors of War

In May 1864 Union physicians had reason to become personally aware of the shortages of food and other supplies in the Confederacy. During a limited exchange of prisoners of war, a ghastly parade of "living skeletons" disembarked from boats that had ferried them from Belle Isle prison in Richmond. Broadly publicized in northern newspapers, the condition of these men deeply shocked the nation. Sketches of their emaciated bodies printed in illustrated newspapers such as *Harper's Weekly* brought the horror into every household. Members of Congress were outraged, as was the leadership of the Sanitary Commission. No single event so countered the view of war as glorious and honorable as did the sight of these men. It was one thing to celebrate fallen heroes, but quite another to grieve for stricken martyrs. Abraham Lincoln, in his second inaugural address, questioned how both sides could call upon the same God. "Both read the same Bible and pray to the same God, and each invokes His aid against the other."[1] No single facet of the war is as contested in memory as the history of the prison camps and what the treatment of those prisoners revealed about the character of the governments that oversaw them. The atrocities committed on both sides competed in summer and fall 1864 with news of how the war was being executed to call the humanity of Americans as a people into deep question.[2]

Psychologist Steven Pinker has identified a "humanitarian revolution" as beginning in the eighteenth-century Enlightenment and leading to a decline of overt cruelty in everyday life, whether in the treatment of slaves, women, children, animals, or prisoners. He even drew on Lincoln's first inaugural address, titling his book on the topic *The Better Angels of Our Nature*.[3] Mid-nineteenth-century elite Americans, whose very country had been born amid Enlightenment ideas, could countenance honorable death in battle as evidence of a higher calling, but the gaunt bodies of the returned prisoners of war brought the whole of American civilization into question. The discovery of Andersonville prison in southern Georgia with its many "martyred dead" further fed this despair.[4] Clara Barton, who visited Andersonville in 1866, called the site "one of the sanctuaries of the nation."[5] Augustus Hamlin, whose account *Martyria; or, Andersonville Prison* described its horrors, reminded his readers that "as prisoners, they were entitled to the care and treatment acknowledged by the general laws and usages of civilized nations."[6] The overt cruelty of Confederates to the men in their camps bewildered northern observers. As a people, they believed, Americans were better than this.

The U.S. Sanitary Commission became deeply enmeshed in this soul-searching and focused particularly on the horrors of prisoner-of-war camps. It was part of a broader attempt to find some meaning, some justification in the loss of lives as well as destruction of property that the war had wrought. The liberation of millions of blacks from slavery gave the North a righteous advantage but could not entirely salve the collective conscience, which was aware that the Union cause had been won with harsh tactics that betrayed the morality of a Christian people. Northern men of goodwill struggled to justify the cruelties of northern prison camps, the aftermath of northern army depredations upon southern civilians, and above all the massive loss of life. How could a war be just, be God's will, when it killed so many and even led to the assassination of the president? Historian Harry Stout has emphasized that preachers on both sides gave their official Christian blessing to the conflict and found moral arguments to justify the course of each side. Historian George C. Rable likewise described how pervasive Christian thinking was among Americans of the time, including the assumption that God's judgment gave meaning to the war.

It became harder for those finding God's beneficence in the war effort in the last two years of the war, as the wretchedness of prisoner-of-war camps stood revealed in the public press, the Union waged "hard war" to grind down the Confederacy, and southern troops committed atrocities against black troops and their white officers. In the midst of this anguish, thoughtful men, including the leaders of the USSC, felt pushed to find some further meaning, some progress in the general condition of mankind that would reassure them that human civilization was not slipping backward into barbarism.[7] The USSC involved itself first on the side of the rebel prisoners held by the Union, decrying the state of their camps and the needless loss of life to disease. But by 1864 their tone had shifted, and the agency chose political expediency over adherence to their own ideals of Christian behavior. Perhaps in some compensation for that moral failing, the USSC in that same year began its association with the International Red Cross, and after the war USSC men spearheaded the first drive to make the United States a participant in the Geneva Convention. They failed in this, as well. The USSC served as a model of how a civilian relief agency could indeed mitigate the horrors of war and channel the benevolence of a people, but it remained for Clara Barton to put that example to use in the American Red Cross in the early 1880s. The USSC was too weak, and perhaps too flawed, to execute this humanitarian response to war in the immediate aftermath of the American Civil War. This chapter first explores the USSC involvement in the prisoner-of-war controversy and then considers their (unsuccessful) efforts to create a permanent relief agency in the United States.

From the perspective of the USSC's official historian, the creation of the Sanitary Commission was "one of the great glories of the war, and . . . the most comprehensive and successful method of mitigating its horrors known in history."[8] The USSC was nothing less than "the organized sympathy and care of the American people for those who suffered in their cause," Charles Stillé rhapsodized. "[It] stands out alone in its every fresh beauty from the dark back-ground of civil strife, and must always and everywhere call forth the homage and admiration of mankind. It is the true glory of our age and our country, one of the most shining monuments of its civilization." Stillé's history closed with these ringing words: "May it ever prove a beacon to warn, to guide and to encourage those who, in future ages, and other countries may be afflicted with the dire calamity of War!"[9]

Stillé's emphasis on this theme revealed a shift in the overall purpose of the Sanitary Commission. In the beginning, the group sought to educate the tyro surgeon in the ways of camp hygiene and the methods of preventing disease. Its impulses merged with the women's aid societies' desire to supply food, clothing, and bedding to the sick and wounded, and later to the army in general. The USSC was the organized expression of the people's benevolence toward the army. But toward the end of the war, USSC communications took on a loftier tone, for now the USSC represented all that was hopeful, uplifting, and moral in the midst of horror and chaos. Its role was indeed to mitigate the horrors of war and, perhaps, to remind human beings of the way of peace. Not denying the justness of the war's aims, the USSC served also to remind the army and the nation of the Christian character of the United States. Led by a minister, the organization sought to direct the Union toward Christian charity when such ideals risked being overwhelmed by military necessity.

Prisoners of War

Speaking to the heart of national character and self-justification, the treatment of prisoners of war during the American conflict generated outrage on both sides. The discourse featured words such as fairness, honor, civilization, and Christian behavior. Both sides shared the same moral code and the same awareness that their governments would be judged as moral or immoral, barbarous or civilized, by the court of international opinion. The very Declaration of Independence sacred to North and South included in its justifying first line the need to exhibit a "decent respect to the opinions of mankind," and during the American Civil War that need to appear just and honorable, especially to Europeans, dominated many discussions about the treatment of prisoners. This was perhaps most obvious in summer and fall 1862, when European recognition of the Confederacy appeared imminent and Lincoln issued the Emancipation Proclamation in part to regain the moral high ground in England, where slavery had been abolished decades earlier. But the Confederates were still concerned about their foreign reputation in March 1865, when they issued a formal refutation of northern claims about southern camps, for "they deemed it a sacred duty, without delay, to present to the Confederate Congress and people, and to the public eye of the enlightened world, a vindication of their country."[10] It was not sufficient that either side feel justified in its actions; the European court of public opinion served as an ever-present judge of national behavior and honor.

The two sides held similar views of what constituted proper treatment of prisoners of war. Northern philosopher Francis Lieber codified proper behavior of soldiers during war, and the Union officially adopted the rules. The southern authorities ridiculed Lieber's rules as not representing actual Union actions, and they found loopholes in his prescriptions, but they did not disagree with the basic outline for honorable engagement and treatment of prisoners. First of all, once a man indicated he was surrendering, he was no longer an enemy combatant and should not be harmed. "It is against the usage of modern war to resolve, in hatred and revenge, to give no quarter," said Lieber's code.[11] Thus, when Nathan Bedford Forrest's troops slaughtered black soldiers who had surrendered at Fort Pillow, the war propagandists in the North trumpeted his shame, and the southern press, far from arguing that the black troops were a special category, instead claimed that the slaughter had never happened or that the men had not fully surrendered.[12] Although the Confederate government did argue that captured Union soldiers who had once been slaves should be treated not as soldiers but as runaways—and their officers should be prosecuted for theft—in general they at least officially shared the code of honor that dictated the proper treatment of prisoners of war.

Lieber went further, in his list of rules, to dictate what care was due to prisoners. They should "be fed upon plain and wholesome food whenever practicable and treated with humanity." Wounded men were to be "medically treated according to the ability of the medical staff"; the proper care of men who became sick while in the hands of the enemy was not described.[13] The Sanitary Commission report on prisoners of war issued in November 1864 expanded on this description. At stake was the very definition of how a Christian, civilized nation should behave and how its actions revealed its character. The USSC pointed out, "In these days of civilized warfare, the cowardly and barbarous usage no longer prevails of maltreating prisoners of war." Then follows the organization's description of proper treatment. "The surgeon, with the high sense of professional duty . . . goes equally to all. The prisoners . . . are made as comfortable as the arrangements necessary for their safe keeping will permit. They are sheltered, warmed, fed and clothed" and, all told, treated as part of "the military family of their enemy." The USSC concluded, "Such is the high principle, and noble usage, which prevails in modern warfare. The perfection of its arrangements is a matter of pride and honor among soldiers, and the proper boast of every Christian government."[14] Protesting about how rebel prisoners were treated in the North, the Confederate government made it clear that it held similar ideals about the handling of prisoners.

Yet neither side came close to providing equal or even comfortable care for the prisoners held in their camps. In northern camps, 25,976 Confederate men died, for a mortality rate of 12.1 percent; in the South, 30,218 Union men died, for a mortality rate of 15.5 percent.[15] Conditions were worst at southern prisons such as Belle Isle, Salisbury, and Andersonville, but men died in nearly equal numbers at northern camps in Elmira, New York, and Chicago and on the mid-Atlantic coast. The condition of prisoners became propaganda fodder for both sides, and publicity about their fate stoked the fires of hatred in both North and South. Controversy has surrounded no other Civil War stories as persistently as those of Andersonville or Elmira.[16]

It is hard to see either side as noble when both were so overtly cruel to their prisoners of war. Historian Charles W. Sanders Jr. lays blame at both doors. "Both of the belligerent powers deliberately and systematically mistreated the captives they held, and the depth of their guilt was such that even before the guns fell silent each was furiously constructing elaborate explanations for and justifications of their actions."[17]

The Sanitary Commission put itself in the middle of these controversies with its publication of *Narrative of Privations and Sufferings of United States Officers and Soldiers While Prisoners of War in the Hands of the Rebel Authorities*, published in fall 1864. The report presented the suffering of Union soldiers with great feeling but argued disingenuously that rebel prisoners in Union hands were well fed and comfortable. Yet early in the war, USSC spokesmen had protested the poor treatment of prisoners in northern camps. In general, as we have seen, the USSC did not hesitate to criticize the federal government's actions, and by 1864 their enemies Secretary of War William Stanton and Surgeon General Joseph Barnes were triumphant in Lincoln's administration. The Union officer in charge of prisoner-of-war camps, William Hoffman, was rude and uncooperative to USSC representatives. With this history, and their avowed commitment to humanity and justice, why did the USSC cover up conditions in northern prison camps?

The USSC and the Prisoners of War

It was after the western campaign in spring 1862 that the Union first began to accumulate prisoners of war in significant numbers. William Hoffman received his appointment as commissary-general of prisoners with command of all Union prisoner-of-war camps and hastily ordered that camps be established in the west at Chicago (Camp Douglas), Indianapolis (Camp Morton), and St. Louis (Gratiot St. Prison). With the expectation that prisoners would be held only a short time before exchange, the camps were constructed shoddily and little attention was paid to maintaining life and health over the long term. As Confederates were jammed into these hastily assembled pens after the battles of Shiloh and Donelson in spring 1862, conditions worsened rapidly and epidemic disease erupted. The Sanitary Commission's president, Henry W. Bellows, happened to be visiting Chicago when he learned of the local camp's growing pestiferous reputation. He asked Hoffman and received permission to inspect the premises in June 1862. Bellows was appalled by the conditions at Camp Douglas, which had opened in September

1861 for training and recruitment and not converted to its prisoner-of-war function until February 1862.[18]

Bellows railed against the "amount of standing water, of un-policed grounds, of foul sinks, of unventilated and crowded barracks, of general disorder, of soil reeking with miasmatic accretions, or rotten bones and the emptyings of camp-kettles." He continued, "[It] is enough to drive a sanitarian to despair." He had only a single recommendation: "The absolute abandonment of the spot seems the only judicious course. I do not believe that any amount of drainage would purge that soil loaded with accumulated filth, or those barracks fetid with two stories of vermin and animal exhalations." Bellows advised that a new camp be established "in some gravelly region" and suggested, "If in the pressure of your engagements you choose to call on the Sanitary Commission for any plan for the camp of 10,000 men or a proper and economical style of barracks I shall be most happy to send a plan and even an architect at the expense of the Commission to aid your purpose."[19]

Bellows's indignation might as well have characterized his assessment of a regimental camp for Union troops. Humans were living in an unsanitary space. This needed to be remedied. That these men were the enemy apparently did not enter his thinking. Granted, at this point Union soldiers were billeted in part of the camp, and local citizens were concerned that epidemics would spread from prisoners to the city, but still, the tone toward the bodies in question is neutral. Hoffman was likely taken aback by this effrontery and did not, in fact, call on the USSC for advice on doing his job. He told Quartermaster General M. C. Meigs a few days later that he disagreed with Bellows on the hopelessness of the site and that he had instituted a cleanup at the camp and ordered some structural improvements. He asked Meigs in particular for authorization to build a sewer for the camp. Meigs refused the funds for sewer construction but approved of plans for remodeling the barracks.[20] While Hoffman explicitly told Bellows that he could release his report, Bellows never did so. Discussions of the status of prisoners of war, in North or South, do not appear in the USSC press until 1864. With prisoner exchanges actively ongoing in summer and fall 1862, the prisoner population of Camp Douglas fell to zero in September, and Hoffman no longer had any jurisdiction over the camp.[21] An empty camp meant no need for sanitary reform.

However by Christmas 1862, the exchange agreement between North and South had so broken down that prisoners had again begun to accumulate.

The battles of Antietam, Perryville, Fredericksburg, and Stones River all produced new ranks of prisoners on both sides, and several thousand of them were housed in Camp Douglas. In spring 1863, the Sanitary Commission sent two inspectors to examine the hospital at the camp and also to evaluate the hospital associated with the Gratiot St. Prison in St. Louis. They reported that the hospitals were in "deplorable condition," a situation so desperate as to be "disgraceful to us as a Christian people." The improvements promised by Hoffman the summer before had not been made, and the prison was in worse shape than ever. Rebel inmates were dying of disease so quickly in Chicago that the "rate of mortality . . . would secure their total extermination in about 320 days," the USSC report declared. The prisoners in the hospital were "without change of clothing, [and] covered in vermin," and many lay on cots without mattresses or at best with rags for blankets. The situation in St. Louis was as bad: poorly ventilated barracks were crowded with sick men who had yet even to be admitted to the filthy hospital quarters.[22]

In a letter to Secretary of War Stanton, USSC physician William Van Buren quoted the report of the USSC inspectors (Drs. Thomas Hun and Mason Cogswell) that described this abysmal setting. "In these prisons and hospitals a condition exists which is discreditable to a Christian people. It surely is not the intention of our Government to place these prisoners in a position which will secure their extermination by pestilence in less than a year." They believed, "This state of things cannot be known to those who have the power to cause it to cease." Disturbed by their findings, the Sanitary Commission inspectors wrote in the report to their own commissioners, "From the circumstances under which we were admitted we feel that we have not the right to speak publicly of what we have seen." The USSC leadership alerted Surgeon General Hammond to these foul conditions but also addressed Secretary of War Stanton directly, because they believed Hammond lacked power to correct the evils. William Van Buren, the USSC spokesman, urged Stanton and the president to "issue such orders as will secure humane and proper treatment for the sick prisoners." Although he took a chiding, Christian stand in his rhetoric, Van Buren did not issue a public call for reform.[23]

Apparently the USSC sent the same report directly to Hoffman, for he answered it in a letter to Stanton eleven days later. While he did not accuse the USSC inspectors of incompetence, he did paint a very different picture of the hospital conditions at these two camps. His answer included claims that the situation was much better than reported, had been improved since the

USSC inspectors were there, or was not his fault because he had so much trouble communicating with the ever-shifting officers at the camps. He closed with a request for the appointment of a medical inspector of the camps, a post subsequently filled by A. M. Clark, MD, of New York. From the content of the letter it is evident that Hoffman agreed, at least at this point, that the hospitals should be clean and orderly. And the request for an inspector indicates that perhaps he really did not know how bad conditions had become in Chicago.[24]

There are no further records of Sanitary Commission involvement with northern prisoner-of-war camps until July 1863, when the USSC sent an inspector to the Fort Delaware camp, on an island in the Delaware River, near Philadelphia. The inspector was S. Weir Mitchell, a prominent Philadelphia physician serving at the Turner's Lane Hospital, who specialized in gunshot wounds and nerve injuries. Mitchell "represented to us that the hospital there is in a disgraceful condition," reported corresponding secretary Frederick Law Olmsted to the head of the Philadelphia branch of the USSC, Robert M. Lewis. He told Lewis to send USSC supplies to the camp, in accordance with Mitchell's requisition, to relieve conditions at the prison-camp hospital. Olmstead justified this use of USSC goods to help the rebel prisoners as ultimately benefiting the Union soldier. "It may be well to say that the Commission proceeds upon the ground that every rebel whose life is saved will increase the inducement presented to the rebel authorities to treat carefully all Union men who fall into their hands," Olmsted argued. "Every Union man whose life is saved by them buys back to them a man of their own, whose life we have saved."[25] Olmsted did not appeal to either humanity or Christianity, but rather to a strict quid pro quo. Although the prisoner exchange program was dormant at this point, there was hope that such exchanges might soon be resumed.

The USSC had reason to see the status of prisoners of war in southern hands from a deeply personal perspective in July 1863. Their own agents, Dr. Alexander MacDonald and three others, had been captured at the Battle of Gettysburg and were being held in Confederate prisons in Richmond. The Confederates justified holding these civilians on the grounds that the Union had imprisoned rebel nurses (although, as Bellows pointed out in a letter to a newspaper, the Confederate nurses were also enlisted men).[26] The Confederates demanded that the USSC prisoners demonstrate that they had ministered to rebel wounded, which had in fact been the case. Sanitary commission physicians caring for the Confederate wounded left at Gettysburg aided MacDonald's cause by having captured rebel surgeons sign a petition for the release of the USSC prisoners.[27] The Confederacy let MacDonald go in September, but another agent, Alfred Brengle, was still under negotiation as late as December 1863.[28]

In November the USSC sent an inspector to evaluate the conditions of Hammond Hospital and an adjacent prisoner-of-war camp at Point Lookout, in Maryland. It is not clear how they received permission to view the camp; it is likely that officers were so accustomed to USSC men showing up to inspect regular camps and hospitals that no one questioned their right to see how the prisoners lived. Point Lookout was located on a peninsula formed by Chesapeake Bay and the Potomac River. Now a pleasant state park, the camp was then one of the hell-holes of the North. Opened in August 1863, the prison held more than seven thousand men by the end of October, with only ragged tents for shelter. Many of the men sent there had come from other camps and brought smallpox and scurvy with them. As the freezing temperatures of winter arrived, the imprisoned Confederates lacked sufficient clothing, blankets, and shelter. Indignant reports appeared in southern newspapers, and presumably rumors about the camp reached the USSC offices.[29]

The USSC sent Dr. W. F. Swalm, a New York physician who had himself been captured at the first battle of Bull Run and who had ministered to Union prisoners in Richmond in fall 1861. One captured officer remembered Swalm as a dedicated physician to the Union troops in Richmond, "Dr. Swalm was . . . indefatigable; night & day he was among the wounded until he was very near brot down with a fever himself."[30] Swalm began his account of the Point Lookout inspection with the statement, "In compliance with orders received from the central office to proceed to Point Lookout, Md., and inquire into the condition, &c. of the rebel prisoners there confined, and also the sanitary condition of the encampment and its inmates, I hereby submit the following report." Would this former prisoner sympathize with the inmates, recalling his own experience, or exercise a hatred toward all Confederates acquired from his own grueling imprisonment?

Swalm described the wretched conditions in camp—and blamed a fair bit of it on the rebels themselves. The hospital inside the camp was filthy, but the prisoners made no efforts at cleanliness and the rebel nurses attending the sick "paid little or no attention to their sick comrades," even though six of these nurses were medical school graduates. Other factors were beyond the prisoners' control. No attempt was made to separate cases, so that the

Point Lookout, Maryland. In the foreground is Hammond Hospital, a pavilion hospital arranged with a central building and radiating wards, in the shape of an aster. Heading up the shore on the right, the fenced-in square enclosure is the prisoner-of-war camp. Library of Congress, Prints and Photographs Collection, LC-DIG-cph.3a05130.

wounded and those with diarrhea or erysipelas lay side by side. The hospital had no stove, and Swalm reported, "The men complain greatly of cold, and I must admit that for the poor emaciated creatures suffering from diarrhea one single blanket is not sufficient." He told the prisoners that there were bricks aplenty and that they should build a stove for the hospital, as indeed the prisoners had done for their own tents. No one died in the prison hospital, for the very ill were moved outside the camp to the general hospital a short distance away. The hospital dispensary had "little or nothing but a few empty bottles," and the few medicine bottles present were covered in dust. The patients had adequate food, but the hospital and grounds around it were filthy.[31]

Swalm laid much blame at the feet of the surgeon in charge and more generally targeted the camp commander for not maintaining discipline. He noted, "A great amount of the misery experienced in the hospital and throughout the camp might be obviated if a little more energy was displayed by the surgeon in charge, Dr. Bunton. There is a lack of system and want of discipline, neither of which (with all due respect to the doctor) do I think he is possessed of." The camp reeked of the human excrement and animal offal that lay in fetid piles near the quarters. The men lacked sufficient clothing, blankets, and shelter. Swalm suggested that the USSC send medicine and clothing to the camp. "I know they are our enemies, and bitter ones, and what we give them they will use against us, but now they are within our power and are suffering." Even though comparing their situation to the North's own men would be, in his words, "hardly . . . adequate to express our indignation," still he believed, "You would be doing right and . . . it might prove beneficial to us."[32]

Swalm closed with a report on Hammond Hospital, a general hospital near the camp that cared for Union soldiers and rebel prisoners. He found the hospital generally in good order but said the rebels were much dirtier in their behavior than the Union patients—they were prone, for example, to spitting on the floor—so that the rebel wards were in consequence less tidy. All told, he took every opportunity to blame the Confederates for their own dirt but could not hide the fact that they were being held in conditions of great deprivation.[33]

Frederick Knapp, the associate secretary of the USSC, forwarded Swalm's report on to Hoffman, with a brief note saying only, "Allow me to transmit to you for your consideration a copy of the report."[34] Hoffman was not pleased. Apparently in a separate letter Knapp had requested that Alexander MacDonald, so lately a prisoner of war himself, be allowed to visit other prison camps in the North for the purpose of inspecting them. In a reply that still drips sarcasm a century and a half later, Hoffman began, "I have the pleasure of acknowledging the receipt of your note of yesterday, and with it the interesting report of the condition of the prisoners at Point Lookout . . . for which permit me to return you my thanks." Just in case Knapp believed these sentiments, Hoffman then snapped, "In reply to your request that permit may be granted to Doctor McDonald to visit the prison camps for the purpose of inspection, I beg to say that medical inspectors of the army make frequent inspections of the camps referred to, and it is therefore not thought necessary to impose this labor on the Sanitary Commission."[35] In other words, the USSC should mind its own business.

Hoffman did not ignore the Swalm report, however. He forwarded it to the prison's commander, Brigadier General Gilman Marston, and demanded a reply. Marston's response to the report was indignation. "I have to remark that one more disingenuous and false could not well have been made," he began. "It is surprising that the commission should employ agents so stupid or dishonest as the author of this report." In contrast to Swalm's gloomy picture, Marston insisted, "The prisoners are treated as prisoners of war ought to be by a civilized people, and they and their friends are content. They have shelter, clothing, and wholesome food sufficient to insure vigorous health." He admitted, "They are a dirty, lousy set," but claimed that "every facility for cleanliness" was offered to them and that the prisoners were too lazy to take advantage of them.[36] Note that Marston did not question the basic standards for "a civilized people" to meet; he just disputed Swalm's claims that his camp did not meet them.

This answer must not have satisfied Hoffman, for he ordered his medical inspector of prisoners of war, A. M. Clark, to the camp twelve days later. Like Swalm, he found Hammond Hospital to be well run and in good shape. The day he inspected Point Lookout camp, there were 8,764 prisoners incarcerated, nearly a quarter of them sick in the hospital. Apparently some changes had been made since Swalm saw the place. The prisoners' tents had been moved; the hospital now had a stove; the hospital patients had a new issue of clothing; and an isolation hospital for smallpox and other contagious diseases had been constructed away from the main camp hospital. Rather than the hapless Dr. Bunton, who had been in charge when Swalm visited, another surgeon was overseeing the camp hospital, and Bunton was reassigned to the contagious disease hospital. Bunton was not up to even this less-demanding job. Clark judged him entirely incompetent but reported that Bunton's commanding officer was about to replace him. Clark otherwise found the camp in fairly good shape. The men each had a blanket—at least so he was informed, the tents had brick fireplaces, the men had access to wood, and vegetables were issued with the rations. Clark did not report seeing scurvy, that ever-telling sign of inadequacy in diet. It is quite possible that Marston, in response to the Swalm exposé, had made as many improvements as he could and went out of his way to give Clark a good impression. Clark did have some recommendations in regard to drainage and camp sanitation, but overall his comments allowed Hoffman to consider the USSC's charges answered.[37]

Hoffman sent a copy of Marston's reply, the one that called Swalm "stupid" and "dishonest," to Knapp at the USSC. Hoffman told Knapp, "Your inspector has strangely been led into many errors in his report." He requested that the USSC not publish the findings, for "their publication cannot but do much harm to our cause by exciting the friendly sympathies of the people for those who seem to be treated with unnecessary harshness and neglect." Not only would publication damage general northern morale; it would give "the rebel authorities an apparent excuse for the cruel treatment which they have heaped upon those of our people who have been so unfortunate to fall into their hands." The USSC did not mention the Point Lookout findings in any of their official publications or further refer to the findings in archival documents.[38]

Throughout the war the USSC looked for ways to provide relief to Union prisoners in rebel hands and to those recently released. USSC depots supplied bedding, food, and clothing to the steamers that carried exchanged prisoners home to the North. At times USSC doctors and nurses attended these weak and sick men, many of whom had to be carried on board and were in need of immediate medical care. The USSC also attempted to send packages of food, clothing, and blankets to prisoners held in the South. Some of these got through, and some released prisoners reported that the only blankets they received while in prison came from the USSC. But there were other reports that the Confederates took these supplies for themselves or that rebel officers sold them at a profit. Union officers at Libby Prison told of seeing stacks of crates marked USSC in view but out of reach, or of being taunted by having the crates broken open before them and the contents destroyed as the hungry men looked on. Such stories fueled northern anger and countered the southern claims that there just was not enough food and other supplies to support the prisoners in comfort. If allowed by rebel authorities, the USSC and other Union supply sources stood ready to feed and clothe the Union prisoners in southern camps.[39]

A year after the Point Lookout exposé, the USSC published a book on the sufferings of prisoners in war camps, but now its primary focus was the fate of Union soldiers in rebel camps. Although the full-scale exchange program did not resume until the winter of 1864–65, the two governments agreed to limited prisoner exchange in spring 1864. Sick prisoners were to be exchanged, especially those who were unlikely to be well enough to serve their respective armies anytime soon. On May 4 a load of such prisoners arrived from the camps in Richmond to a Union receiving hospital in Annapolis, Maryland. As mentioned earlier, their condition set off a firestorm of northern indignation and became a major factor in the coming elections.

Summer and fall 1864 were pivotal in the Civil War. U. S. Grant's army in Virginia fought battle after battle that killed tens of thousands of men but seemed to lead only to stalemate. W. T. Sherman appeared likewise stuck outside of Atlanta, unable to grasp that objective. Northern morale reached a low ebb by the end of the summer, and support for the Copperhead Democratic ticket was growing. Its presidential candidate, George McClellan, campaigned on a platform of ending the war by negotiating peace with the Confederacy. Lincoln openly worried in August that the election was lost, and it might well have been had Sherman not taken Atlanta in September.[40] Republicans struck back by emphasizing southern atrocities, hoping to arouse northern hatred of the Confederacy and strengthen the Union backbone to fight on. The Radical Republicans who populated the congressional Committee on the Conduct of the War first publicized the massacre of surrendered black Union troops at Fort Pillow and then trumpeted the horrible conditions of returned Union prisoners of war in their summary report on returned prisoners. The committee printed a pamphlet with their investigatory reports of these two atrocities and circulated it widely, including to northern newspapers, which gave the contents much play.[41]

The report on the returned prisoners was particularly inflammatory because it included pictures of the sickest men, men who indeed looked like skeletons draped in skin. These sketches made from photographs of nearly naked, starving men conveyed a record of abuse that words could not begin to express. The committee ordered the report printed only five days after the men arrived in Annapolis. Souvenir cartes de visites were likewise created from the images and sold to the indignant, who goggled at the misery of nearly starved men.

The Sanitary Commission responded ten days later by sending a committee of its own to determine "the true physical condition of prisoners, recently discharged . . . from confinement at Richmond" and whether "they did, in fact, during such confinement, suffer materially for want of food . . . or from other privations and sources of disease." The committee was particularly charged to discover "whether their privations and sufferings were designedly inflicted on them by military or other authority of the Rebel Government, or were due to causes which such authorities could not control." As an afterthought the USSC leaders also directed the committee to examine "some of the Rebel prisoners, recently captured, with reference to the question whether they have, while in the Confederate service, suffered like privations to those experienced by the Federal captives." The committee appointees included leading medical, religious, and commercial leaders from New York and Philadelphia.[42] The committee proceeded at once to Annapolis, where they interviewed men who had returned from prisons in Richmond. They also interviewed surgeons for their opinions about their patients. Finally, committee members solicited information from Quartermaster Meigs and William Hoffman about the treatment of rebel prisoners in Union hands and inspected prisoner-of-war camps at Fort Delaware (south of Wilmington on the Delaware River) and David's Island, New York (in Long Island Sound, near New York City).

The resulting report, the *Narrative of Privations and Sufferings*, attempted to convey an air of objectivity, but outrage and blame came through loud and clear. The report repeatedly used judgmental language to lay down the standards dictated by honor, Christianity, and humanity and called the Confederates to task for failing to meet this universal code. The committee recognized that in the early years of the war there were problems owing to poor preparation, but three years' experience should have removed this excuse from the table. The men who were exchanged in May 1864 were in ghastly shape, and the report demanded that the crimes of the guilty be laid before mankind for judgment.

"Months ago, we heard reports that our men were starving and freezing in Southern prisons," began their narrative. In the most recent exchange, "boatloads of half-naked living skeletons, foul with filth, and covered with vermin," were landed at Annapolis and Baltimore. For this delivery of men who were "diseased and dying, or physically ruined for life," the Union sent home southern soldiers who were, on the contrary, "well fed, well clothed, and well sheltered" during their captivity. Having heard all this, the USSC committee resolved to find out whether this state of affairs was due to unavoidable shortages in the Confederacy (whether Confederate soldiers fared just as badly), or whether it was all due to "a horrible and predetermined scheme."[43]

The committee members viewed and at times spoke with some of the more than three thousand men who had come to Annapolis and Baltimore from prisons near Richmond. (Andersonville, the most infamous southern prison camp, had only begun to fill in spring 1864.) Committee members were horrified by what they saw. "The best picture cannot convey the reality, nor create that startling and sickening sensation which is felt at the sign of a human skeleton, with the skin drawn tightly over its skull, and ribs, and limbs, weakly turning and moving itself, as if still a living man!"[44] Many of the men had "lost their reason," an outcome likely due not only to the traumas endured but also to vitamin deficiencies. The men examined testified to a litany of abuses: inadequate food, clothing, and shelter; physical abuse from guards for petty offenses, for example, being shot for daring to look out the windows at Libby Prison in Richmond; theft of the prisoners' personal ef-

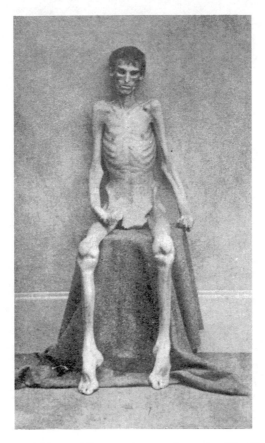

Photograph of a Union soldier and former prisoner of war, one of the "living skeletons" released from the Belle Isle Prison camp in Richmond and received in Annapolis, Maryland, in May 1864. This is one of several similar images that shocked the nation and moved the USSC to action. Library of Congress, Prints and Photographs, LC-DIG-cph.3g07966.

fects; and having relief packages sent from the North appropriated by prison staff.[45]

The question remained: was this cruel treatment deliberate or inevitable given the state of supply in Virginia? The returned prisoners reported seeing stores of food, clothing, bedding, and tents in Richmond, countering the claim that the treatment of prisoners was due to actual shortages. Many of the returned prisoners could not walk, and none of them were in shape to fight a battle. If the rebel army indeed received rations similar to those given the prisoners, then they should be likewise incapacitated—and were

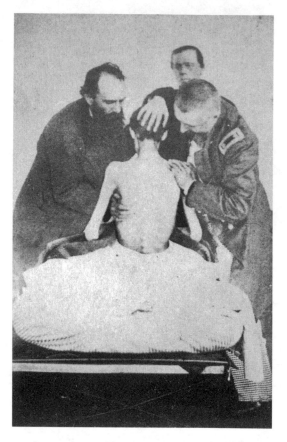

A Union surgeon and two other gentlemen examine a released Union prisoner of war in May 1864. Unknown photographer. This image is similar to one of the photographs published in the USSC's *Narrative of Privations,* which were copied from those taken by the Committee on the Conduct of the War during their inspection of Union prisoners exchanged in May 1864 from Belle Isle Prison, Richmond. It is thus very likely from that collection, although not published by the committee or the USSC. Library of Congress, Prints and Photographs Division, LC-DIG-cph.3g07961.

not. The guards watching the prisoners appeared hale and hearty in their charges' eyes.[46] To the explanation that the returned Union soldiers were suffering from illness, not starvation, the committee responded that their physicians found that the only treatment needed to restore the men who were salvageable was food and stimulation in the form of alcoholic beverages. Yes, the men often had chronic diarrhea, but it resolved as the men put on weight and regained their strength. Most of them weighed less than one hundred pounds on deliverance to Union lines. They had to be fed a liquid diet at first, because they were so unaccustomed to solid food.[47] Having reviewed and refuted any defense of the rebel behavior, the committee concluded, "The excuse and explanation are swept away. There is nothing now between the Northern people and the dreadful reality." To the committee's collective mind, "The conclusion is inevitable. It was in the power of the rebel authorities to feed sufficiently, and to clothe, whenever necessary, their prisoners of war."[48]

The committee drew a strong contrast between this abuse and the treatment of rebel soldiers in Union hands. Citing their own inspection tour as well as the official information received from Hoffman, they pronounced the northern prisoners to be well fed, well housed, and well cared for. Nothing in their description recalls the critical inspections of prior years. They take Hoffman at his word on the quantity of food given to the prisoners, even though it was well known that the ration had been dropped July 1 in direct retaliation for the treatment of Union prisoners of war in the South. Unable to deny that Confederate soldiers were dying in the Union pens, the committee claimed that such excess mortality was due either to the wounds and diseases the men brought into the camps or to the unavoidable ravages of smallpox.[49] For the USSC committee, the whole story was evident at the moment when the two lots of prisoners were exchanged. "The contrast must have been overwhelming at the point to which this narrative has now come. When the flag-of-truce boat landed within the rebel lines, the two systems confronted each other." The committee laid out the grim scene. "On one side, hundreds of feeble, emaciated men, ragged, filthy, hungry, diseased, and dying; on the other an equal number of strong and hearty men, clad in the army clothing of the Government against which they had fought, having been humanely sheltered, fed, cleansed of dirt, cured of wounds and disease, and now honorably returned in a condition to fight that Government again."[50]

Others presented a very different image of the rebel soldiers dying in Union camps from hunger, scurvy, and exposure. One prisoner at Fort Delaware recalled in his memoirs, "What a grand chance the United States Sanitary Commission missed in not having a photograph made of the survivors of Secretary Stanton's brutality. What a grand contrast our photo would have made with those photos alleged to have been made at Andersonville and other southern prisons," he lamented nearly fifty years after the end of the war. "It is a pity, indeed, those loyal souls who were ever anxious to stir the Northern heart did not have taken, for distribution in the North, our photos," the old soldier said. "Our condition would have brought the blush of shame to every Northern cheek, and made even Edwin M. Stanton turn pale at the sight of the victims of his brutality."[51] Given the high rates of scurvy in the camp and the many reports of prisoner deprivations, it is clear that the USSC account of Fort Delaware was overly rosy. One historian argues that the USSC men never went there and instead relied on the reports of others. But if one takes the USSC claim of eyewitness inspection as truthful, then their positive description of the camp was deliberately deceptive.[52]

In their report the USSC committee explicitly laid before the court of public opinion what they had found. They quoted a recent Confederate appeal to the European powers, which cried, "*We commit our cause to the enlightened judgment of the world, to the sober reflections of our adversaries themselves, and to the solemn and righteous arbitrament of heaven.*"[53] In response the committee appealed to the southern conscience to "remember that the judgment of Heaven is on the side of humanity, and against cruelty and oppression." They evoked the words of Jesus, "I was a stranger, and ye took Me not in: naked and ye clothed Me not," and called upon the Confederacy to act as a Christian, honorable, civilized nation.[54] If the Confederacy wished "to be taken for what they claim[ed] to be, . . . a mighty government, and a 'superior race,' first in civilization, in culture, and in courage, distinguished for all that is magnanimous, chivalric, humane, and noble," then they must abide by the standards of modern, civilized warfare and treat their prisoners decently.[55]

The Confederate government took these charges seriously, and while the war effort was crumbling around them, a Confederate committee responded with a report of their own in March 1865. To the USSC's *Narrative of Privations* and the Committee on the Conduct of the War's report on returned prisoners, they offered explanations and counterattacks but did not dispute the standard of proper care. They claimed that the skeletal men exchanged at Annapolis were sick, not starved; such malnourishment that the prisoners did suffer came from unavoidable food shortages, and northern camps were deadly as well. The rebel report sought to shift blame to the Union, which by ravaging southern agriculture had diminished the food supply. Casting the Bible back on their opponents, the Confederate committee urged the USSC to attend to problems in their own camps. "The cruelties inflicted on our prisoners at the North may well justify us in applying to the 'sanitary commission' the stern words of the Divine Teacher," the Confederate report cried. "Thou hypocrite, first cast out the beam out of thine own eye, and then shalt thou see clearly to cast out the mote out of thy brother's eye."[56]

Although much of the Confederate report contains out-and-out falsehoods, some aspects of it rang true. Northern prison camps were full of disease and death, and the USSC men probably knew this, at least to some extent. Who, then, made the decision to go to press with a glowing account of the northern camps? And why? The authorship of the text is unknown, and it is hard to identify who was in charge and making decisions at the USSC headquarters at the time. The *Narrative of Privations* was in print by early October 1864, and the last dated testimony it contains was from July 5. So it probably went to the printer by the end of July at the latest.[57] Henry Bellows was in California from May through October 1864, raising money for the USSC and tending to the Unitarian congregation of his great friend Thomas Starr King, who had died in March. His only contact with the home office in Washington was by telegraph or very slow mails. (The transcontinental railroad was not completed until 1866.) Olmsted was in California, too, having resigned from the USSC in September 1863. If there are archival documents relating to the publication decisions about the *Narrative of Privations*, they have proved elusive.[58]

When Bellows returned home in October, he discovered operations in the Washington office of the USSC in disarray. "I find things considerably mixed up during my absence—several feuds & much snarling," he wrote his wife. "I think my absence has been very injurious to our interests as I was the *lubricator* & balance wheel of the concern. I hope to get things back—but it looks *stormy*."[59] The main problems may have involved accounting records in the office, for F. N. Knapp, the corresponding secretary, resigned a few weeks later amid charges and countercharges of financial impropriety. Bellows found the Standing Committee members at odds but did not describe the topics of discord in his letters.[60]

Issues surrounding the prisoners of war immediately occupied Bellows's professional attention. "I was imperatively called to Genl Grants Headquarters on Friday last," Bellows told his wife. "The condition of our prisoners in Rebel Hands, is the subject of my important business with the Commander-in-Chief. I go with the greatest reluctance but the Commission will not let me off, and the most crying humanity demands my conference with him. *I may even see Gen. Lee!* altho that is doubtful." Bellows closed by reporting, "I have had a conference with Mr. Seward to-day & have an appointment with him & the President at 12 o'clock tomorrow."[61] Bellows did go down to City Point in the following weeks to inspect the USSC station that took in the released prisoners and prepared them for the trip north toward Union hospitals and home. The prisoner-of-war issue had assumed prominence on the national stage, and the USSC's humanitarian role in the care of these debilitated men had become an important aspect of their charge.

The only mention Bellows made of the *Narrative of Privations* amid this flurry of prisoner-of-war activity was in a letter to his son Russell, who had taken leave from his undergraduate studies at Harvard and traveled to California with his father. Russell had stayed in San Francisco to assist the local branch of the USSC. Bellows told him, "Our Report (which I send along to the California office) on the Sufferings of our Prisoners, has excited a profound feeling & is doing good at home & abroad."[62] What sort of "doing good" was Bellows talking about? Most likely Bellows was referring to the impending presidential election. He might have wished his nemesis Stanton and his underling Joseph Barnes to go to the devil, but overall he was in favor of the Lincoln presidency. Lincoln had it right on the prosecution of the war, the eradication of slavery, and the preservation of the government, Bellows believed, and he was pleased that the *Narrative of Privations* document might shore up the Lincoln candidacy. McClellan's Democrats were for ending the war and making peace with the South; the *Narrative of Privations* portrayed the southern leadership as too evil for honorable negotiations to be possible. At the time he wrote his son, Bellows was giving campaign speeches for the Lincoln ticket.[63]

Here, then, is one explanation for the obvious biases in the *Narrative of Privations*. The USSC did not hesitate to criticize the Medical Department or Hoffman's administration of the prisons. That was their accustomed duty. But summer 1864 brought a higher calling, the saving of Lincoln's presidency and the preservation of the Union. Whatever problems the USSC had with various members of Lincoln's fractious administration, they were sure that right and justice fought on the side of the North and that the Union must prevail. They also had to preserve their own public-relations image, as fund-raising remained an important source of income for their relief efforts. The *Narrative of Privations* was a campaign document, and one rather quickly forgotten at that.[64] In his official history of the USSC published in 1867, Charles Stillé mentioned the *Narrative of Privations* only in a footnote to the 553-page text. There Stillé wrote, "The facts stated in this report in regard to the cruel treatment received by our men have never been successfully controverted."[65] This language sounds rather defensive and also does not address the question of whether the book accurately described the northern camps. Was this a backhand acknowledgment that the rosy picture of Fort Delaware was easily controvertible? The historian can only speculate, but the strong political bias of the *Narrative of Privations* is hard to ignore.

Beyond publishing the *Narrative of Privations,* the Sanitary Commission had little to say about the prisoner-of-war camps in the North. It was planned that the sanitary history of the war, designed and compiled in 1865 and 1866, would include a chapter on northern camps, but such a chapter was never published. According to the handwritten prospectus in the USSC papers, there were to be five volumes; the prospectus printed for distribution had pared them down to four, which were to include detailed histories of individual hospitals and campaigns, an account of the ambulance system, and a review of the prisoner-of-war camps. Ultimately, only two of these volumes were published. (A third, written by B. A. Gould on anthropometric data from soldiers, bears the USSC label but was not described in the original prospectus.) Was the chapter titled "Hygienic History of Military Prisons of the Late War" deliberately suppressed? One could imagine that in the retaliatory mood of Reconstruction, such a chapter chronicling Union cruelty to prisoners would be unpopular; perhaps cutting this work was a political, or emotional, decision. But the evidence does not support that conclusion.[66]

In December 1865, Dr. A. M. Clark agreed to write the essay on rebel prisoners in three months, for a compensation of $120.[67] Clark had been the Union medical inspector of prison camps, and the USSC leaders probably knew of his reputation for bluntness and accuracy. His reports on Union camps pulled no punches; he did not hesitate to tell William Hoffman in his report on Camp Morton, for example, "This camp is a disgrace to the name of military prison."[68] By December 11 he had sent an outline to Elisha Harris of the USSC, but already four days later he was complaining that the work was not as straightforward as he had expected. "I find it rather slow work, tabulating my records," he told Harris. "I am anxious to inject all that is of importance, and yet allow no *dispensable* matter to creep in."[69] At this point in the correspondence, an experienced editor might well begin to worry that this writer would not be able to meet his deadline. By April 4, 1866 (the essay was now several weeks late), he wrote for help in receiving papers and information from certain physicians who had served in northern camps. "I do hate the idea of bringing my work to a close without making it as complete as possible," he stalled, and then he added flattery to sweeten the tardy report:

"This 'History' is not to be an ephemeral work of a day, but a monument to the *United States Sanitary Commission* for all time."[70]

Did he ever finish it? On April 10 he wrote Harris that he had been sick, and while some parts of the manuscript were written, he had not yet made a copy, since he was waiting for some promised information to arrive from another source. The copied partial manuscript would be there in two weeks, he promised. The last entry in this correspondence has Clark telling Harris on May 15, "[A] packet [was] sent by express yesterday."[71] This essay does not appear in the cataloged or microfilmed papers of the USSC, and Clark never published it independently. At this late date we cannot know whether he sent the completed manuscript in May, or just the portion mentioned on April 10. Clark's letters contain responses familiar to those with writer's block. Although he churned out reports steadily enough during the war, he did not publish in medical journals and may have been stymied by the public nature of this product.[72]

Why did the USSC not include an essay on northern prisoner-of-war camps in its medical history of the war? If Clark did finish the essay, the planned volume where it was placed in the prospectus never materialized, so that may explain its nonappearance. It is not clear why these planned volumes were canceled, but the most likely reason relates to the cost of publication and the difficulty of eliciting material from war-weary physicians. With the war over, the USSC had little fund-raising power at its command. But the USSC did find 172 pages for Confederate physician Joseph Jones's report on Andersonville in the *Medical Memoirs* volume.[73] One suspects that if the USSC had a completed essay on northern camps at hand, it could have published it there, as well. The USSC had clearly wanted the essay as late as December 1865, wanted it enough to pay for it, so it is likely that the agency's principal reason for not publishing it was that Clark could never bring himself to complete it.

It is unlikely that the members of the USSC committee that generated the *Narrative of Privations,* and members of the USSC governing committees such as Henry Bellows and Elisha Harris, remained untouched by the powerful emotions elicited by the returned soldiers of Belle Isle and Libby Prisons. Three years of war had brought so much death, had generated so much grief, had instilled so much hatred, that basic feelings of humanity and Christian charity were lost in the maelstrom. One historian has dubbed this anger "war psychosis" and used the concept to explain the cruelty to prisoners of war. "Apparently an inevitable concomitant of armed warfare is the hatred engendered in the minds of the contestants by the conflict," wrote William Hesseltine in 1930. "The spirit of patriotism which inspires men to answer the call of their country in its hour of need breeds within those men the fiercest antagonism toward that country's enemies." As a result, "the enemy becomes a thing to be hated; he does not share the common virtues"; such men "appear defective in all principles which are held dear by that people."[74] Even though the opponents in America's Civil War by and large shared the same skin color, religion, national traditions, and ethnicity, it was possible with enough goading for men to see the other side as significantly different. Remember Dr. Swalm, the inspector who found the southerners to be deliberately dirty; even though he saw the men as suffering from deprivations, he found a way to both separate their experiences from his own and blame them for their own condition.

Other physicians faced with the prisoner-of-war experience likewise hardened their views by 1864. Eugene F. Sanger, MD, physician to the Union camp at Elmira, New York, offers one study in transformation. Early on in his appointment, he tried to relieve epidemic diarrhea in the camp by draining a foul pond that bifurcated the stockade, and he called for a change in rations when he diagnosed 793 cases of scurvy among the prisoners. But after several months at the camp, he was desperate to receive a new posting. He wrote to an official of his home state, Maine, "I now have charge of 10000 Rebels[,] a very worthy occupation for a patriot . . . but I think I have done my duty having relieved 386 of them of all earthly sorrow in one month." Two weeks later he wrote the same official that he had "served in every medical capacity and butchered by the cart load."[75] The prisoners of war were helpless targets for the distilled hatred of both sides in this most uncivil of conflicts. It is understandable that even the men of the Sanitary Commission, men who carried the banner of beneficence toward all, should have yielded to an overwhelming hatred toward the enemy. And perhaps awareness of such hatred generated a tide of regret that led them to seek redemption in an international movement to bring humanity to the battlefield.

The American Red Cross

The 1850s and 1860s saw brutal wars in Europe and America but also elevated the voices of men and women calling for the more humane conduct of those wars. Florence Nightingale, appalled by the conditions of the British wounded in the Crimea, led a revolution in hospital nursing and sparked a massive volunteer effort to lessen the horrors of battle. Her work directly influenced the founders of the USSC, and the USSC's work in turn demonstrated to Europeans what an organized national relief effort could do to channel a nation's benevolence to its army. Henry Dunant, a Swiss civilian, witnessed the aftermath of the Battle of Solferino (June 1859), which pitted the French and the Sardinians against the Austrians and left thousands of casualties scattered on the field. After he organized what impromptu relief he could for the many men left helpless without medical attention, Dunant resolved to do something to prevent such nightmarish scenes in the future. Although he did not learn the details of the USSC's work until later, it was just this sort of national, voluntary relief agency that Dunant envisioned as the proper source of emergency battlefield relief.

In 1862 Dunant wrote a memoir of his experiences at Solferino, which he distributed to leaders across Europe, carrying his message of battlefield relief and humanitarian aid. In 1863 he called a conference in Geneva to discuss what could be done; that group met again in 1864 and generated the first set of recommendations of what became known as the Geneva Treaty. The treaty, quickly approved by most European governments, called for the humane treatment of wounded soldiers and declared the neutrality of medical providers, ambulances, and hospitals. The convention had adopted a reverse Swiss flag as its emblem, and the treaty declared that medical personnel should display the mark of the medically neutral, the red cross, as a signifier to protect them. The treaty made initial provisions for prisoners of war, and it declared that civilians should likewise be seen as neutral and not harassed. The convention urged each country to create its own national society to be ready to provide battlefield relief should the need arise. And such national societies could, in times of crisis, offer aid to those of other countries as well.[76]

The sort of national volunteer relief organizations that the Geneva gathering called for was very like that of the USSC, which had by this point three years of experience in doing just what was envisioned. There were two unofficial delegates at the meeting from the United States, Charles Bowles, USSC agent to Europe, and George P. Fogg, the U.S. ambassador to Switzerland. Secretary of State William Seward instructed Fogg to attend, informally (but to leave if a Confederate delegate showed up). Bowles was somewhat miffed that the delegates showed so little interest in learning about the USSC's activities; he felt that little credit was given by Dunant and others to what was, in effect, the prototype of the national Red Cross organization.[77]

In December 1865, Henry Dunant sent a letter to Bellows, asking him to organize an American branch of the Committee of Relief for Wounded Soldiers (Comité International de Secours aux Militaires Blessés), as the International Red Cross was first known. Bellows called a meeting in January and invited several USSC men as well as other prominent citizens, including Theodore Roosevelt, whose son later became president.[78] Taking on the imposing title Association for the Relief of the Misery of Battle Fields (ARMB), the group's main goal was to urge Congress to make the United States a signatory of the Geneva treaty. It was embarrassing that by 1866 all European powers had signed on, with the exceptions only of Austria and Turkey, but the United States remained outside the fold of civilized lands. The U.S. Government had turned away the first appeal, which came from the French government in 1864, saying that the United States already voluntarily followed the Geneva rules. From its founding in 1866, the ARMB lobbied Congress and sought administrative support. Success seemed near when Secretary of State William Seward granted his support, but ultimately the effort failed. The main reason for the opposition, according to the minutes of the American branch, was on the "general ground of not wishing to take part in any European Complications."[79] Perhaps general war-weariness had a role, too. Considering the Geneva Treaty meant looking ahead to the next war, an activity that could not have had many enthusiasts so soon after the Civil War's carnage.

Bellows's group took up other causes during its brief existence. With Europe roiled by war in the 1860s and 1870s, the ARMB raised battlefield relief money, especially from among ethnic French and Germans in the United States. In 1870, with France now at war with Germany, it issued an appeal that was nonpartisan in regard to the warring factions, focused simply on raising money to aid the wounded soldiers of both sides. Clara Barton later claimed that these efforts, which included gathering food, clothing, and bedding to send to Europe, were ill conceived because the American groups lacked European agents to efficiently distribute the humanitarian goods. In any event, the ARMB disbanded in 1870, and this remnant of the USSC's vision of relieving the horrors of war came to an end.

The American Red Cross did not take official form until the 1880s, under very different leadership. Clara Barton, a patent-office worker when the Civil War began, made a reputation during the war as a one-woman relief agency. She served at various sites in Virginia as a nurse and organizer of nurses. At the end of the war, she went to Andersonville to secure an accurate list of the names of the men who had died there and, where possible, to mark their graves. Barton was traveling in Europe in the 1870s when she learned of the Geneva Convention and its efforts. Now carrying the Red Cross label, the international organization still sought to expand its influence and gain new signatories for the basic treaty. Barton took the treaty back with her to the United States. Unsuccessful with the Hayes administration, she had more luck with presidents James Garfield and Chester Arthur and with their secretary of state, James G. Blaine. The United States finally signed the Geneva Treaty in 1882. A year earlier Barton successfully founded the U.S. branch of the international committee, the American Red Cross.[80]

Conclusion

Although it is undeniable that the USSC did important work of battlefield relief, hospital provisioning, and transport of the wounded, it was much less successful in its mission to bring a higher, Christian consciousness to the conduct of war. Its members, like others in the American states, were affected in the decades following the war by a complex dance of retaliation, reconciliation, and memorial reconstructions. If the Union was again to exist as a Union, the war and all its horrors was better forgotten than faced. While the USSC was eager to publish essays that drew on the medical knowledge learned in the war, it never fully revealed the ignoble cruelty of the prison camps. Early efforts to reform Union prisoner-of-war camps resulted in little effective reform, and the USSC never publicized these reports. The USSC leaders no doubt feared a drop in donations if they took up the unpopular cause of rebel prisoners. In 1864, at the peak of Union-Confederate hate propaganda, the USSC followed the path of political expediency, hiding the true conditions of Union camps in order to paint the Confederates as unilateral villains in the court of public opinion. Perhaps if Clark had been able to complete his article, some balance might have been brought to this story, but for whatever reason, Clark's essay never appeared. Bellows and his USSC colleagues failed to convince Congress and the president to formally accept the Geneva Convention, ending the career of the USSC. Although the Sanitary Commission's moral authority faded with this ignominious defeat, its influence in sanitary and public health persisted into the following decades' revolution in hygiene and disease control. The preacher failed, but the teacher proved triumphant in the long term.

CHAPTER TEN

A Public Health Legacy

Elisha Harris probably would not have phrased it this way, but he and the cause he held most dear had a very good war. In the 1850s Harris had been frustrated by failures to institute sanitary reforms in the United States. Through his work with the USSC during the war, however, he saw the goals of sanitation broadcast nationwide, and he participated in the flowering of public health institutions in the years following the end of hostilities. Harris had been the quarantine physician in New York's Staten Island from 1855 to 1860 and had helped organize the first national quarantine conferences in the late 1850s. After the war he participated in the founding of the New York City Metropolitan Board of Health, and he helped organize the American Public Health Association in 1872 (and served as its president in 1878). In his career and interests, he embodied the change in public health institutions and actions that was given great impetus by the Civil War.

This chapter looks at the impact of the war's medical crisis on postwar medical institutions and the response to that crisis. Sixty years ago historian Howard Kramer described the influence of the war on public health reform, concentrating on concepts of hygiene and how the war spread such ideas throughout the medical and civilian leadership. A Civil War camp was a city, he wrote, with a population density of "about 85,000 persons to the square mile, compared to London's 50,000 or Philadelphia's 45,000." Just as in cities, drainage, sewerage, water supply, and food provision were central to the health of the inhabitants. The U.S. Sanitary Commission's inspectors drove these points home, and their many informative pamphlets reinforced the message. Camp physicians left the war educated in the basics of community hygiene; officers likewise went home imbued with knowledge that proved valuable in crafting public health legislation. Many physicians who later became public health leaders acquired a passion for public health in the Civil War setting. The war provided a vast school in the importance of general sanitation and the potential of disease to destroy the strength of a community.[1]

U.S. public health reform was stalled in the decades before the war. While legislation in England, France, and Germany responded to the health evils of urbanization with new efforts toward sanitation by midcentury, in America public health advocates made little headway. John Griscom and Lemuel Shattuck produced damning reports on the health of New York and Boston, respectively, but their pleas fell on deaf legislative ears.[2] Many physicians called for the collection of vital statistics to document the health of the community. But much public discourse on health and disease prevention was marked by bickering over the appropriate role of sanitation versus that of quarantine, and there was little evidence to back up the sanitarians' claim that cleaner streets and purer water would improve the health of the population.[3]

The war helped to break this impasse. By widely disseminating the idea of a specific and transportable poison as the cause of major epidemics such as cholera and yellow fever, and by demonstrating the ability to neutralize the poison with disinfectants, physicians such as Elisha Harris of the USSC offered a new, direct path for public health action that appealed to common sense. The mandatory use of smallpox vaccination among troops prompted a vigorous debate over its safety and efficacy, both the strengths and weaknesses of this tool, and had a mixed influence on government-sponsored vaccination after the war. Kramer was right in emphasizing the importance of the war in furthering general concepts of sanitation, but he did not go far enough in recognizing the specific ways in which the ideas spread during the conflict gave physicians new power to take on disease. Public-health-minded physicians such as Elisha Harris left the war thinking that there was much physicians could do to prevent disease and that the government had the right and the duty to assist them in the endeavor. Further, the war strengthened the concept of the expert, and of the official government expert, in combating disease and regulating the environment and behavior.

Harris was particularly interested in the promotion of vital statistics, and if nothing else, the war accustomed physicians to filling out forms. Those who were unfamiliar with the official ways that diseases were categorized learned how to write diagnoses on a line of a form and to keep count of the sick and the dead. For the first time, the federal government, in the form of the army's medical staff, tallied cases of sickness by cause and created tables of morbidity and mortality on a massive scale. The army of clerks engaged to keep track of these numbers in turn generated the hundreds of pages of tables reporting on disease and death in the war. As urban areas began to collect vital statistics in earnest in the postwar years, medical veterans took up a now-familiar task. And for municipal leaders who had served as officers, and who likewise had had to keep count of their men, supplies, and arms, the paperwork for vital statistics may have seemed both routine and straightforward. The war had taught all of them about the bureaucratic machinery of keeping count as a means of discovering the state of affairs and as a basis for taking action to deal with problems.[4]

Although yellow fever and cholera were of minimal importance during the war, they quickly reemerged at the center of public health controversies at war's end, cholera in 1866 and yellow fever in 1867. Physicians applied lessons learned during the war to these familiar enemies, approaching them with new confidence. Disinfectants, sanitation, and isolation had limited the number of infections during the war. Girded with this knowledge, physicians were ready to take on these major threats to the nation's health and commerce.

Yellow Fever

The understanding that yellow fever was spread by a mosquito was four decades away when the Civil War erupted. In 1860 the cause of yellow fever was still hotly debated, with one side saying that the disease arose from rotting animal filth and another claiming that the disease was contagious and imported. A third faction had emerged in the decade before the war, which held that the disease was not contagious but that the specific poison of yellow fever could be carried from place to place, perhaps on infected clothing and bedding or in the holds of ships that had captured the infected air of an epidemic site. Such contaminated, disease-carrying entities were referred to as fomites. USSC physician Elisha Harris typically explained, "The initial and all-essential fact in the etiology and geography of yellow fever, is that which relates to the implantation of the epidemic germs or the *fomites,* in a locality where temperature, humidity, and personal conditions favor their pestilential propagation."[5] This emphasis on the introduction of the germ into an appropriately receptive environment became common in the 1870s; at that time the language of seed and soil was often used. The word *germ* here, like the use of *virus* elsewhere, should not be taken to mean a living microorganism. Harris did not know what it was, just that it could be transported and contained and that the epidemic would not grow if the ground was suitably cleansed.

As noted in chapter 3, yellow fever was not a major problem during the Civil War. There were scattered epidemics on the East Coast, most likely sparked by the activity of blockade runners, and the number of cases in New Orleans was greatly diminished compared to the prevalence of yellow fever in the prewar decade. Physicians drew different lessons from these outbreaks. Some, such as William Wragg of Wilmington, persisted in their beliefs that yellow fever emerged spontaneously from the filth of a city, especially in its dockside areas.[6] Others agreed with Elisha Harris that the seed of an epidemic had to imported from elsewhere, even if the affected locale's sanitary condition was still critical in the eruption of an epidemic. The case of New Orleans attracted the most commentary, as the military government had full rein to enforce quarantine, hire cleanup crews, and in general pursue the goals of cleanliness and isolation that many sanitarians sought but lacked the power or means to achieve. Harris saw the small number of yellow fever cases in New Orleans as a vindication of these methods.[7]

In 1870 Stanford E. Chaillé, MD, of New Orleans took issue with the conclusions of Harris and others that the minimal number of yellow fever cases in New Orleans during the war was due to "an impregnable system of Quarantine" and the enforcement of "excellent sanitary regulations." It was not true, he maintained, that "so clean a city had never before been seen upon the continent." Granted, the more public commercial sections and the affluent neighborhoods were cleaner than usual, but the impoverished districts were just as dirty as usual. The history of New Orleans was, he granted, "one long, disgusting story of stagnant drainage, foul sewerage, environing swamps, ill- and un-paved streets, no sanitary regulations, and filth, endless filth every where." Chaillé challenged Harris's conclusions that strict sanitary police had restricted the spread of yellow fever, since it did not actually happen. (Chaillé not only was there at the time; he could interview other New Orleans physicians for their opinions, something the New York–based Harris could not do so easily.)[8]

Chaillé was likewise unimpressed with the supposedly rigid quarantine. He noted (as did Harris) that many cases of yellow fever did appear, so if they were introduced from abroad, then they were prima facie evidence that the quarantine had failed. What was important was that the epidemic did not erupt into a major outbreak, in spite of a city full of susceptible northern troops. Chaillé contended that Harris had set up a false dichotomy by comparing New Orleans in 1862–65 to the very worst epidemic years before the war, 1853–55. If Harris had instead looked at 1857–59, he would have found little difference, because yellow fever was not epidemic in those years, either. Chaillé further demonstrated that the death rates in 1862–65 were high in New Orleans, even if those deaths were not from yellow fever. The environmental cause of multiple fevers was pervasive and deadly, belying the argument that the city was significantly cleaner in its atmosphere than before. It would require significant work on the swamps around New Orleans as well as the creation of a citywide sewage system to make any dent in the pestiferous atmosphere that both killed citizens year in and year out and fed the yellow fever epidemics when they occurred. Chaillé reviewed prevailing theories of the cause of yellow fever and concluded that the evidence was not conclusive for any of them. He believed that the poison of yellow fever could arise spontaneously in the appropriate conditions and spread as an epidemic, again in the appropriate conditions. And he remained sure that cleanliness could prevent both actions but that the quagmire that was New Orleans required more than a little street sweeping in the commercial areas to ameliorate its deadly atmosphere.[9]

Chaillé's detailed analysis of mortality statistics and disease patterns failed to impress most physicians. His article presented no clear path for action and was too complex for readers to appreciate his argument. Most medical authors in the 1870s continued to tout the safety of New Orleans during the war and point out the yellow fever epidemic of 1867 that followed it, as signs that available techniques, properly and rigorously applied, could prevent yellow fever outbreaks. After yellow fever again scourged the South in the early 1870s, Congress called for an investigation of southern quarantines and public health activity. The surgeon general put this project in the hands of Harvey Brown, who had fought yellow fever in the army for the previous decade. Brown traveled throughout the South, discussing yellow fever with physician groups and health officers. His report was published in 1873.[10]

Brown stated, "A majority, especially of the younger physicians, believe decidedly in the transmissibility of the yellow-fever poison; but they hesitate openly to express their views, partly because by so doing they would find themselves in opposition to men who may be said to lead the profession at the South, and who, with the people at large, are considered to speak almost with infallibility." Further, he concluded, "In the vast majority of [yellow fever] epidemics, if not all, that have occurred in the United States, the germinal principle of the disease was imported from elsewhere, and was not due to local causes." Brown was unsure about the identity of this germinal principle: "The fundamental deduction arrived at by recent observers is, that every kind of contagion consists of extremely minute particles of organic character, capable of growth and multiplication to an indefinite extent; and that the contagious fevers are due to the introduction of such particles into the blood." He then reviewed various theories, including the idea that the causative agent might be a fungus, but he drew no definitive conclusions on that score. He was clear about what needed to be done. Rigid quarantines should exclude the "germinal principle," and rigorous sanitation should deny the disease seed a nurturing environment. Included in the quarantine recommendation was a call not just to have ships wait under observation. They were to be vigorously fumigated, cleansed, and disinfected.[11]

Although southern boards of health were slow to take up Brown's suggestions, by the early 1880s a quarantine based on disinfectant principles was being applied at all vigilant southern ports. The National Board of Health, a short-lived entity formed in the wake of the disastrous 1878 yellow fever epidemic, fumigated railroad cars and applied the powers of disinfectants to limit the spread of yellow fever by ship and rail. By 1880 the use of the word *germ* to describe the cause of yellow fever was pervasive, although researchers were not much closer to identifying the nature of the germ. As more and more infectious diseases were ascribed to particular microbes, the certainty grew that yellow fever's cause would soon be found. If, as for cholera, a bacterial agent had been implicated, there would have been no major conceptual break between using disinfectants to destroy an "active organic agent" and using disinfectants to destroy the modern, specific bacterium. Only when yellow fever was shown to be carried by a mosquito were public health officials challenged to abandon a whole system of belief and action for a novel new approach.[12]

Cholera

Cholera was less of a threat than yellow fever during the war, but it broke out among army troops in 1865 and 1866 as the third great cholera pandemic swept through the United States. It also penetrated the refugee camps of freed people and displaced whites during those years, threatening widespread mortality.[13] Using familiar language about poison, physicians applied disinfectants to neutralize it while isolating patients to prevent them from becoming a source of poison that could engender disease in others.

The connection between Civil War experience and postwar public health was perhaps most evident in New York City. There a group of reform-minded physicians called for a sanitary survey of the city in 1864, and Elisha Harris wrote its report. Harris applied his knowledge of disinfectants and disease poisons directly to the foul conditions of New York City, and when cholera struck again in 1866, his ideas prevailed: cholera patients were isolated and their excretions sanitized with disinfectants. These efforts were organized by the nascent Metropolitan Board of Health, for which Harris worked as registrar of record. Historian Charles Rosenberg said of these events, "In the history of public health in the United States, there is no date more important than 1866, no event more significant than the organization of the Metropolitan Board of Health. For the first time, an American community had successfully organized itself to conquer an epidemic." And they used "the tools and concepts of an urban industrial society." The "tools and concepts" in question were isolation and disinfection. Harris instructed the board that the causative agent of cholera passed through the stools and that disinfectants could render them safe. Although he had not fought a cholera epidemic during the war, his disinfectant work for the U.S. Sanitary Commission was finding immediate application in civilian life.[14]

The relationship between disinfectant use in the war and the response to the cholera epidemic becomes more convincing when compared to the British response to the epidemic in the same year. English and Scottish physicians had discussed the possibility that the "rice water" discharges of cholera patients might spread the disease, particularly after having some time to transform under the influence of heat and moisture. William Budd, known for his theories about typhoid spreading through contaminated water, particularly urged this response to cholera, as mentioned in works commenting on the 1854 epidemic in England. But in 1866 the "disinfectant" plan was not widely used in England; if Budd is to be believed, this measure was applied only in the city of Bristol. Writing in 1871, he was still pleading for the disinfectant approach—prevent the specific poison of cholera from moving between source patient and new victim by disinfecting all that surrounded and came out of the patient—with an air of exasperation and futility. As late as 1871, Budd believed that the Bristol experience presented "the only example of a great city, eminently disposed . . . by the character and crowded state of its population, to the propagation of contagious disease, succeeding in protecting itself against cholera by measures especially designed for that object." Meanwhile, the disinfectant method had become common practice in the United States, a fact that Budd apparently did not know; at least, he didn't mention it.[15]

Rosenberg's account of cholera in 1866 in New York City has become familiar to a host of historians and students through his widely read book *The Cholera Years*. Less well known are the many cholera epidemics that occurred in military camps in 1865–66 and the major epidemic that afflicted the Mississippi Valley in 1873. But these epidemics, and the public health response to them, familiarized more physicians, and politicians, with the power of disinfectants and public health than the New York City case did.

Cholera made its U.S. debut at Fort Columbus on Governer's Island in New York Harbor in 1866. It subsequently afflicted multiple cities across the country and particularly attacked military encampments. An 1867 military report on its path summarized, "Cholera spread over the country during the year 1866, extending as far westward as Forts Leavenworth, Riley, and Gibson; and in the southwest as far as Texas. In its progress the disease followed the lines of travel rather than any general westward course, and, in the case of the army, it especially followed the movements of bodies of recruits." Although the war had ended in 1865, many men remained in uniform. Some were sent to Texas to wage peace there and guard against an uprising in Mexico. Others were scattered throughout out the South or to the western reaches of the country to guard against American Indian hostility. This mobility fueled cholera outbreaks. More than 1,200 soldiers died of cholera in 1866; the army was proud to hold this number down to 230 in 1867.[16]

Surgeon General Joseph Barnes circulated orders detailing the steps to be taken in response to the epidemic. "In addition to the strictest hygienic police, enforcement of personal cleanliness and thorough disinfection, attention should be paid to the quality of water used for drinking and cooking purposes. When pure rain water cannot be procured in sufficient quantities, and the spring or river water contains organic impurities, it should be purified by distillation, or the noxious matter precipitated by permanganate of potash." This message was conveyed to all physicians in the army, and district medical commanders reiterated its message in orders to their subordinates. For example, physician Ebeneezer Swift at Jefferson Barracks in St. Louis told the physicians in his command on July 17, 1867, that "they should disinfect all slop vessels, drains, privies, etc., by means of chloride of lime or zinc, sulphate of iron, or carbolic acid. Strict cleanliness should be enforced in dormitories, kitchens, etc., and former should be fumigated with fumes of burning sulphur."[17] When the army's cholera mortality in 1867 was a fifth of that in 1866, the army physicians applauded their successful application of these measures. These physicians acquired impressive experience in the process. They learned that clean drinking water was essential to health and that disinfectants can stop epidemic diseases.

One army correspondent described the power of disinfectants to change a fearsome disease into a manageable entity. The setting was Fort Harker, in Kansas. Dr. J. W. Brewer noted in July 1867 that few hospital workers acquired the disease, even though there were a large number of cholera patients. "To the free use of disinfectants in the cholera tents and sinks, I attribute the immunity from the disease enjoyed by the nurses and attendants," he wrote. "No case of cholera occurred among them after I took charge. . . . The nurses and attendants in the cholera wards were almost all of them negroes. I was very much surprised at their fidelity; they placed great faith in disinfectants, and after their use they seemed to have no fear of the disease."[18] Army protocol had reduced the caring for cholera patients to a simple routine: isolate the cholera patient, disinfect the diarrheal discharges, rinse soiled bedding and clothing in disinfectants before laundering, and use the liberal disinfectant application in the latrines.

These lessons were put to extensive use when cholera returned in 1873. Whereas previous epidemics had ravaged the cities of the eastern seaboard, this epidemic was largely confined to the vast network that was the Mississippi and Ohio valleys. The official death count was 7,356, but one chronicler of the outbreak put the likely death toll much higher. So little has been written about this epidemic that it is difficult to know whether it spared the East Coast because of vigilant and reactive public health officials or because of the presence in these cities of so many people who had acquired immunity during the epidemic eight years before. What is clear is that the epidemic demonstrated the capacity of the river transport system to spread disease, a capacity that became even more evident during the yellow fever outbreak of 1878.

Congress was concerned about this return of cholera and its widespread dissemination. It commissioned a report, written by the new director of the Marine Hospital Service, John M. Woodworth. The Marine Hospital Service had hospitals in many of the port cities of the Mississippi and Ohio valleys, and Woodworth could call upon the physicians in such institutions for information. Woodworth's conclusions illustrate the same sort of transformation in ideas that was simultaneously occurring for yellow fever. He argued, "Malignant cholera is caused by the access of a specific organic poison to the alimentary canal; which poison is developed spontaneously only in certain parts of India." So cholera could not arise de novo in the United States. He further posited that the cholera poison was contained in the stools, vomit, and urine of the diseased persons. But when freshly deposited, it was benign; only after it had time to decompose in an alkaline environment did the poison become dangerous. This latter idea came from German scientist Max von Pettenkofer and had been widely discussed from the mid-1850s on. Far from believing that bacteria caused cholera, Woodworth thought bacteria appeared at end of the period of incubation (if, say, stools were allowed to sit around) and disappeared later. The dejecta were most poisonous when bacteria were present. "It is not meant by this that the bacteria so found are the cholera-poison, since they differ in no appreciable manner from bacteria found in a variety of other fluids." If the stools, urine, or vomit were treated with appropriate acid disinfectants, the resulting mess became innocuous.[19]

By the early 1870s, physicians had reached something of a consensus on the cause and prevention of yellow fever and cholera. That consensus was revealed as well as created, in part, by official government pronouncements on the infections. Although the United States did not yet have a national public health service, and most states had no state board of health, there were still national spokesmen who tried to summarize and disseminate the best knowledge of the time. One source of authoritative material was the surgeon general of the army, or his delegates, who gathered information from multiple physicians in the service and in turn used military channels to communicate their findings. The reports by Woodworth and Harvey Brown in the 1870s were both in response to congressional requests for information, summonses that marked the beginning of a nonmilitary federal role in public health. The consensus that emerged argued that both diseases were spread by a specific poison, that the poison could travel from place to place, and that it could be blocked by proper measures of isolation and disinfection. When the formal theories of microbiology emerged in the 1880s, little about this understanding had to change. The cholera vibrio replaced the cholera poison; the yellow fever poison became the yellow fever germ. Conveniently, both were still to be controlled by isolation of patients and disinfection of the environment surrounding them.

The president of Columbia University, F. A. P. Barnard, summarized the status of the "germ theory of disease and its relations to hygiene" in an address before the nascent American Public Health Association in 1872. Barnard was not a physician but a scientist with broad interests in biology, mathematics, and astronomy. His friends were many of the public health leaders in New York, and he had read widely in the medical and biological literature on such topics as cellular anatomy, spontaneous generation, and the various theories on the causes of epidemic diseases. He argued that there were mainly two regnant theories. "The chemical theory is founded on a presumed analogy between the propagation of disease in living organisms, and the process of fermentation in certain forms of organic matter without life." He went on to explain, "This theory assumes a ferment to be an organized substance in a certain state of decay, which possesses the property of exciting the same decay in other organic substances with which it is in contact." Barnard contrasted this idea with "the opposing theory [which] presumes that the diseased person is suffering from an invasion of his system by microscopic algoid or fungoid vegetative forms having the property of rapid self-multiplication, and that the spores which proceed from these fungi or the cells of the algae are wafted in like manner by the air from person to person." Barnard found the evidence for the latter, animalcular theory, to be inconclusive. He argued that when researchers saw tiny life-forms in the blood or other tissues of the ill patient, they were seeing the product of the infectious process, not its cause. Although he cited Casimir Davaine's work on anthrax and a few other authors of works on bacteria as causal organisms, Barnard found that most proponents of an animalcular theory proposed fungi or algae as the agents of disease.[20]

Having considered the debate over causation in detail, Barnard turned to the prevention of infectious diseases. In a remarkable passage, he concluded that it did not matter whether one accepted the chemical or the germ theory of disease, since the steps against them were the same. "Pure air, pure water, wholesome food, thorough drainage, rigidly enforced cleanliness, the severe exclusion from towns and cities of industries which contaminate the air . . . [and] the prevention of overcrowding in dwellings" were the first order of business for "the guardians of the public health." Further, attention was to be paid to the individual case, with "the prompt and complete disinfection of every spot where pestilence may lift its head, and in every article and substance, including the dejecta of the sick, which may serve as a vehicle of disease." Barnard was happy to conclude that "the champions of conflicting theories, however freely they may splinter lances in the arena of controversy, are always found, in the field of actual warfare and in the face of the common enemy, marching harmoniously side by side."[21]

Smallpox

Smallpox was the exception to public health intervention in the nineteenth-century United States. It was obviously contagious person to person and could be prevented by its own distinctive method, vaccination. Whatever the specific cause of the disease was, it could be conveyed by the tip of a scalpel or dried and carried in an envelope. From the earliest efforts of Benjamin Waterhouse to popularize vaccination in the early republic, vaccination was a singular preventive technique. But not until the Civil War did the government step in to impose vaccination on a population. Vaccination had been a private affair, something for individuals to negotiate with their physicians, not a government-sponsored intervention. The Civil War highlighted the public importance of this dread disease and created the impetus for laws imposing vaccination on the populace, particularly the school-aged populace.

It became quickly evident during the Civil War that smallpox could and would spread through an unvaccinated army. George Washington learned that lesson a century earlier, when he ordered the inoculation of his army to prevent its utter dissolution. The British and Hessian troops, drawn from urban areas where smallpox was endemic and seasoned by long time in service, were mostly immune to smallpox. But Washington's rural men were vulnerable, as smallpox was an intermittent problem in the American colonies and children did not routinely acquire the disease. The process of inoculation, in which vesicular lymph from smallpox patients was scratched into the arm of a well recipient, transferred a mild case of smallpox and immunized the treated patient. Edward Jenner improved on this process in the late eighteenth century, when he began using fluid from cowpox instead of smallpox lesions. By the mid-nineteenth century, the vaccine fluid was transferred from human arm to human arm, ideally using infants as incubators, since they were unlikely to have other diseases such as syphilis. Scabs were dried and then reconstituted with water when the vaccine matter was to be applied. The Union Army required that men be vaccinated on admission, or show a vaccine scar or evidence of previous smallpox. This rule was rarely observed, at least in the enthusiastic opening days of the war.

Smallpox spread through the Union troops and from there into the Confederate Army. According to some reports, there was no smallpox in the Confederacy until summer 1862, when Confederate troops pushed into Maryland and made contact with the disease. Smallpox was most prevalent in cities crowded with troops, especially Washington, DC, so it is not surprising that while in its vicinity the Confederates may have had contact with an infected person.[22] As white and black southerners fled the fighting and slavery, refugee camps on the borders between Union and Confederate domains provided particularly fertile grounds for smallpox amplification. Prisoner-of-war camps likewise provided an environment in which every nonimmune person was likely to be exposed and infected. Armies on both sides tried to block this outbreak with mass vaccination, but the process often failed. Vaccine scabs were unavailable or inactive, and arm-to-arm vaccination between adult men, many of them poorly nourished or ill with other diseases, at times led to gangrene or the transfer of syphilis. The southern press charged that the Union deliberately infected southern prisoners of war, especially at prison camps in the Midwest. Both sides created isolation hospitals for eruptive fevers, including smallpox, measles, and chicken pox. These hospitals often succeeded in cross-infecting the patients unfortunate enough to be sent to them.

Smallpox was a nasty, frightening disease that caused significant controversy precisely because it was commonly understood to be preventable. Stories of failed vaccination raised the possibility that vaccination itself was ineffective, but most believed instead that the failure was due to poor procedure. In 1867 Elisha Harris summarized the lessons learned during the war. He pointed out that in Europe infant vaccination was common and that all troops were systematically revaccinated upon enrollment in the army. Such was the practice of the Prussian army, for example, and as a result smallpox was very rare among those troops. In contrast, there were no "adequate regulations for encouraging or enforcing vaccination [in the United States] and in but few States did the local military authorities succeed in vaccinating their volunteers previous to departure for the field." It was true that army regulations required vaccination, but the regulations were not enforced. Only after smallpox erupted, with major epidemics in the winter of 1863–64, did vaccination become a major priority in both the Union and the Confederacy.[23]

There were many stories of vaccination disasters during the war, especially among Union troops west of the Appalachians. The vaccine procedure itself opened a wound that could become infected, particularly in men whose bodies and clothing were dirty, who were malnourished, and who were weakened by other diseases such as chronic diarrhea. Such men healed poorly because of vitamin deficiencies, and they had weakened immune systems. Thus the vaccine process was dangerous to them, and they reaped less benefit. The vaccine supplied by the medical purveyor was frequently inactive, many army surgeons reported, based on the lack of classic vesicle formation. Given the outbreaks of smallpox and general fear of its effects, men took to vaccinating themselves or vaccinating arm-to-arm from a soldier with an active vaccine lesion.

Surgeon Ira Russell described the ill effects that could follow. He was in charge of a military hospital in St. Louis in the winter of 1863–64 when smallpox broke out in Benton Barracks, nearby. A surgeon at his hospital directed the vaccination of a ward from the vaccine vesicle on one soldier's arm. The first day, the vesicle was opened, drained, and the material distributed to the initial ward via scalpel. The second day the same vesicle was opened again, and a second ward received the fluid. "On the third day the same man was taken into another ward and lent his inflamed and now purulent vaccine sore to the patients there." This action was repeated for several more days. The men who received the vaccine on day one were protected from smallpox and healed well. But the men who received the material from the soldier's arm on day two and beyond developed nasty ulcers and skin infections at the site and were not protected from smallpox.[24]

Even more casual vaccination protocols no doubt prevailed where vaccine was in short supply or the men decided to take matters into their own hands. Dr. Owen M. Long reported such an event in an Illinois company stationed near Vicksburg. "In August 1863 J. B. a private Co. B 11th Illinois Infantry, whilst under an attack of syphilis, was vaccinated with genuine vaccine virus furnished by the Medical Purveyor at Vicksburg, Miss. . . .The vaccine pustule presented all the usual characteristics of a healthy eruption." So far, so good, but J. B. decided to share his good fortune. "One fine morning whilst the men were engaged in washing themselves he exposed his arm to view, the pustule was seen and much admired by one, in the ranks who at home had been a country doctor. Without any consultation on this point with either myself or my assistant [he] undertook to and did vaccinate some twenty five men of said company." Long reported that most of these men had already been vaccinated once, near the start of the war. "Their arms became immediately very sore, with enlarged glands in the axilla, so much as to disable most of them from duty. . . . I found most of these men labouring under well developed chancres at the points of puncture with enlarged axillary glands."[25] These men had apparently inoculated themselves with J.B.'s syphilis as well as his vaccinia virus. At other times and places, men used rusty nails, pocket knives, or hat pins to scratch their skin and smear liquid from a friend's vesicle into the wound. The outcome was fairly predictable.[26]

After detailing all the ways in which smallpox vaccination went wrong during the war, Elisha Harris was still confident in his 1867 summary about smallpox that properly applied vaccine from a healthy host, given to a man who was himself otherwise in good health, was highly effective at preventing this deadly disease. "The experience of more than two millions of American soldiers in the war of the rebellion has reaffirmed the great doctrines of Jenner," Harris concluded. It has "demonstrated anew, and upon a gigantic scale, both the importance and the correctness of his rules for procuring the full benefit of vaccinia and transmitting it to others in all its original and pure prophylactic power." There was no doubt, Harris trumpeted, that "Genuine vaccination was an absolute safeguard against small-pox."[27] The success of vaccination, done properly, to protect the army was manifest from the Civil War experience. But many men also went home with nasty scars from purulent vaccine sites, and that memory likewise had a lasting influence on the popular perceptions of vaccination in the United States.

Conditions were apparently even worse in the Confederacy. Many fewer citizens, slaves, and soldiers had ever been vaccinated, and from reports published after the war, it appears that the circulating vaccine virus in the South was often ineffective. "Untoward results of vaccination appear to have been at one period the rule rather than the exception among civilians as well as soldiers within the Confederate lines," one commentator noted, "so much so that for some time after the war the people, and in some instances even physicians, manifested a fear of resorting to this protective measure."[28] Many Confederate soldiers endured purulent ulcers and painful axillary lymphadenopathy following vaccination, and yet they were not protected from smallpox when it circulated.

The official medical history of the war put the white Union soldier smallpox case and mortality numbers at 12,236 and 4,717; for the black troops the count was 6,716 and 2,341. Since the black troops made up only about one-tenth of the Union Army but contributed one-half of the smallpox cases and deaths, their increased susceptibility to the disease was evident, and it most likely reflected the lack of previous vaccination.[29] Some ten thousand Confederate prisoners of war contracted the virus in northern camps, and about a quarter of them died of it, according to the *Medical and Surgical History of the War;* it is hard to know whether these numbers are accurate or an underestimate.[30] Doctors at the Andersonville prison camp attempted to vaccinate the Union soldiers incarcerated there, only to see the vaccine site fester and suppurate, with many men requiring subsequent amputation. The starved, malnourished, filthy men could not endure the wound with infection, and Union prosecutors initially accused the rebel surgeons of deliberatively spreading poison. Joseph Jones, in his review of the causes of mortality at the prison, exonerated the rebel surgeons and argued that the vaccine was likely inactive but not poisonous.[31]

So, did the war heighten awareness of smallpox, spread the disease to new locales, and generally engender an urgent belief in the need for government to enforce compulsory vaccination to stop this dread disease? Or did the many instances of vaccination gone bad during the war create a fear of vaccination and disdain for its protective properties? The war's influence was mixed; it both broadened knowledge of vaccination (while spreading the disease and creating a need for vaccination) and created widespread awareness of the procedure's potential for damage. Many Americans would have been immune by the end of the war, as a result either of mass vaccination campaigns or of exposure to the natural disease. But in the late 1860s, a new smallpox outbreak spread across the United States, beginning in California. The West Coast would have been the least affected by the civilian smallpox epidemics of the war, so it is perhaps no surprise that the flame first caught there. It traveled over the next few years to many cities, finding nonimmunes among recent immigrants and young children in the cities of the Midwest, the South, and the Northeast.[32]

When the new epidemic arrived in New York City, its fledgling board of health took immediate and thorough action. Physicians were recruited to go house to house immunizing the occupants, especially in the tenement areas. When they found a case of smallpox, the patient was taken to an isolation hospital and the domicile treated generously with disinfectants. The visiting physician promptly vaccinated all contacts. The city had established a general recommendation that infants be immunized at birth and again upon entering school, although as yet these recommendations had not been turned into laws (or, at least, laws that were enforced). Some three hundred thousand New Yorkers received the vaccine in 1871, and a board of health official believed that through the publicity for this campaign, "the public mind was in a great measure disabused of the opposition hitherto manifested towards this method of applying preventive medicine."[33]

Boards of health were less successful in selling vaccination in other cities. In Milwaukee, when smallpox erupted in 1868, the medical profession was divided on the propriety of the procedure. As historian Judith Walzer Leavitt has noted, this left the citizenry confused. One wrote to the local paper, "On the one hand we have been informed by certain physicians that vaccination not only does no good but is really injurious. On the other hand the Board of Health have advised *all* to get vaccinated. . . . By this disagreement of the doctors a great many people are undecided what to do."[34] This split in the medical community was not just between regulars and sectarians, as had been argued elsewhere by another historian; it included prominent orthodox physicians on both sides.[35] Resistance was particularly strong among Milwaukee's German community; this was perhaps the first instance of the strong immigrant influence on the spread of smallpox and the vaccination debate that became more prominent in the last three decades of the nineteenth century.

In the summary of the laws regarding vaccination that Elisha Harris presented to the American Public Health Association in 1875, he reiterated the importance of education and persuasion and urged public health officials to compel compliance only as a last resort. He also emphasized the importance of providing pure vaccine matter and the state's role in ensuring that vaccinating practitioners were trained in their jobs. This procedure, he believed, would prevent the "loathsome and dreaded contagion of small-pox." Harris did not refer directly to his Civil War experience in describing campaigns against smallpox, but his awareness of how such efforts could go wrong was evident in comments that recognized how singular a public health measure vaccination was. "The prevention of small-pox is secured only by a kind of interference with the individual which must first be justified," he wrote. The state had a duty to protect the safety and welfare of the people in general, but it must also see that "the laws are so applied as to do the individual no harm."[36]

While the legacy of the war in terms of vaccination was mixed, it did inspire one Boston physician to reform the vaccine supply in the United States. Confederate surgeons had attempted to obtain vaccine matter from cows or horses, but without clear success.[37] Henry Austin Martin, an 1845 graduate of Harvard Medical School and surgeon-in-chief to the second corps of the Army of the Potomac, was disgusted with the many problems caused by impure vaccine and poor technique during the war. He argued that the only way to guarantee a safe and consistent vaccine supply was to grow it on animals, producing the vaccine through animal inoculation in farm factories maintained for that purpose. By the 1890s the result of his efforts and those of like-minded physicians, often channeled through the bully pulpit of the American Medical Association, was that most vaccine used during the major smallpox outbreaks of that decade came from animal production. Vaccination still had its opponents, but at least there was no shortage of vaccine material when these epidemics once again created a major public health crisis in American cities.[38]

Conclusion

Events of the American Civil War energized and educated physicians and the ruling classes to improve the health of the American population through sanitary measures. Although the National Board of Health was itself short-lived, and the USSC vision of a nationalized public health effort failed with it, the public health movement did achieve a solid local footing in many communities in the decades following the war, especially in the more prosperous North. The military model of medical efficiency influenced the reorganization of the Marine Hospital Service in 1872, and that agency ultimately evolved into the U.S. Public Health Service, uniforms and all. Concepts of specific poisons and the disinfectant response laid the groundwork for the bacteriological revolution in public health which followed from the 1890s on. Public health continued to be contested terrain, as the troubled history of vaccination illustrates, but events of the war made sure that the health domain had a persistent place on the map of American political discourse.

During the 1870s and 1880s, the germ theory became more and more accepted as an organizing concept for disease causation. Robert Koch's methodology, including his invention of microbial growth media, supplied the means for demonstrating the connection between particular microorganisms and particular diseases, opening the door to public health measures specifically targeting each microbe and its transmission. The work of Koch and his colleagues finally meshed with that of John Snow and William Budd in England to illustrate that water was a common carrier of disease, especially intestinal diseases. Public health advocates in the late nineteenth century accordingly turned from fretting about impure air to focusing on securing a pure water supply through proper sewerage and water-purification systems. This reform in turn had a major impact on infant mortality and the public's health in general by decreasing exposure to the many intestinal organisms—salmonella, shigella, and cholera particularly—that plagued Americans.[39]

This chapter has highlighted the influence of the war on public health debates surrounding yellow fever, cholera, and smallpox, all diseases central to discussions of governmental responsibility to control disease and questions about the contagiousness of these dangerous entities. Wound infections were a fourth general category of infectious disease that received much attention during the war and sparked continuing debate in years to come. Chapter 11 explores the connection between surgery, disinfectants, and the germ theory and demonstrates that the war brought a new level of interest in these questions, preparing the way for the growth of surgery as a field in the late nineteenth century and the maturation of American hospitals.

CHAPTER ELEVEN

Medicine in Postwar America

How did the war influence medicine in postwar America? Chapter 10 considered how understandings of the causation and prevention of infectious disease directly affected public health. This chapter follows other threads of the narrative into the world of postwar medicine, exploring influence, change, and sometimes stasis. Historians of American history have long debated the effect of the war on the trajectory of such trends as industrialization and modernization. As historian Morton Keller has noted, "This was the ambiguous inheritance of the Civil War: aside from the end of slavery and the defeat of secession, no profound alteration of American society; yet for many of that generation, a sense of vast and sudden change." Keller concluded, "The contradiction inherent in this legacy set the tone of the nation's postwar public life."[1] We have seen that the war radically relocated the sickroom from home to hospital, exposed physicians to a wide range of medical practices and ideas, and brought women into medicine. At the same time, the war destroyed southern institutions, liberated slaves only to allow many to die of disease, and brought countless other alterations to American life. Through this chapter I hope to encourage other historians to approach medicine in the late-nineteenth-century United States with an eye to the war's influence even as they explore more familiar foreign sources of change in American medicine.

Coming Home

By fall 1865, most soldiers had been discharged, and the sick and wounded who could travel had gone home. They were back where they belonged, at home, where they picked up their "normal" lives. Women resumed their traditional roles as caregivers within families, especially when their men were persistently weak or ill. Certainly there was no longer the need for the war's huge hospitals, which the army rapidly dismantled. The boards taken from Chimborazo's pavilions no doubt assisted in the rebuilding of downtown Richmond. By the 1880s the acres of Satterlee Hospital in Philadelphia had become the neighborhood of Spruce Hill, with a small corner persisting as Clark Park. Hospital buildings in Charlottesville reverted to the university, and seminaries, female academies, warehouses, and other structures resumed their civilian roles. The war had created the hospital's purpose, and at war's end that purpose returned to the private household. There were "soldier's homes," to be sure, institutions designed to care for the disabled who had nowhere to go, and hospitals continued to serve the troops still in uniform, such as those sent to Texas in summer 1865. Freedmen's hospitals persisted as well and often incorporated black troops not yet well enough for release.

The return to normality could not be complete, of course. As David Hacker has recently calculated, "1 in 10 [Union] men of military age in 1860 died as a result of the war and 200,000 white women were widowed." Southern losses were proportionally greater, with 22.6 percent of southern-born white males aged 20–24 in 1860 losing their lives in the war.[2] Overall, among men of military age, 6 percent of northerners died in the war and 18 percent of southerners died. Many of the survivors came home maimed and unable to perform the work that had sustained them and their families before the war. Even before the war ended, large numbers of men were discharged home when it became clear that they were not likely to recover from their illnesses. These men had not been counted as official casualties of the war until Hacker's research used the 1870 census to estimate domestic postdischarge excess deaths that can be attributed to the war's health effects. Those who lived made an impact at home that could be almost as severe as the loss of loved ones who died. Some men came home deeply disturbed by the horrors they had seen and never recovered the mental stability needed for daily life. Jeffrey McClurkin's study of postwar Lynchburg, Virginia, has revealed, for example, how many Confederate veterans spent time in the local insane asylum after the war. As is probably true after every war, reentry into everyday life was not easy, even for those with intact bodies.[3]

Soldiers return home not just as survivors but as vectors of disease. Those who had contracted tuberculosis or typhoid brought those infections to their families. Hugh Donnelly, a relative of mine, mentioned in the acknowledgments, came home from the war chronically ill and died in 1868. Two of his brother Peter's young children died in the following four years. Did their illnesses originate in their uncle's wartime exposure? Did Hugh Donnelly bring the typhoid bacterium home to Anoka, Minnesota? Or was it tuberculosis? It seems likely that a war that so effectively spread typhoid (the Union list included 27,056 cases, likely an undercounting) also created abundant new carriers who carried the disease back to their communities.[4] The official Union lists indicate that 5,286 men died of consumption, the contemporary label for pulmonary tuberculosis.[5] Many more would have been discharged home on disability when this diagnosis became apparent. Returning New Englanders brought vivax malaria back to the Connecticut River valley. And smallpox spread horrifically among African Americans and American Indians in the South and the West.

Thousands of men exited the war with body parts missing. One estimate puts the number of Union amputations at sixty thousand, with three-quarters of the patients surviving the surgery. Of the amputations performed on Union troops, more than twelve thousand involved surgery on the lower extremity, significantly impairing mobility. In 1862 the U.S. Congress granted war-related amputees fifty dollars for a prosthetic arm and seventy-five dollars for a leg.[6] Former Confederates were not eligible for federal money under this program, but southern states did what they could to help rebel veterans. In South Carolina, government records show that nearly 1,250 men applied to a program that paid for prosthetic legs.[7] In North Carolina the state laid out more than eighty-one thousand dollars from 1866 to 1871 to buy prostheses for veterans, and other southern states had similar programs or, in the case of Louisiana, a private charity that raised money for the cause.[8] Cultural historians have latched onto the disabled Civil War "body" as an analytic focus for understanding issues of nationalism, sacrifice, and commemoration of the great war. However construed, there can be no doubt that these disabled men were a persistent reminder of the war that had been and the enduring costs of the conflict.[9]

Innovation

The war brought few dramatic elements of progress to American medicine, at least as compared to the changes during twentieth-century wars. The yellow fever mosquito was not discovered, penicillin was not developed, and vaccines were not tested. Instead, the changes brought by the war were at the level of education and the development of ideas about what in modern parlance is called "best practices." First, there was the dissemination of knowledge to physicians who had, many of them, received the barest minimum of medical education. Second, military physicians received hands-on experience in surgery and anatomy such as they would likely never have achieved in their whole careers. Third, the idea that the best-educated physicians should work to educate their brethren and elicit research to generate new knowledge was boosted by the medical hierarchy of the war. Fourth, the war brought notions of standardization to the forefront, a new universalism emphasizing that there was a best way to do something—a particular surgery or the choice of a drug to treat a given illness. The war allowed less focus on the patient as a distinct individual and more emphasis on matching diagnosis with therapy. All of these effects prepared at least Union physicians to receive the marked changes that emerged from the microbiology, physiology, and pharmacology laboratories of Europe in the late nineteenth century.

Southern physicians' wartime experiences influenced them in many of the same ways, but the destruction of their homeland left them scrambling for survival after the war. The few antebellum centers of medical learning in the southern states—New Orleans, Richmond, Charleston, Nashville—slowly revived in the 1870s; medical journals resumed publication, and medical school classes reconvened. Because of the persistent poverty of the region, it had no medical institutions on a par with those in the North until the twentieth century. The southern population was plagued with diseases of both tropicality (malaria, yellow fever, hookworm) and poverty (pellagra, hookworm, typhoid). The South was slow to institute the sort of public health reforms that characterized northern cities; it remained a depressed, diseased region well into World War II. For the South it is difficult to argue that anything positive came of the war (at least from the perspective of white elites), with the exception of medical leaders, such as J. Lawrence Cabell and Joseph Jones, who drew on medical experiences during the war to build postwar careers in public medicine. By and large it was in the North that the war's impact on medical innovation was most obvious.

As for the Union, several specific advances can be appreciated. As noted in chapter 6, the exhibit at the Philadelphia Centennial Exhibition of 1876—meant to illustrate the most exciting medical progress—was a model hospital room, complete with illustrative manikins demonstrating splinting devices. The display also included photographs of amputations, a case of medical and surgical curiosities, trays of surgical and medical supplies, cases of artificial limbs, and even a medical mess room complete with silverware.[10]

The exhibition highlighted order, modern tools, and the power of surgery to alter medical outcomes.

In that vein, one more feature of that Philadelphia Centennial medical exhibit is noteworthy. Hanging on the wall was Thomas Eakins's painting *The Gross Clinic,* which he completed in 1875 to honor the great surgeon Samuel Gross and the importance of medicine in Philadelphia. The painting depicts an amphitheater operating-room scene, with the surgeon pausing in his work as students look on eagerly. Later, Eakins painted D. Hayes Agnew in a similar setting. Agnew was another famous Philadelphia surgeon, and this painting was commissioned as a gift to be presented on Agnew's retirement in 1889. The paintings have often been compared, as they demonstrate the change in surgical garb (from street clothes to surgical gowns) and the presence of a female nursing attendant in the later painting. A third painting of that era shows the first operation that was conducted under ether anesthesia. It was completed in 1894, fifty years after the landmark surgery. All three paintings were perceived as too gory for public display; indeed, *The Gross Clinic* was moved from the art-exhibit portion of the Philadelphia fair to the medical ward. But the three paintings do present the surgeons as powerful figures, the center of attention of rapt students, men of action and capability. Perhaps it is best to say that this is how postwar surgeons saw themselves, rather than reflecting a queasy public's perception of surgical power. Certainly Luke Fildes's *The Doctor* (1891), in which a thoughtful physician sits pensively at the bedside of a sick child, had more appeal to the Victorian public.[11] Would the many men who went under the surgeon's knife in the Civil War have praised his efforts in memory?

They might have been more grateful for improvements in the ambulance system. Historians have pointed to accomplishments in the efficient transport of battlefield casualties as a major outcome of the war.[12] Jonathan Letterman's ambulance system came to refer not just to the efficient movement of the wounded on litters or in wheeled vehicles from the battlefield to the first dressing station, but more broadly to the smooth transfer of the wounded all the way down the line, from battlefield to dressing station to field hospital to general hospital (if needed); the system was important in concept and often in execution. Army surgeons learned to apply the concept of triage, although that word did not come into common use in military medicine until the twentieth century. Triage sorted the wounded into three rough categories—those who were going to die, no matter what the surgeon did; those who might recover with rapid intervention; and those mildly injured who could wait for medical attention. Men with major abdominal or chest wounds were treated with opiates and made as comfortable as possible but otherwise abandoned as dying. Men with major crushing wounds to arms or legs were the first priority, as rapid amputation could save lives. Finally, men with flesh wounds, dehydration, sunstroke, or simple fractures were seen to last. Whether this strategy saved lives during the war is impossible to gauge. It is impressive that one Union surgeon reported in June 1865 that over the course of the previous six months, some 69,200 men had been returned to active service from the field and depot hospitals of southern Virginia during the spring campaigns. He is full of praise for the "efficiency of the medical department."[13]

The most obvious medical innovation during the war was not the creation of new knowledge but the diffusion of ideas and management techniques. George Fredrickson criticized the members of the U.S. Sanitary Commission for being elitist in their attitudes toward the ignorant general surgeon and overly controlling in the imposition of their ideas. Fredrickson wrote in the mid-1960s, when criticizing elites was the fashion.[14] I admit his charge but see the USSC elitism as a positive force for the improvement in medical practice and camp hygiene. An alternative point of view is to see the war as offering opportunities for widespread education of the physicians—most of them poorly trained—who practiced in antebellum America. Even in an era lacking the powerful drugs and techniques of twenty-first-century medicine, the spectrum in the quality of care offered by physicians was broad. The best American doctors had studied at the elite eastern schools, and many had supplemented their training abroad in the wards and dissecting rooms of hospitals in Edinburgh or Paris. They had learned anatomy and surgical techniques and knew how to administer the new anesthetics safely. They had read Florence Nightingale and other writers on the proper construction of hospitals and had absorbed the sanitary gospel that filth caused disease. And many had come to understand the lesson that less drugging was usually better than more, that relying on a course of therapy that strengthened the body's own "healing power of nature" afforded the best outcomes in medical care. It was these lessons that the best American physicians could and did spread to the medical profession during the war.

The USSC particularly broadcast the message of camp and hospital hygiene and promoted the role of nutrition in the care of the sick (while supplying the nutrients they recommended). If they sounded like a scolding nanny while doing it, so be it. Sometimes fussy women were just what the doctor ordered. Their advice about separating fecal output from oral input was wise. Reducing the bacterial load in the water consumed by troops could only be for the good, and that was the immediate effect of "policing" the camp to see that the men's feces were kept separate from their water sources. Teaching line officers and surgeons about the importance of "anti-scorbutics" and, further, supplying them with barrels of onions, pickles, potatoes, and lemons prevented scurvy and assisted healing in those who were sick or wounded. No surgeon who served in the Civil War could have left it unfamiliar with the importance of fresh fruits and vegetables in preserving health, even if he might persist in worrying that too much fresh fruit could cause illness.[15]

But the USSC was on the periphery of the army medical system, and in the South similar state organizations were even more peripheral; the army's medical hierarchy was the overwhelmingly dominant presence in the lives of Civil War surgeons, in both the North and the South. Most practitioners had been in independent settings before the war. Other than their time of apprenticeship or a short stint as a house officer, they had been in solo practice, making decisions without supervision and without interference from other practitioners. Physicians who served as army surgeons suddenly had "superiors" who told them what drugs were available for order and use, commanded them to supply written reports, and brought them up on charges if their behavior was found wanting. Although at the beginning of the war, the medical leadership in the North was drawn from the seniority system of the army and was not in fact particularly elite or well trained, this quickly changed. The new leadership, on both sides, saw their army surgeons as badly in need of "elevation" and education and set about providing it with books and pamphlets. They put hierarchical systems in place, so that only the best surgeons were the "cutters" after major battles, while lesser men were assigned as wound dressers, first responders, or overseers of hospitalized medical patients.

The distributed literature carried several messages about the choice of drugs. Hammond caused a major stir when he withdrew calomel and tartar emetic from the supply table, and even if the surgeons were angry, they were all exposed simultaneously to the idea that an elite regular physician, not just sectarian quacks, was pushing for the diminution in use of these drugs.[16] Likewise, Hammond's protégé Joseph Woodward's book on camp diseases, which the Union Army distributed to its surgeons, acknowledged that "mercurials [had] been very generally employed in the treatment of enteritis [diarrheal illness] in civil life" but stated, "The remedy is certainly objectionable in the treatment of this disease in the army."[17] With such language Woodward avoided criticizing a surgeon's prior practice while urging him to change it in the military setting. Woodward generally avoided mercury, recommending instead opiates, rest, appropriate food, and quinine for the malarial fevers. Southerners received a similar message, albeit with the added problem that many standard remedies were less readily available owing to drug shortages and the blockade. Necessity pushed them to rely on botanical remedies, a circumstance that confused the distinction between allopath and botanical practitioner. The southern leadership urged practitioners to try botanical products and report back, or it carried out research in hospitals on native plant remedies. All together, both sides attempted to regulate prescribing habits and diffuse a concept of best practices that diminished the use of heroic remedies.

Such supervision and education continued during the war in the general hospitals of both sides. Specialists from local medical schools walked the wards, advising on cases and teaching the surgeons much as specialists do today, making rounds with interns, residents, and fellows. If a surgeon proposed surgery for a patient, a committee decided whether it was appropriate, what the best approach was, and who should perform the operation. Surgeons in charge of hospitals looked over the shoulders of the doctor in charge of each ward, and if the care was found wanting, the man's work was examined. He might be offered further education, be discharged for incompetence, or even be court-martialed. In St. Louis Ira Russell noticed that black patients in some wards did better than others, and he talked to their physicians. When one surgeon claimed that black men were just too difficult to treat and that his patients were sicker than the others, Russell switched the assignments and demonstrated that the better doctor's techniques worked no matter which ward he served in.[18] The war did not necessarily make all of its surgeons into great doctors, but it offered opportunities for learning that were unprecedented in American medicine.

No major new surgical techniques developed during the war, but the spread of surgical knowledge was phenomenal. Both sides distributed books with illustrations describing amputation methods and the control of bleeding and infection. The hierarchy of surgeons created by both the Union and the Confederacy put the better-trained surgeon in a teaching role over the tyro, to the benefit of both operator and patient. In the general hospitals, surgeons conferred over the wisdom of further surgical procedures, exposing some physicians for the first time to the experience of "teaching rounds" and modeling the process of dispute, discussion, and decision. Surgeons were encouraged to report interesting cases, whether to local medical "societies" that formed in the vicinity of troops in winter quarters, or to more formal medical groups such as the one Moore organized in Richmond, or to the Army Medical Museum in Washington, DC. Although there were local and national medical societies before the war, such encounters reinforced the value of hierarchical learning and group interchange in medical progress. If there were best practices to be shared and learned, such structures allowed for the ready transmission of knowledge.

The Surgical Profession

Gauging the influence of Civil War surgical experiences on the direction of surgery as a profession later in the century is challenging. The Civil War operation was largely determined by trauma and the demand for immediate intervention. Such needs persisted after the war but with much less frequency. One history of surgery commented that during the 1870s major hospitals in the Northeast might do one operation a week.[19] Still, surgeons began to push the boundaries of the possible and to explore new surgical procedures. In 1880 Samuel Gross founded the American Surgical Association (ASA), whose list of initial members contains many familiar names from the leadership rosters of the Confederate and Union armies. The American Medical Association was three decades old by this time, but the founding of the ASA may point to one impact of the war on the subsequent practice decisions of these men. Did they now consider themselves to be more surgeons than physicians, along the classic distinctions of an earlier era in Europe, where practitioners were one or the other? Not, at least, at first. Gross argued that the ASA was not a challenge to the AMA, and there was much overlap in their memberships. Gross had been president of the AMA, and some of the early members of the ASA, such as James L. Cabell and Stephen Smith, had major careers in public health work. Gross saw the work of the ASA as "the cultivation and improvement of the art and sciences of surgery, and the promotion of the interests not only of its Fellows, but of the medical profession at large."[20] All physicians assigned to either army during the war had been called surgeons; being a surgeon became an integral part of what it meant to be a doctor in America. Gross was not creating a separate specialty with his new organization but rather continuing the process of educating general physicians in a core skill of their trade.[21]

In a memoir about the founding of the association, published twenty-eight years later, James E. Mears recalled that Gross "designed it to be a school for mutual instruction and improvement, a court of supreme authority into which the great questions of Surgery should be brought for discussion and judgment, a gathering in social intercourse of the individual workers in surgical science." The ASA included men who subspecialized in various branches of surgery, including ophthalmology, gynecology, orthopedics, and genitourinary problems, but rather than seeking to separate surgeons from the rest of the profession, the ASA sought to continue the elite role of education and supervision that had emerged during the war. Both Samuel Gross and D. Hayes Agnew had served in the role of ward consultant in the Philadelphia general hospitals during the war. Hunter Holmes McGuire, John Brinton, Robert Kinloch, John Packard, and others had held major positions in the medical hierarchy of the two armies. These men wanted to see a continuation of what they regarded as the progress gained during the war in advancing American surgery through education, research, and dissemination of knowledge.[22]

Mears's reference to the "court of supreme authority" echoes the concept promoted during the war that there was a best way to treat a given disease, to perform a particular operation, to design a hospital, or to maintain a hygienic camp. John Harley Warner has emphasized a transformation in American medicine over the nineteenth century from an emphasis on regional variation in practice to a universalist approach to patients and their problems. He draws particularly on the influence of the study of physiology in bringing about this shift, since if studying the lungs, heart, or kidney in a dog reveals valuable information about humans, the differences among humans fade in comparative significance.[23] Clearly the war also stimulated this transformation. Surgeons treating hundreds of patients did not have much time to tailor their therapies to individuals. Manuals and pamphlets delivering educational advice spoke in terms of universals. An illustrated surgical manual demonstrating a foot amputation did not comment on whether the foot was white, black, male, female, southern, or northern. The regimented pharmaceutical supply kits issued by the military imposed therapeutic standardization. The hierarchical structure of military medicine, especially in the large general hospitals, encouraged uniformity in approach and technique. Of course, variation in method still occurred, but the military experience accelerated the trend toward universality.

This transformation echoed broader transformations in the American polity. Morton Keller has described how the strong national government that emerged from the war led to federal interventions in a variety of areas—agriculture, railroads, banking—that increasingly drew on the interstate commerce clause to impose national standards.[24] In 1879 Congress created the National Board of Health (NBH) to fight the yellow fever epidemics ravaging the South and the Midwest. It was a board staffed by men who had all served in the war and shared a common point of view about the value of public hygiene and the power of disinfectants. They created a set of interstate standards for the transport of goods from diseased areas. Joseph Jones, an unreconstructed Confederate who now headed the Louisiana State Board of Health, may have railed against them and called forth the ghost of states' rights, but his neighbors welcomed the NBH's rules because they brought uniformity and reliable practices.[25] The various regional and national medical associations formed in the 1870s all had, at least in part, the intention of promoting their areas of interest, finding the best techniques, and pursuing new knowledge. The efforts of the USSC leadership after the war to bring the International Red Cross to the United States, complete with its universal regulations for the treatment of civilians and battlefield casualties, echoed this trend. If the Progressive Era was marked by a reliance on experts, in medicine the concept was well begun during the Civil War and promoted by the various national organizations that followed in the 1870s and 1880s.

Certainly the American Surgical Association embodied the elitist ethos surrounding the leadership of American medicine. Membership by election was initially limited to one hundred. Many of these physicians held teaching positions in medical schools, and the majority resided in the northeastern United States. One historian has noted that this exclusivity made it difficult for promising young surgeons to gain entry and access to the very wisdom that the association claimed to be dispensing.[26] The founding of the ASA occurred amid one of the most heated debates in American medical culture: did sectarian practitioners have the right to both practice legally and to consult with doctors who considered themselves "regulars"? The medical hierarchy that controlled the commissioning of surgeons during the war for the most part demanded that candidates prove their allopathic credentials. While sectarian opponents characterized such men as "gouty old Allopathic bigots," and some sectarians slipped through by hiding their proclivities from examiners, in general the war did put the stamp of government legitimacy on the allopathic profession.[27] In the 1870s a new line of argument was promoted by physicians who claimed that knowledge of science, particularly physiology—and not necessarily choices in therapeutics—should determine medical legitimacy. Coming in the wake of war, in which regular physicians had been urged to decrease the use of heroic remedies and, especially in the South, increase the use of botanical ones, this argument had more power. So often in the war a practitioner's skill was judged by his handling of surgical tools, not the medicines he dosed. Michael Flannery has argued that the sectarian physicians had similar formal education in surgery and the same skills as the allopaths.[28] The time was ripe for a reconsideration of sectarian boundaries and medical identity.

Much happened to change medical practice and professional identity in the four decades following the Civil War. John Harley Warner has emphasized the promise of experimental physiology, pioneered in Germany and France, for making American medicine scientific, in rhetoric if not in reality. Physicians who took up bench research as the beacon for therapeutic progress tended to dismiss concerns about sectarian challenges; instead, they insisted that all claims would be tested by science and that there was no need for legal prohibitions against homeopaths, for example, since all would be equal before the bar of science. Warner and other social historians have focused mostly on medical therapeutics in this period. Accompanying this discussion has been similar attention to the maturation of the American public health movement in the late nineteenth century and the incorporation of the new science of microbiology into its practices. In addition, historians have rightly recognized the growth of specialism, the expansion of hospitals and medical schools, and new models of medical training exemplified by the founding of Johns Hopkins Hospital and Medical School.

The history of surgical practice and surgical specialization in this period has received much less attention (although George Rosen wrote about ophthalmology as long ago as 1944).[29] Yet the most memorable experience of the more than twelve thousand doctors who served in the American Civil War was surgery, often performed under the most primitive of circumstances and in mind-numbing volume. Certainly they learned their craft in the process, even if it was "see one, do one, teach one" taken to the extreme.[30] In the process they learned human anatomy to a depth unknown to earlier American physicians, as they saw every part of the body flayed open by bullets. As Michael Sappol has emphasized, knowledge of anatomy was the hallmark of the well-trained physician in 1860, and these men had ample new exposures to build on during the war.[31] Other laboratory sciences—physiology, pharmacology, microbiology, genetics—were not yet essential for the elite physician, but a solid grounding in anatomy distinguished the best men. Added to that was the experience of preserving specimens for the new Army Medical Museum. This was a war that glorified the surgeon and surgical anatomy and trained a legion of physicians in both the skills of their craft and the knowledge of anesthetic application. Such skills in manipulating the body had little to do with previous faith in mercurials, botanicals, or homeopathic remedies; manual skill and anatomical knowledge were the key factors in success. So how did this experience influence the postwar debates about professional identity?

One controversy turned on something dubbed the "medical ethics clause" of the American Medical Association, which proclaimed that it was unethical for a member to consult with a sectarian practitioner, such as a homeopath or an eclectic. Such consultation was seen as an affirmation of the sectarian's status as a legitimate doctor. Reformers argued that instead of fretting about such distinctions, the profession should turn its attention to physiology and pharmacology, to the emerging bench sciences that would explain the action of medications and offer new approaches to therapeutics. Opponents noted that while such ambitions were laudable, few drugs had yet appeared to justify such optimism.[32] The American Surgical Association adopted the ethics clause as part of its constitution; as a result, at the fourth meeting of the ASA, in 1884, three fellows were asked to resign for violating the consultation clause. James Mears, a witness to these events, noted later, "Since then this clause of the Constitution has been, very wisely, expunged," although he did not give his reasons for the decision.[33] One might well imagine that surgeons seeking patients were more interested in the case (and its potential fees) than what approach the referring clinician took.

The elite of American medicine saw the 1880s as an era of striking novelties in medicine, and their focus was on disseminating that knowledge. They sought to put new ideas to the test in the various "courts of supreme authority" made up by professional associations, scientific medical schools, and an improving cadre of medical journals that brought the best of German, French, and British medical science to American practitioners. These were the major innovations of the day, and the consultation clause and other concerns over sectarians grew less and less important. Surgeons were forging ahead, perhaps overly aggressively, with abdominal and pelvic surgical procedures that their teachers would not have dared to attempt. While American medical students still studied abroad, especially in Germany, the great medical centers in New York, Boston, Philadelphia, and Baltimore increasingly promoted concepts of excellence, research, and the best of medical education. This is a story about the growing acceptance of elite leadership, of institutions that modeled, created, and disseminated new ideas in medicine. The war nurtured this transformation from the Jacksonian democratic medical rhetoric of the 1830s to a medical system that could foster and respond to the Flexner Report (1910), which set standards for medical schools that were accepted across the country. The Flexner Report and the similar efforts that preceded it represented a reform effort that assumed there was a best way and that elite institutions should serve as beacons for those in woeful need of guidance.[34]

Many factors brought about and shaped changes in American medicine in the second half of the nineteenth century, and of course not all of them were tied to the war. Eight eye surgeons met in New York City in 1864 to found the American Ophthalmological Society; nowhere in the initial discussion of the rationale or goals of the organization was the ongoing war mentioned. Perhaps this is because, of the eight founding members, only one served in the army. All of the doctors who met were from New York and Boston, a distribution no doubt determined partially by the association of specialization with urban areas but likely also due to the limitations of wartime travel.[35] Still, it seems likely that this organization occurred in spite of the war, not because of it. It is certainly reasonable to ask in what ways the war limited progress in medical education and organization. Would the new organizations that followed in the 1870s—the American Public Health Association, the American Neurological Association, the American Gynecological Society, the American Dermatological Association, and the American Laryngological Association—have formed sooner? Or later?[36] On the one hand, the process of specialization was well under way in Europe, and the best-educated American surgeons did not need the war to be pushed in that direction. On the other hand, the war had demonstrated that physicians working in groups could stimulate the dissemination and testing of new ideas.

The large numbers of wounded and ill promoted the concept of specialty care for separate diseases in some instances. Surgeon General Hammond, who became a specialist in neurology after the war, allowed S. Weir Mitchell to create a separate institution for the treatment of nervous diseases in Philadelphia. Mitchell culled men with nerve injuries from other military institutions and summarized his findings first in an 1864 publication and later in multiple articles and a landmark work on nerve injury.[37] Union physicians likewise transferred cardiac patients to a separate ward of the hospital, where

Jacob da Costa studied the entity he dubbed "the irritable heart of soldiers," a war-related cardiac disorder characterized by palpitations and weakness.[38] Other specialty hospitals focused on orthopedic injuries, or they isolated infectious patients, especially those with smallpox. The specialty hospitals no doubt encouraged specialization, as they gave practitioners intensive experience with distinct categories of cases.

The Germ Theory: Transformation and Dissemination

Nothing in medicine altered more dramatically in the four decades following the Civil War than the understanding of infectious diseases. Rather than medical physiology, as reformers might claim, the greatest changes came in the introduction and acceptance of the germ theory of disease. Throughout my career I have been fascinated by this transition, how physicians came to believe that microscopic creatures could do anyone any harm, especially when it seemed so obvious to them that filth and its bad odors caused disease. Many historians have approached this question, of course, and none so thoughtfully as Charles Rosenberg. He identified many of the pertinent factors. The model of contagion already existed, in the peculiar case of smallpox. Whatever caused this disease clearly passed from person to person and could be transmitted by a small bit of pus taken from a sick person's arm. The idea of specific cause and specific disease needed time to evolve; as long as physicians believed that there was really just one fever with local modification by climate, gender, time of year, diet, constitution, and so on, then each case of disease was obviously caused by many factors.

But certain diseases were "great teachers," as historian C. E. A. Winslow noted of bubonic plague in his chapter on the history of infectious diseases published in 1943.[39] This epidemic was unfamiliar, came in great waves, and moved perceptibly across the landscape. In the nineteenth century, cholera and yellow fever were instructive, prompting great debates over whether they were caused by poisons that were somehow transported or arose spontaneously under appropriate conditions of filth, temperature, and moisture. By midcentury many physicians had come to believe that these were specific diseases, with specific causes, and that those causes were poisons or even ferments of some sort. As discussed in earlier chapters, a simultaneous development was the awareness that various disinfectants seemed able to counter certain infectious diseases. When physicians believed bad smells from rotting animals or festering feces caused disease, they observed that certain disinfectants poured on the offensive substance took the smell, and hence the danger, away.

During the Civil War, concepts of disinfection crossed with ideas about putrefaction, as discussed in chapter 3. Men were reminded of the contagiousness of smallpox, and men like Ira Russell in St Louis, witnessing outbreaks of meningitis and pneumonia, began to at least question whether other diseases moved the same way. Confederate surgeon Joseph Jones pondered the absence of typhus in the hellholes of Civil War prison camps (where surely there was ample filth and susceptible starving bodies) and concluded that some sort of specific poison must be missing from the mix. For the connections to finally be made, it remained only for the concept of putrefaction, as caused by a chemical process, to be transformed by substituting "infusoria" for "chemical process" or poison. Before the war the two diseases at the center of the controversies over contagionism versus anticontagionism were yellow fever and cholera, because these two diseases seemed to travel (and hence raise questions of quarantine) and repeatedly threatened European and American cities. Plague might have been the great teacher, but it was not a major western problem in the nineteenth century. While the war brought outbreaks of measles, smallpox, pneumonia, and meningitis to the metaphorical table, the most important new topic for conversation was the cause and treatment of wound infections. Did some cause, some special poison, travel from person to person in the hospital, as the patients lay recovering from wounds? How did this question link up with others about cholera and yellow fever, and how were all three tied to the use of disinfectants and the development of what in shorthand can be called the germ theory of disease?

In an otherwise excellent analysis of American attitudes toward the germ theory in the 1870s, Nancy Tomes neglects to consider the impact of Civil War experiences on the American medical profession.[40] Prewar observations of cholera and yellow fever had convinced many that some sort of poison traveled in ships or from person to person spreading these infections. The war taught every surgeon who confronted it that hospital gangrene could spread rapidly through a ward unless it was contained by strict measures of cleanliness and disinfectants, or unless infected men were removed and isolated in tents that were open for ample ventilation. The power of disinfectants during the war was immediately translated afterward into a public health measure during the cholera epidemic of 1866. Disinfectants played an important role in controlling epidemics, and a new disinfectant suddenly came to center stage in the decade after war's end in a discussion of a new surgical procedure. In the debates over and reception of Joseph Lister's ideas, it is worth keeping in mind that the arena was the surgical theater; the problem was wound infection. Physicians in Britain, France, Germany, and America who had served in military conflicts were peculiarly prepared to understand the stakes regarding wound infection and peculiarly familiar with the many arguments over how to prevent and treat them. Cholera and yellow fever containment aside, this topic—wound care—was the central stage for discussions of the germ theory before Robert Koch and colleagues redirected the discourse to such diseases as tuberculosis, diphtheria, pneumonia, and cholera.[41]

It would be handy to be able to ask discharged army surgeons how they responded to the brief reports of Lister's first papers in late 1867 and to compare their answers to the responses of those who did not serve, but of course that is impossible. Nevertheless, the point should be made. For every surgeon who served in the war, there is a story, and any consideration of the physicians' careers and accomplishments after the war should ask about the influence of wartime experiences on their subsequent choices and ideas. For example, Walter Kempster served as a hospital steward during the war, with the Tenth New York Cavalry. It is not known whether he interrupted medical training to enter the army, but in 1864 he graduated with the MD degree from Long Island College Hospital. Shortly thereafter he was treating patients at the U.S. Army's Patterson Park General Hospital in Baltimore. There he and his surgical colleagues confronted an epidemic of hospital gangrene in fall 1864. Their treatment included moving the men to hospital tents located on the hospital grounds, to guard against the spread of the disease and ensure adequate ventilation. The wounds were cleansed and debrided with nitric acid, because Kempster had found that other agents, such as "turpentine, bromine, permanganate of potash, Labarraque's solution, sugar, &c." worked less well. Kempster noted that "all proper precautions were taken to guard against the spread of the disease, each patient being supplied with his own sponge, bowl, and towel; the bandages and other appliances were immediately burned after their removal." Kempster did not specify what agent was involved in the potential spread of the disease, although he did conclude that "a powerful blood poison" was at work.[42] There is nothing in Kempster's account to indicate that these ideas were particularly novel, just that his team was quite thorough in applying them.

It is likely that Kempster was familiar the various publications on wound infections that were distributed widely during the war (and discussed in chapter 3). It is evident that he, and probably many other war surgeons, were thus primed by their war experience to appreciate the publication of Joseph Lister that appeared in 1867 on the use carbolic acid–soaked pads in the treatment of compound fractures. Kempster read about Lister's work in October 1867 and published his research on its use in the following year. He did not see Lister's work as a marked departure; rather, he called the acid a "comparatively new antiseptic and disinfectant" whose "powers have doubtless been exaggerated, [but which] nevertheless . . . stands in advance of any other article of its class both for efficacy and variety of application." Kempster used it in a variety of ways in his current role as superintendent at the State Lunatic Asylum in Utica, New York. He applied it to infected skin wounds; in cases of scarlatina, he prescribed it as a gargle; for sinus infections he had patients inhale carbolic acid steam; and for gastrointestinal complaints brought on by "yeasty stomach," he dosed by the spoonful. He speaks of the power of carbolic acid in killing organisms such as flies, maggots, crickets, rodents, and "the lower forms of organic life [such as] microscopic infusoria and cheese mites."[43] In his writing it is fairly easy to see the concept of a disinfectant (carbolic acid or otherwise) moving from "neutralizing a poison" to "killing an organism." He was not quite there—Kempster did not yet clearly see that disinfectants healed wound infections because the infections were *caused* by microscopic infusoria, but the process of this cognitive transition had clearly begun.[44]

George Derby did take the next step. In an article that historian Thomas Gariepy has identified as the earliest report on carbolic acid use by an American physician in response to Lister, Derby described using the substance in the treatment of a compound fracture of the femur in a 9-year-old boy who had fallen from a tree. After probing the wound with his finger, Derby swabbed it inside and out with carbolic acid and dressed the wound with a carbolic acid–soaked pad, which remained in place, with fresh applications of the solution, for the next four weeks. The boy's bone was realigned with traction, and he made a complete recovery.[45] The standard of care in the case should have been to amputate the leg, which, depending on the location of the fracture, might have put the stump dangerously close to the hip socket. Derby was taking a chance, and if his choice had gone the wrong way, the boy might well have died.

Derby was a Harvard-trained physician who had served in the war as surgeon to the Twenty-Third Massachusetts Regiment, a regiment that saw action in North Carolina and Virginia, including at Cold Harbor and the siege of Petersburg. He would have been exposed to all of the same literature as Kempster and would have had experience with far worse injuries than that suffered by this child. And he would have seen the power of disinfectants to prevent and heal infected wounds. It is highly likely that this background readied him to appreciate the ideas of Pasteur and Lister. "The use of carbolic acid in surgery is based directly upon the investigations of M. Pasteur on putrefaction, in which he shows the relation of infusoria to the process," Derby proclaimed. These infusoria "are everywhere part of the atmosphere," and the process of preventing wound infection was the process of keeping these "vibrios and bacteria" out of the wound and, by initial cleansing, to kill those that had already entered. "The application of these ideas to practical surgery was first made by Mr. Lyster [sic], a surgeon of Glasgow, and an account of his experiments has been recently published in the London *Lancet.*"[46]

Derby's decision not to amputate the young boy's leg is particularly telling because he had learned during the war that early amputation of complicated fractures was essential to saving lives. He had been in the thick of things as a surgeon in North Carolina, where his unit had seen action, as reported in a published letter of 1862. "The sum of my day's work was, in capital operations," he reported, "amputation of the thigh, 3; of the leg, 1; of the arm, 1; at shoulder joint, 1. Next day, another arm."[47] When Derby summarized the lessons of the war for the medical profession, he listed the value of early amputation to save lives, the "very near perfection" the ambulance system had acquired in the last two years of the war, and the fact that disease was the "direct and logical consequence of the rules of hygienic science as applied to war." Where officers and medical men enforced camp cleanliness and well-ventilated, hygienic hospitals, men suffered much less from sickness. Derby carried these lessons into his work as secretary of the Massachusetts State Board of Health, an appointment he held from 1868 until his death in 1874. Derby also lectured on hygiene and sanitary science at the Harvard Medical School. General cleanliness and the power of disinfectants to limit epidemics like cholera were powerful lessons that he taught throughout the rest of his career.[48] Like James Lawrence Cabell, he combined interest in surgery with fervor for public health reform, a duality of passions that might seem unlikely in later years but was a natural outcome for the physicians who served in the war.[49]

The influence of Civil War hospital practices was also felt in claims for American exceptionalism that entered the debate over Listerian methods. By the 1870s American surgeons could note with pride that mortality rates in

American hospitals were much lower than those of Europe and the British Isles, a result, they claimed, of the greater cleanliness and ventilation of the American institutions. Thus the Listerian process and dressing was unnecessary, since the air in the hospitals was purer and less prone to encourage wound infections. Surgeons had supervised the building of numerous hospitals during the war, had read Hammond and Nightingale on proper hospital design, and had put these ideas into practice upon returning to their home institutions.[50] The debate over Listerian surgical techniques lasted into the 1890s, when such methods became commonplace and gradually transformed from antiseptic practices to aseptic ones, in which the focus came to be on preventing the introduction of microorganisms instead of killing the ones already there. Thus resulted the shift from washing and dressing the surgical wound with antiseptics to the operating team dressing in special garb (gown, mask, gloves) and working in a space made as germ-free as possible before the patient was introduced into it.

This brief account can only suggest the role of Civil War influences and experiences on the transformation in surgical practices and the broader acceptance of the germ theory in the United States. At the same time the American Surgical Association was debating Listerism, the National Board of Health was disinfecting railroad cars thought to be carrying the yellow fever germ. The assault on the "special poison" of cholera with disinfectants in 1866 continued when cholera returned in 1873; and when Koch announced the discovery of the cholera vibrio in 1883, physicians could redefine the poison as a microbe without altering their disinfectant approach. The Civil War very much helped prepare physicians for this transformation, and the scholar of American medicine would do well to ask of the late-nineteenth-century actors under consideration: "What did he (or she) do in the war, and what influence did those experiences have on the postwar development of American medicine?"

Afterword

Imagine that on July 21, 1861, there you stand, an observer near Manassas Junction, Virginia. Out of the smoke and noise staggers a man, clutching his arm, calling out in pain as blood drips onto his boots. He falls on the bare ground at your feet. What can you do for him? Tear up your petticoat to make a tourniquet to stop the bleeding? Cup some water in your hands from the nearby stream, for his dry lips? Search in vain for a doctor, for an ambulance, for any sort of official medical person? This is how it began. The calls for help were ultimately met by men and women on both sides who struggled to care for their soldiers, while at the same time taking advantage of the war's exigencies to promote agendas of reform, research, and progress.

Themes of gender have run throughout this narrative. Health care has always been a predominantly female activity, even if men held the higher professional positions that brought prestige (at least until recently). Women knew how to *care* for the sick, and when war put men in hospitals instead of home sickrooms, those women pushed to assume their competent role at the bedside, either directly or through surrogates. The U.S. Sanitary Commission both "channeled" this feminine force by organizing the conveyance of needed goods to the men and acted as a forceful nanny in teaching the men how to be clean and preserve their health. The best of Civil War hospitals, drawing on the work of Florence Nightingale, were healing environments precisely to the extent that the home sickroom could be recreated. Nutritious food and cheerful nursing were at least as important as the actions of physicians, and these essentials differentiate northern from southern success in healing the men.

Historians of medicine have long been aware of the antebellum debates about reform within American medicine. Physicians pushed for the adequate collection of vital statistics so that the health of the people could be ascertained; sanitary conventions argued over proper measures to prevent disease; professional associations sought to elevate the profession through improved education and research; medical students went abroad to have fulfilling learning experiences in foreign hospitals; women sought equal access to medical schools. What is less well known is how the war transformed these various reform impulses, much less how those changes played out in the decades following war's end. Medical historians need to "put the war back in" to the history of nineteenth-century medicine in the United States. One way to approach this agenda would be through group biography. Of the 2,432 women who were listed as physicians in the 1880 U.S. census (see chapter 2), how many had served in some capacity during the war? The same question could be asked of the (male) leadership of medical societies and schools and boards of public health. American medicine changed rapidly in the years after the war, and it is highly likely that the leaders of that transformation had themselves been transformed by their wartime experiences.

Germs love war, and this conflict was a field day for the parasites that prey upon humans. It is clear that certain microorganisms took advantage of all those young men gathered in filthy camps, to spread abundantly. Their story has yet to be told in all its glory. The tools of the "spatial humanities" would be helpful here, using geographic information systems to show how, for example, smallpox traveled over space and time. Like many other topics that I originally intended to cover in this volume, that one deserves, instead, a book-length treatment of its own. Elizabeth Fenn showed not only that smallpox was a major challenge for George Washington during the Revolutionary War, but that the war's effects so amplified the epidemic that it reached the western Plains Indians and Mexico City. In a similar way, smallpox broke out among freed people in 1865 North Carolina and reached the Cherokee by 1868. Other diseases that need mapping are malaria, typhoid, and tuberculosis. I suspect that each of these diseases not only took their toll among soldiers but carried over into civilian epidemics.[1]

In this time of sesquicentennial "celebrations," we would do well to remember this: The Civil War was a disaster that killed more than a million Americans. That some responded so grandly to the humanitarian challenge is small comfort amid such carnage. We live, still, with the war's legacy.

Abbreviations

CSMSJ	*Confederate States Medical and Surgical Journal,* 1864–65, reprinted with an introduction by William D. Sharpe (Metuchen, NJ: Scarecrow Press, 1976).
Duke SC	David M. Rubenstein Rare Book & Manuscript Library, Duke University, Durham, NC.
MOC	Eleanor S. Brockenbrough Archives, Museum of the Confederacy, Richmond, VA.
MSHW	*Medical and Surgical History of the War of the Rebellion (1861–1865),* 12 vols. and 3 index vols. (Washington, DC: Government Printing Office, 1870–1883; facsimile repr., Wilmington, NC: Broadfoot, 1990). Volumes 1–6 of this reprint reproduce volume 1 (medical) in the original; volumes 7–12 reproduce volume 2 (surgical). As the reprint is a facsimile, it reproduces the original page numbers exactly. Hence the original volume 1, which spanned pages 1–966, is spread out over the first six volumes of the reproduction, with the same page numbers. The exact translation for citations of this source can be found in the reproduction index, 1:xv. I have used the reproduction volume numbers throughout.
NARA	National Archives and Records Administration. All NARA documents reviewed for this book are located at the National Archives building on the Mall in Washington, DC.
NYPL	Manuscripts and Archives Division, Astor, Lenox, and Tilden Foundations, New York Public Library, New York.
OR	*The War of the Rebellion: A Compilation of the Official Records of the Union and Confederate Armies* (Washington, DC: Government Printing Office, 1880–1901), series I, 1–53; series II, 1–8; series III, 1–5; series IV, 1–4. Now available and searchable online at Making of America, http://cdl.library.cornell.edu/moa/browse/monographs/waro.html. These volumes consist of the reproduction of various official documents saved from the war and reprinted together in the 1880s and 1890s.
SCSA	South Carolina State Archives, Columbia, SC.
UNC SHC	Southern Historical Collection, Wilson Library, University of North Carolina–Chapel Hill.
USSC Microfilm Papers	Microfilmed collection of the United States Sanitary Commission Records, series 1, Medical Committee Archives, 1861–1865 (Wilmington, DE: Scholarly Resources, 1998). The originals are held in the Manuscripts and Archives Division, Astor, Lenox, and Tilden Foundations, New York Public Library, New York. Cited by reel and frame numbers; the images on each reel are numbered sequentially.

Introduction · Call and Response

1. For the latest estimate of Civil War dead, see J. David Hacker, "A Census-Based Account of the Civil War Dead," *Civil War History* 57 (2011): 307–48. Hacker puts the military death count at over seven hundred thousand; here I've added estimates of the deaths of freed people and slaves from war-related violence and disease. See Jim Downs, *Sick from Freedom: African-American Illness and Suffering during the Civil War and Reconstruction* (New York: Oxford University Press, 2012).

2. Nina Silber, *Gender and the Sectional Conflict: The Steven Brose Lectures in the Civil War Era* (Chapel Hill: University of North Carolina Press, 2008), xii. Silber in turn refers to LeeAnn Whites, "The Civil War as a Crisis in Gender," in *Divided Houses: Gender and the Civil War*, ed. Catherine Clinton and Nina Silber (New York: Oxford University Press, 1992), 3–21. Silber's 2008 volume, especially in the notes to the preface, is a good source for works on gender and the Civil War. See also Lyde Cullen Sizer's review article "Mapping the Spaces of Women's Civil War History," *Journal of the Civil War Era* 1 (2011): 536–48.

3. Walt Whitman, "The Real War Will Never Get in the Books," *Specimen Days*, in his *Prose Works* (Philadelphia: David McKay, 1892), facsimile reproduction at Bartleby.com, www.bartleby.com/229/1101.html (accessed July 11, 2012).

4. Margaret Mitchell, *Gone with the Wind* (New York: Scribner, 1936), 109.

5. Steven Pinker, *The Better Angels of Our Nature: Why Violence Has Declined* (New York: Viking, 2011), 183.

6. Lorien Foote, *The Gentlemen and the Roughs: Manhood, Honor, and Violence in the Union Army* (New York: New York University Press, 2010); for the southern perspective, see Stephen W. Berry II, *All That Makes a Man: Love and Ambition in the Civil War South* (New York: Oxford University Press, 2003).

7. Lynn Hunt describes the history of the ideal of human rights in her *Inventing Human Rights: A History* (New York: Norton, 2007), although she does not consider the American Civil War in her discussion.

8. David W. Blight, *Race and Reunion: The Civil War in American Memory* (Cambridge, MA: Harvard University Press, 2001).

9. William Faulkner, *Intruder in the Dust* (New York: Random House, 1948), 194–95.

10. On the monument that makes this claim (speaking particularly of the point furthest north achieved by Confederate troops at Gettysburg), see Gettysburg, National Park Service, www.nps.gov/archive/gett/getttour/tstops/tstd3-20.htm (accessed Sept. 13, 2010). The battle in general can be referred to in these terms, as the peak of Confederate efforts. In an imagined alternate universe, General Robert E. Lee might have turned and headed south to capture Washington, DC, or marched northeast to Philadelphia. Instead his troops retreated quickly to Virginia.

11. Faulkner, *Intruder in the Dust*, 194–95.

12. Pat Conroy, *My Reading Life* (New York: Doubleday, 2010), 18.

13. See chapter 2 for references on women and the war. For sources on African Americans and the war, see Thavolia Glymph, *Out of the House of Bondage: The Transformation of the Plantation Household* (Cambridge: Cambridge University Press, 2008); Margaret

Humphreys, *Intensely Human: The Health of the Black Soldier in the American Civil War* (Baltimore: Johns Hopkins University Press, 2008). For one example of "home front" analysis, see Matthew Gallman, *Mastering Wartime: A Social History of Philadelphia during the Civil War* (Cambridge, Cambridge University Press, 1990).

14. Drew Gilpin Faust, *This Republic of Suffering: Death and the American Civil War* (New York: Knopf, 2008); George C. Rable, *God's Almost Chosen Peoples: A Religious History of the American Civil War* (Chapel Hill: University of North Carolina Press, 2010).

15. Maris A. Vinovskis, "Have Social Historians Lost the Civil War? Some Preliminary Demographic Speculations," in *Toward a Social History of the American Civil War: Exploratory Essays*, ed. Maris Vinovskis (Cambridge: Cambridge University Press, 1990), 1–29.

16. George Worthington Adams, *Doctors in Blue: The Medical History of the Union Army in the Civil War* (Baton Rouge: Louisiana State University Press, 1952), 228, mentions the end of the medical middle ages; he may have been quoting William Hammond, but the source is not documented. On the Haitian situation, see Deborah Sontag, "Doctors Haunted by Haitians They Couldn't Help," *New York Times*, Feb. 12, 2010, www.nytimes .com/2010/02/13/world/americas/13doctors.html?_r=1 (accessed Sept. 13, 2010). George Wunderlich spoke against the label "Civil War medicine" as applied to Haiti in "Learning from History," Feb. 1, 2010, http://gwunderlich.wordpress.com/tag/george-wunderlich/ (accessed Sept. 13, 2010). Wunderlich rightly emphasized that the triage and transport system that was devised in the Civil War was much more efficient than the chaos of Haiti.

17. Karen Buhler-Wilkinson, *No Place Like Home: A History of Nursing and Home Care in the United States* (Baltimore: Johns Hopkins University Press, 2001), 1–13.

18. Lasalle Corbell Pickett, "My Soldier," *McClure's Magazine*, March 30, 1908, 563–71, 569.

19. As, for example, in James M. McPherson, *Drawn with the Sword: Reflections on the American Civil War* (New York: Oxford University Press, 1996), 115. In *The Mighty Scourge: Perspectives on the Civil War* (New York: Oxford University Press, 2007), 63, McPherson rendered it as "I always thought the Union army had something to do with it," without attribution. George C. Rable noted these various versions in *God's Almost Chosen Peoples*, 397.

20. Edward L. Ayers, Gary W. Gallagher, and John L. Nau III, "Fighting and Freedom: United States Military Forces and the Geography of Emancipation," paper presented at the Annual Meeting of the Society of Civil War Historians, Richmond, VA, June 17–19, 2010. At the date of this writing, Gallagher has not published this paper, although he does devote a chapter to emancipation in his book *The Union War* (Cambridge, MA: Harvard University Press, 2011), 75–118.

21. Military historians have been slow to recognize the importance of health and medicine in war, although the publication of John Keegan's work *The Face of Battle: A Study of Agincourt, Waterloo, and the Somme* (London: Jonathan Cape, 1976) pushed them to consider this factor. Keegan reiterates the point in his *The American Civil War: A Military History* (New York: Knopf, 2009), 77–78. Historians of other conflicts, especially historians at British institutions who study the twentieth century, have been quicker to take up this topic than American historians and have often approached it from the cultural-history perspective. See Roger Cooter, Mark Harrison, and Steve Sturdy, eds., *Medicine and Modern Warfare* (Amsterdam: Rodopi, 1999), esp. the introduction, 1–27; Roger Cooter, Mark

Harrison, and Steve Sturdy, eds., *War, Medicine, and Modernity* (Stroud, UK: Sutton, 1998); and Roger Cooter, "War and Modern Medicine," in *Companion Encyclopedia of the History of Medicine*, 2 vols., ed. W. F. Bynum and Roy Porter (London: Routledge, 1993), 2:1536–73.

22. Cooter, "War and Modern Medicine."

23. Victoria A. Harden, *Inventing the NIH: Federal Biomedical Research Policy, 1887–1937* (Baltimore: Johns Hopkins University Press, 1986). Even though the NIH was a civilian agency, its creation was influenced by the support for federal research that characterized World War II.

24. John Harley Warner, *The Therapeutic Perspective: Medical Practice, Knowledge, and Identity in America, 1820–1885* (Cambridge, MA: Harvard University Press, 1986); Norman Gevitz, *Other Healers: Unorthodox Medicine in America* (Baltimore: Johns Hopkins University Press, 1988).

25. Ross Thomson, "The Continuity of Innovation: The Civil War Experience," *Enterprise and Society* 11 (2010): 128–65.

26. Charles E. Rosenberg, *The Care of Strangers: The Rise of America's Hospital System* (New York: Basic Books, 1987), 18–22.

27. Perpenduum, http://perpenduum.com/2008/01/a-brief-history-of-the-drinking-straw-part-1-from-sumer-to-stone/ (accessed July 28, 2011).

28. Esther Hawk, *A Woman Doctor's Civil War: Esther Hill Hawk's Diary*, ed. Gerald Schwartz (Columbia: University of South Carolina Press, 1984).

29. Reid Mitchell, *The Vacant Chair: The Northern Soldier Leaves Home* (New York: Oxford University Press, 1993).

30. Florence Nightingale, *Notes on Nursing: What It Is, and What It Is Not* (London: Harrison, 1859), 1; Jane E. Schultz, *Women at the Front: Hospital Workers in Civil War America* (Chapel Hill: University of North Carolina Press, 2004).

31. Bobby A. Wintermute, *Public Health and the U.S. Military: A History of the Army Medical Department, 1818–1917* (New York: Routledge, 2011), chap. 1.

32. Humphreys, *Intensely Human.*

33. Margaret Humphreys, "A Stranger to Our Camps: Typhus in American History," *Bulletin of the History of Medicine* 80 (2006): 269–90.

34. Jeanie Attie, *Patriotic Toil: Northern Women and the American Civil War* (Ithaca, NY: Cornell University Press, 1998); Judith A. Giesberg, *Civil War Sisterhood: The U.S. Sanitary Commission and Women's Politics in Transition* (Boston: Northeastern University Press, 2000); Schultz, *Women at the Front;* Jane E. Schultz, ed., *This Birth Place of Souls: The Civil War Nursing Diary of Harriet Eaton* (London: Oxford University Press, 2011); Michael A. Flannery, *Civil War Pharmacy: A History of Drugs, Drug Supply, and Provision and Therapeutics for the Union and Confederacy* (New York: Haworth Press, 2004); Andrew McIlwaine Bell, *Mosquito Soldiers: Malaria, Yellow Fever, and the Course of the American Civil War* (Baton Rouge: Louisiana State University Press, 2010).

35. James Downs, "Diagnosing Reconstruction: Sickness, Dependency, and the Medical Division of the Freedmen's Bureau, 1861–1870" (PhD diss., Columbia University, 2005); and Downs, *Sick from Freedom;* Michael Flannery, *Well Satisfied with My Position: The Civil War Diary of Spencer Bonsall* (Carbondale: Southern Illinois University Press,

2007); Richard M. Reid, ed., *Practicing Medicine in a Black Regiment: The Civil War Diary of Burt Wilder, 55th Massachusetts* (Amherst: University of Massachusetts Press, 2010).

36. Richard H. Shryock, "A Medical Perspective on the Civil War," *American Quarterly* 14 (1962): 161–73; Adams, *Doctors in Blue;* Horace H. Cunningham, *Doctors in Gray: The Confederate Medical Service* (Baton Rouge: Louisiana State University Press, 1958).

37. Alfred Jay Bollet, *Civil War Medicine: Challenges and Triumphs* (Tucson: Galen Press, 2002); Frank Freemon, *Gangrene and Glory: Medical Care during the American Civil War* (Chicago: University of Illinois Press, 2001); Ira Rutkow, *Bleeding Blue and Gray: Civil War Surgery and the Evolution of American Medicine* (New York: Random House, 2005).

38. J. H. Salisbury, "Remarks on Fungi," *Boston Medical and Surgical Journal* 65 (1862): 509–15, 532–36.

39. Ira Russell to Medical Committee, July 15, 1865, reel 1, frame 969, USSC Microfilm Papers.

40. Reid, *Practicing Medicine.*

41. Shauna Devine's doctoral dissertation (University of Western Ontario, 2010, not available at the time of this writing) and forthcoming book on the Army Medical Museum will expand our understanding of this important aspect of Civil War medicine.

42. Wayne E. Lee, "Mind and Matter—Cultural Analysis in American Military History: A Look at the State of the Field," *Journal of American History* 93 (2007): 1116–42, quotations on 1117 and 1138.

43. David Goldfield, *Still Fighting the Civil War* (Baton Rouge: Louisiana State University Press, 2002).

44. Gallagher wrote: "Any historian who argues that the Confederate people demonstrated robust devotion to their slave-based republic, possessed feelings of national community, and sacrificed more than any other segment of white society in United States history runs the risk of being labeled a neo-Confederate. As a native of Los Angeles who grew up on a farm in southern Colorado, I can claim complete freedom from any pro-Confederate special pleading during my formative years." Gary W. Gallagher, *The Confederate War* (Cambridge, MA: Harvard University Press, 1997), 13.

Chapter 1 · Understanding Civil War Medicine

1. Reid Mitchell, *The Vacant Chair: The Northern Soldier Leaves Home* (New York: Oxford University Press, 1993).

2. Kathryn S. Meier, in her "'No Place for the Sick': Nature's War on Civil War Soldier Mental and Physical Health in the 1862 Peninsula and Shenandoah Valley Campaigns," *Journal of the Civil War Era* 1 (2011): 176–206, http://muse.jhu.edu/journals/journal_of _the_civil_war_era/summary/v001/1.2.meier.html, emphasizes the importance of the soldier learning "self care" and finding ways to counter the surrounding hostile environment.

3. Nancy Cott, *The Bonds of Womanhood: "Woman's Sphere" in New England, 1780–1835* (New Haven, CT: Yale University Press, 1977).

4. Jane E. Schultz, *Women at the Front: Hospital Workers in Civil War America* (Chapel Hill: University of North Carolina Press, 2004).

5. Harriet Eaton, *This Birth Place of Souls: The Civil War Nursing Diary of Harriet Eaton,* ed. Jane E. Schultz (New York: Oxford University Press, 2011), 75.

6. D. A. B[uie] to Kate McGeachy, Fort Caswell, NC, Dec. 31, 1861, Buie, Jane (McGeachy) Papers 1859–1861, Catherine McGeachy Buie Papers, 1819–1899, Duke SC.

7. Jonathan Letterman, *Medical Recollections of the Army of the Potomac* (New York: Appleton, 1866), 47.

8. Edward H. Dunster, "The Comparative Mortality of Armies from Wounds and Disease," in *Contributions Relating to the Causation and Prevention of Disease and to Camp Diseases; Together with a Report of the Diseases, etc., among the Prisoners at Andersonville, GA,* ed. Austin Flint (New York: Hurd and Houghton, 1867), 169–92, quotation on 182. Charles Smart, writing two decades later, likewise compared these statistics in *MSHW,* 5:1–17, although he cautioned that the environmental conditions of the various campaigns might also account for some of the disparities.

9. Roberts Bartholow, "The Various Influences Affecting the Physical Endurance, the Power of Resisting Disease, etc., of the Men Composing the Volunteer Armies of the United States," in Flint, *Contributions Relating to Disease,* 3–41, quotation on 3.

10. Samuel Cartwright, "Diseases and Peculiarities of the Negro Race," *DeBow's Review* 11 (1851), reproduced online at Africans in America, www.pbs.org/wgbh/aia/part4/4h3106t .html (accessed July 12, 2012).

11. For a fuller explication of nostalgia and its understanding in the French military context, see Lisa O'Sullivan, "The Time and Place of Nostalgia: Re-situating a French Disease," *Journal of the History of Medicine and Allied Sciences* 67 (2012): 626–49.

12. Bartholow, "The Various Influences," 21.

13. George Andrew to Elisha Harris, July 17, 1865, reel 1, frame 418, USSC Microfilm Papers.

14. Nostalgia among black troops puzzled Union physicians, because so many black soldiers had been slaves on plantations, and it was thought such a life was unlikely to generate longings. The physicians failed to realize that most black men lacked that common connection to loved ones, the written letter, since so few former slaves could read.

15. On the often inadequate health care of black troops, see Margaret Humphreys, *Intensely Human: The Health of the Black Soldier in the American Civil War* (Baltimore: Johns Hopkins University Press, 2008).

16. Martin S. Pernick, *A Calculus of Suffering: Pain, Professionalism, and Anesthesia in Nineteenth-Century America* (New York: Columbia University Press, 1985). While Pernick alludes to the Civil War usage of anesthesia, he does not analyze the impact on subsequent surgical practice of the widespread use of chloroform and ether in the war.

17. On the French influence on American medicine, see John Harley Warner, *Against the Spirit of System: The French Impulse in Nineteenth-Century American Medicine* (Princeton, NJ: Princeton University Press, 1998).

18. For the figures on alcohol prescription, see the list, in *MSHW,* 6:966, of drugs purchased or produced by the Union Army. On the use of alcohol as a therapeutic agent and the growth of its acceptance in medicine, see John Harley Warner, "'The Nature-Trusting Heresy': American Physicians and the Concept of the Healing Power of Nature in the 1850's and 1860's," *Perspectives in American History* 11 (1977–78): 291–324; John Harley

Warner, "Physiological Theory and Therapeutic Explanation in the 1860s: The British Debate on the Medical Use of Alcohol," *Bulletin of the History of Medicine* 54 (1980): 235–57.

19. John Harley Warner, *The Therapeutic Perspective: Medical Practice, Knowledge, and Identity in America, 1820–1885* (Cambridge, MA: Harvard University Press, 1986).

20. *MSHW*, vol. 3, devotes 482 pages to this topic. Morbidity and mortality statistics for all reported diseases in the war are contained in vol. 1.

21. C. C. Hanks to Dear Mother, Oct. 18, 1862, Carver Hospital, Constant C. Hanks Papers, 1861–65, Duke SC.

22. Joseph Janvier Woodward, *Outlines of the Chief Camp Diseases of the United States Armies as Observed during the Present War: A Practical Contribution to Military Medicine* (Philadelphia: Lippincott, 1863; repr., New York: Hafner, 1964), 209–66 (page references are to the 1964 edition).

23. *MSHW*, 6:966. On the calomel controversy, see Bonnie Ellen Blustein, *Preserve your Love for Science: Life of William A. Hammond, American Neurologist* (Cambridge: Cambridge University Press, 1991), 84–86; Warner, *Therapeutic Perspective*, 220–24.

24. [Satterlee] *Hospital Register*, March 19, 1864, 2, 119. The original source (which includes the poem with minor variations) is Benjamin Colby, *A Guide to Health, Being an Exposition of the Principles of the Thomsonian System of Practice, and Their Mode of Application in the Cure of Every Form of Disease* (Milford, NH: John Burns, 1846). The Thomsonians used only botanical remedies.

25. Michael Flannery discusses the use of calomel in his *Civil War Pharmacy: A History of Drugs, Drug Supply and Provision, and Therapeutics for the Union and Confederacy* (New York: Haworth Press, 2004), 143–55. He argues for the persistence of calomel as a commonly used drug well into the 1880s, in M. A. Flannery, "What Did Doctors Really Do? In Search of a Therapeutic Perspective of American Medicine," *Journal of Clinical Pharmacy and Therapeutics* 24 (1999): 151–56. In their 1932 movie *Horsefeathers*, the Marx brothers have a joke about calomel, an indication that the drug persisted at least in popular memory.

26. *MSHW*, 6:966.

27. Flannery, *Civil War Pharmacy*, 156–68; Andrew McIlwaine Bell, *Mosquito Soldiers: Malaria, Yellow Fever, and the Course of the American Civil War* (Baton Rouge: Louisiana State University Press, 2010); Meier, "No Place for the Sick."

28. See, for example, John H. Packard, *A Manual of Minor Surgery* (Philadelphia: Lippincott, 1863; repr., San Francisco: Norman, 1990), 28–30, 76–84.

29. Humphreys, *Intensely Human*.

30. Elisha Harris, "Yellow Fever on the Atlantic Coast and at the South during the War," in Flint, *Contributions Relating to Disease*, 236–68.

31. The following works on military surgery were all published during the Civil War, and many were distributed to the army surgeons as field manuals. Ira M. Rutkow edited a series of facsimile reproductions of these volumes and wrote introductions that included biographical material on the authors. The reproductions were published by Norman Publishing, San Francisco. Confederate works: Felix Formento Jr., *Notes and Observations on Army Surgery* (New Orleans: L. E. Marchand, 1863; repr., 1990); Moritz Shuppert, *A Treatise on Gun-Shot Wounds Written for and Dedicated to the Surgeons of the Confederate States Army*

(New Orleans: Bulletin Book and Job Office, 1861; repr., 1990), bound with Formento; Edward Warren, *An Epitome of Practical Surgery for Field and Hospital* (Richmond: West and Johnston, 1863; repr., 1989); Office of the Surgeon General [Samuel Preston Moore], *A Manual of Military Surgery Prepared for the Use of the Confederate States Army* (Richmond: Ayres and Wade, 1862; repr., 1989); J[ohn] Julian Chisolm, *A Manual of Military Surgery for the Use of Surgeons in the Confederate Army; with an Appendix of the Rules and Regulations of the Medical Department of the Confederate Army* (Richmond: West and Johnston, 1861; repr., 1989). Union works: Frank Hastings Hamilton, *A Practical Treatise on Military Surgery* (New York: Baillière Brothers, 1861; repr., 1989); S[amuel] D. Gross, *A Manual of Military Surgery; or, Hints on the Emergencies of Field, Camp, and Hospital Practice* (Philadelphia: Lippincott, 1861; repr., 1988); Packard, *Manual of Minor Surgery;* John Ordronaux, *Hints on the Preservation of Health in Armies, for the Use of Volunteer Officers and Soldiers* (New York: Appleton, 1861; repr., 1990); Stephen Smith, *Hand-Book of Surgical Operations* (New York: Baillière Brothers, 1862; repr., 1990); S. Weir Mitchell, George R. Morehouse, and William W. Keen, *Gunshot Wounds and Other Injuries of Nerves* (Philadelphia: Lippincott, 1864; repr., 1989). For more information on the history of surgery during the American Civil War, see Ira M. Rutkow, *Bleeding Blue and Gray: Civil War Surgery and the Evolution of American Medicine* (New York: Random House, 2005).

32. Ken Burns, *The Civil War*, PBS documentary, 1990, now available on DVD; Dan Piraro's "Bizarro" cartoons for March 1, 2011, and Sept. 18, 2009 (available at *Bizarro Blog*, www.bizarrocomics.com, accessed June 14, 2011), depend for their humor on the assumption that viewers associate amputation with the Civil War.

33. See references in note 31.

34. *MSHW*, 12:878.

35. *MSHW*, 12:876.

36. *MSHW*, 12:874.

37. *MSHW*, 12:866–81; statistics on 877.

38. Mary C. Gillett, *The Army Medical Department, 1818–1865* (Washington, DC: Center of Military History, U.S. Army, 1987), 193, comments that the terminology used to refer to various levels of hospitals in the Civil War was "confused and confusing."

39. John H. Fahey, "Bernard John Dowling Irwin and the Development of the Field Hospital at Shiloh," *Military Medicine* 171 (2006): 345–51, quotation on 345.

40. Burt G. Wilder, *Practicing Medicine in a Black Regiment: The Civil War Diary of Burt G. Wilder, 55th Massachusetts*, ed. Richard M. Reid (Amherst: University of Massachusetts Press, 2010), 66.

41. Fahey, "Bernard John Dowling Irwin."

42. Charles S. Tripler, "Report of Surgeon Charles S. Tripler, U.S. Army Medical Director, Army of the Potomac, of Operations March 17–July 3 [, 1862]," *OR*, series 1, vol. 11, part 1, 177–96.

43. Tripler to Brigadier General S[eth] Williams, May 29, 1862, in *OR*, series 1, vol. 11, part 1, 206–7; quotation on 207.

44. William A Hammond to Jonathan Letterman, June 19, 1862, Jonathan Letterman Correspondence and Diary, 1860–1864, MS C 96, National Library of Medicine, Bethesda, MD.

45. Witold Rybcznski, *A Clearing in the Distance: Frederick Law Olmsted and America in the Nineteenth Century* (New York: Scribner, 1999), 212.

46. Alfred Jay Bollet, *Civil War Medicine: Challenges and Triumphs* (Tucson: Galen Press, 2002), 97. On the history of army ambulances and their design, see John S. Haller Jr., *Farmcarts to Fords: A History of the Military Ambulance, 1790–1925* (Carbondale: Southern Illinois University Press, 1992).

47. Letterman to Brigadier General S[eth] Williams, March 1, 1863, Report of Surg. Jonathan Letterman, U.S. Army Medical Director of the Potomac, of Operations from July 4 to Sept. 2, THE PENINSULAR CAMPAIGN, VIRGINIA, March 17–Sept. 2, 1862, in *OR*, series 1, vol. 11, part 1, 210–20, quotation on 213–14.

48. Ibid., 213. Tripler commented on the scurvy too, and he had begun to deal with it when Letterman replaced him. Charles S. Tripler to General R. B. Marcy, June 14, 1862, in *OR*, series 1, vol. 11, part 1, 207–8.

49. Letterman to Williams, March 1, 1863, 213–14.

50. Rutkow, *Bleeding Blue and Gray*, 220–33.

51. Eaton, *Birth Place of Souls*, 67–74.

52. A. N. Read, "Report of Dr. A. N. Read, Oct. 23, 1862," in J. C. Newberry, "Operations of the Sanitary Commission at Perryville, Ky.," Oct. 24, 1862, USSC Doc. 55, in *Documents of the U.S. Sanitary Commission*, 2 vols. (New York: U.S. Sanitary Commission, 1866), vol. 1 (consists of documents numbered and bound together but each paginated separately).

53. C. C. Hanks to Dear Mother, Sept. 26, 1862, Hanks Papers.

54. Frank Haskell to Harrison Haskell, July 16, 1863, reprinted in *Haskell of Gettysburg: His Life and Civil War Papers*, ed. Frank L. Byrne and Andrew T. Weaver (Kent, OH: Kent State University Press, 1989), 190.

55. Ibid., 191.

56. Gregory A. Coco, *A Vast Sea of Misery: A History and Guide to the Union and Confederate Field Hospitals at Gettysburg, July 1–November 20, 1863* (Gettysburg, PA: Thomas, 1988).

57. Roland R. Maust, *Grappling with Death: The Union Second Corps Hospital at Gettysburg* (Dayton, OH: Morningside House, 2001). See also Michael A. Dreese, *The Hospital on Seminary Ridge at the Battle of Gettysburg* (Jefferson, NC: McFarland, 2002).

58. Maust, *Grappling with Death*, 330–35; Coco, *Vast Sea of Misery*. Meade told General Halleck that he could "not delay to pick up the *débris* of the battle-field" in a sentence that came before his report that the wounded would be left at Gettysburg. Although it is tempting to see Meade as callously equating such debris with the wounded, it seems likely that this was not his meaning, that instead he was referring to the battle flags, colors, and arms mentioned in the sentences that preceded this one. See George G. Meade to H. W. Halleck, July 5, 1863, in *OR*, series 1, vol. 27, part 1, 79. Historian Gerard A. Patterson seems to imply the opposite in his book *Debris of Battle: The Wounded at Gettysburg* (Mechanicsburg, PA: Stackpole Books, 1997), esp. 167.

59. Maust, *Grappling with Death*.

60. George Worthington Adams, *Doctors in Blue: The Medical History of the Union Army in the Civil War* (Baton Rouge: Louisiana State University Press, 1952), 84–111.

61. John T. Cumbler, "A Family Goes to War: Sacrifice and Honor for an Abolitionist Family," *Massachusetts Historical Review* 10 (2008): 57–83.

62. Eaton, *Birth Place of Souls*, 156.

63. Sanders Marble, "Forward Surgery and Combat Hospitals: The Origins of the MASH," *Journal of the History of Medicine and Allied Sciences*, doi 10.1093/jhmas/jrs032; first published online May 30, 2012 (print publication to follow).

64. Peter C. English, *Shock, Physiological Surgery, and George Washington Crile: Medical Innovation in the Progressive Era* (Westport, CT: Greenwood Press, 1980); William H. Schneider, "Blood Transfusion between the Wars," *Journal of the History of Medicine and Allied Sciences* 58 (2003): 187–224; Kim Pelis, "Taking Credit: The Canadian Army Medical Corps and the British Conversion to Blood Transfusion in WWI," *Journal of the History of Medicine and Allied Sciences* 56 (2001): 238–77; Bobby A. Wintermute, *Public Health and the U.S. Military: A History of the Army Medical Department, 1818–1917* (New York: Routledge, 2011), 157–88; Vincent Cirillo, *Bullets and Bacilli: The Spanish American War and Military Medicine* (New Brunswick, NJ: Rutgers University Press, 2003).

65. "The Sick Women of Bellevue Hospital, New York, Overrun by Rats," *Harper's Weekly*, May 5, 1860, 273.

66. "The General Hospitals," *MSHW*, 6:964.

67. Drew Gilpin Faust, *This Republic of Suffering: Death and the American Civil War* (New York: Knopf, 2008).

68. On the affluent insane in antebellum America, see Nancy Tomes, *A Generous Confidence: Thomas Story Kirkbride and the Art of Asylum-Keeping, 1840–1883* (Cambridge: Cambridge University Press, 1984); and Lawrence B. Goodheart, *Mad Yankees: The Hartford Retreat for the Insane and Nineteenth-Century Psychiatry* (Amherst: University of Massachusetts Press, 2003).

69. See Ulysses S. Grant National Historic Site, www.nps.gov/ulsg/upload/Winter%20 2008.pdf (accessed Sept. 29, 2009). This site includes a photograph of Churchill.

70. James O. Churchill, "Wounded at Fort Donelson" [a letter to his parents dated April 10, 1862 from a hospital in St. Louis on the corner of Fifth and Chestnut Streets], reproduced in *War Papers and Personal Reminiscences, 1861–1865: Read before the Commandery of the State of Missouri, Military Order of the Loyal Legion of the United States* (St. Louis: Commandery, 1892), 1:146–68, quotations on 165–66.

71. Ibid.

72. John D. Thompson and Grace Goldin, *The Hospital: A Social and Architectural History* (New Haven, CT: Yale University Press, 1975), 118.

73. Ibid., 118–69; Florence Nightingale, *Notes on Hospitals* (London: J. W. Parker, 1859).

74. Charles E. Rosenberg, "Florence Nightingale on Contagion: The Hospital as Moral Universe," in *Healing and History: Essays for George Rosen*, ed. Charles E. Rosenberg (New York: Science History Publications, 1979), 116–36, quotation on 124.

75. There is a large literature on the design of mental asylums in the nineteenth-century United States. See, for example, David J. Rothman, *The Discovery of the Asylum: Social Order and Disease in the New Republic* (Boston: Little, Brown, 1971); and Gerald Grob, *Mental Institutions in America: Social Policy to 1875* (New York: Free Press, 1973).

76. William H. Van Buren and C. R. Agnew, "Report on Military Hospitals in and around Washington, DC," July 31, 1861, Doc 23, pp. 3, 8, *Documents of the U.S. Sanitary Commission*, vol. 1.

77. William Alexander Hammond, *A Treatise on Hygiene: With Special Reference to the Military Service* (Philadelphia: Lippincott, 1863), 305–445.

78. Carol C. Green, *Chimborazo: The Confederacy's Largest Hospital* (Knoxville: University of Tennessee Press, 2004).

79. Charles Stuart Tripler and George Curtis Blackman, *Hand-Book for the Military Surgeon* (1861; repr., San Francisco: Norman, 1989), 1.

Chapter 2 · Women, War, and Medicine

1. Quoted in Anne Scott, *Natural Allies: Women's Associations in American History* (Urbana: University of Illinois Press, 1991), 74. For general background on women and the war, see "We Are Now Very Busy: Women and War," in the same volume, 58–77.

2. Mary Elizabeth Massey, *Bonnet Brigades* (New York: Knopf, 1966), x.

3. Barbara J. Harris, *Beyond Her Sphere: Women and the Professions in American History* (Westport, CT: Greenwood Press, 1978).

4. Judith Giesberg, *Army at Home: Women and the Civil War on the Northern Home Front* (Chapel Hill: University of North Carolina Press, 2009), 10; Thavolia Glymph, *Out of the House of Bondage: The Transformation of the Plantation Household* (Cambridge: Cambridge University Press, 2008); Victoria E. Bynum, *Unruly Women: The Politics of Social and Sexual Control in the Old South* (Chapel Hill: University of North Carolina Press, 1992); Laura Edwards, *Scarlett Doesn't Live Here Anymore: Southern Women in the Civil War Era* (Urbana: University of Illinois Press, 2000).

5. Stephanie McCurry and Thavolia Glymph, comments at the plenary session of the Society of Civil War Historians meeting, Lexington, KY, June 14, 2012; I have paraphrased their messages here, having heard them talk but lacking text.

6. Mary Putnam Jacobi, "Women in Medicine," in *Woman's Work in America,* ed. Anne Nathan Meyer (New York: Henry Holt, 1891), 139–205, quotation on 167. Jacobi cites Blackwell as the authority for the number 300; Thomas Bonner cites an unpublished paper by Edward Atwater as finding 247 women physicians in the antebellum period. Thomas Neville Bonner, *To the Ends of the Earth: Women's Search for Education in Medicine* (Cambridge, MA: Harvard University Press, 1992), 14.

7. Carla Bittel, *Mary Putnam Jacobi and the Politics of Medicine in Nineteenth-Century America* (Chapel Hill: University of North Carolina Press, 2009), 31–49.

8. Jacobi, "Women in Medicine," 197.

9. Jane E. Schultz, ed., *This Birth Place of Souls: The Civil War Nursing Diary of Harriet Eaton* (London: Oxford University Press, 2011), 78.

10. Ibid., 101.

11. Ibid., 109.

12. Jane Schultz, *Women at the Front: Hospital Workers in Civil War America* (Chapel Hill: University of North Carolina Press, 2004), 20–21.

13. Barbara Melosh has argued that nursing never did become a profession, because a key component of profession was autonomy. *"The Physician's Hand": Work Culture and Conflict in American Nursing* (Philadelphia: Temple University Press, 1982). There is no good word to replace "profession," though. "Trained" encompasses some compo-

nents of the concept, but a potential nurse could be "trained" for ten minutes or two years.

14. Bittel, *Mary Putnam Jacobi,* 39.

15. Said in a conversation with me in 2008; the speaker will remain anonymous here.

16. Glymph, *Out of the House of Bondage;* Jim Downs, *Sick from Freedom: African American Illness and Suffering during the Civil War and Reconstruction* (New York: Oxford University Press, 2012).

17. There were 11,515,239 native-born white males and 11,171,105 native-born white females in the 1860 census (344,134 more men than women). J. David Hacker, "A Census-Based Account of the Civil War Dead," *Civil War History* 57 (2011): 307–48, 322. For all races and national origins, there were about 15,135,000 women in 1860 and 15,865,000 men (730,000 more men than women). Introduction to the 1860 Census, www.census .gov/prod/www/abs/decennial/1860.html, xviii. These numbers may reflect differential rates of immigration by sex to the United States, as well as the loss of women's lives in childbirth.

18. Maris A. Vinovskis, ed., *Toward a Social History of the American Civil War: Exploratory Essays* (Cambridge: Cambridge University Press, 1990). And see Theresa McDevitt, *Women and the American Civil War: An Annotated Bibliography* (Westport, CT: Praeger, 2003).

19. Laurel Thatcher Ulrich, *A Midwife's Tale: The Life of Martha Ballard, Based on Her Diary* (New York: Knopf, 1990); Judith Walzer Leavitt, *Brought to Bed: Childbearing in America, 1750–1950* (New York: Oxford University Press, 1986).

20. Mary Roth Walsh, *Doctors Wanted: No Women Need Apply: Sexual Barriers in the Medical Profession, 1835–1875* (New Haven, CT: Yale University Press, 1977); Virginia G. Drachman, *Hospital with a Heart: Women Doctors and the Paradox of Separatism at the New England Hospital, 1862–1969* (Ithaca, NY: Cornell University Press, 1984); Regina Markell Morantz-Sanchez, *Sympathy and Science: Women Physicians in American Medicine* (New York: Oxford University Press, 1985); Ellen Singer More, *Restoring the Balance: Women Physicians and the Profession of Medicine, 1850–1995* (Cambridge, MA: Harvard University Press, 1999); Bonner, *To the Ends of the Earth.*

21. Arleen Tuchman, *Science Has No Sex: The Life of Marie Zakrzewska, M.D.* (Chapel Hill: University of North Carolina Press, 2006), 60.

22. Patricia A. Cunningham, *Reforming Women's Fashion, 1850–1920: Politics, Health, and Art* (Kent, OH: Kent State University Press, 2003).

23. Sharon M. Harris, *Dr. Mary Walker: An American Radical, 1832–1919* (New Brunswick, NJ: Rutgers University Press, 2009); Elizabeth D. Leonard, *Yankee Women: Gender Battles in the Civil War* (New York: Norton, 1994), 105–57; Charles M. Snyder, *Dr. Mary Walker: The Little Lady in Pants* (New York: Vantage Press, 1962); Albert Castel, "Mary Walker: Samaritan or Charlatan?" *Civil War Times Illustrated* 33, no. 2 (1994): 40–43, 62–64; Allen D. Spiegel and Andrea M. Spiegel, "Civil War Doctress Mary: Only Woman to Win Congressional Medal of Honor," *Minerva* 12, no. 3 (1994): 24–34; Mary E. Walker, "Incidents Connected with the Army," n.d., Mary E. Walker Papers, Special Collections, Syracuse University Library, Syracuse, NY.

24. Walker, "Incidents"; and Leonard, *Yankee Women,* 106–23.

25. Walker, "Incidents."

26. Stuart Galishoff, "Bartholow, Roberts," *Dictionary of American Medical Biography,* 2 vols., ed. Martin Kauffman, Stuart Galishoff, and Todd Savitt (Westport, CT: Greenwood Press, 1984), 1:39.

27. Roberts Bartholow, letter to the editor, Jan. 10, 1867, *New York Medical Journal* 5 (1867): 167–70, quotations on 168, 169. Bartholow's letter was prompted by a report that Walker had given a lecture in London about her war work; see "Dr. Mary Walker," *New York Medical Journal* 4 (1867): 314–16.

28. By 1864 the Union Army was desperate for surgeons, any surgeons, and commanders often overrode examining boards in order to fill their rosters. See Margaret Humphreys, *Intensely Human: The Health of the Black Soldier in the American Civil War* (Baltimore: Johns Hopkins University Press, 2008), chap. 4.

29. Nixon B. Stewart, *Dan. McCook's Regiment, 52nd O.V.I.: A History of the Regiment, Its Campaigns and Battles: From 1862–1865* (1900; repr., Huntington, WV: Blue Acorn Press, 1999), 91.

30. Walker, "Incidents"; Harris, *Dr. Mary Walker,* 53–74.

31. B. J. Semmes to Jorantha Semmes, April 12, 1864, Dalton, GA, folder 11, Jan. 1864–April 1864, Benedict J. Semmes Papers, Collection 2333, UNC SHC.

32. Harris, *Dr. Mary Walker;* and other sources in note 23.

33. Tuchman, *Science Has No Sex.*

34. Mercedes Graf, introduction and back cover of *Hit: Essays on Women's Rights,* by Mary E. Walker (1872; repr., Amherst, NY: Humanity Books, 2003).

35. John Raves, "A Woman Surgeon: Eclecticism or Oxymoron?" paper presented at the Southern Association for the History of Medicine and Science meeting, Durham, NC, Feb. 22, 2003. Raves was then in the Department of Surgery, Allegheny General Hospital, Pittsburg, PA.

36. Harris, *Dr. Mary Walker.*

37. Willie Lee Rose, *Rehearsal for Reconstruction: The Port Royal Experiment* (New York: Oxford University Press, 1976); Gerald Schwartz, ed., *A Woman Doctor's Civil War: Esther Hill Hawks Diary* (Columbia: University of South Carolina Press, 1994).

38. Humphreys, *Intensely Human,* chap. 4.

39. Schwartz, *Woman Doctor's Civil War,* 49.

40. Ibid., 50.

41. Ibid., 47, 51–54.

42. Margaret Humphreys, *Malaria: Poverty, Race, and Public Health in the United States* (Baltimore: Johns Hopkins University Press, 2001), 31–36.

43. Mercedes Graf, "Women Physicians in the Civil War," *Prologue: Quarterly of the National Archives and Records Administration* 32 (2000): 86–98.

44. Ibid.

45. Pauline Heald, "Mary F. Thomas, M.D., Richmond, Ind.," *Michigan History Magazine* 6 (1922): 369–73, quotation on 372. See also Schultz, *Women at the Front;* and Graf, "Women Physicians."

46. See "Dr. Mary Walker," 314–16.

47. Frank Moore, *Women of the War: Their Heroism and Self-Sacrifice* (Hartford, CT:

S. C. Scranton, 1866); L. P. Brockett and Mary C. Vaughan, *Woman's Work in the Civil War: A Record of Heroism, Patriotism, and Patience* (Philadelphia: Zeigler, McCurdy, 1867). See also Mary A. Livermore, *My Story of the War: A Woman's Narrative of Four Years Personal Experience* (Hartford, CT: A. D. Worthington, 1889).

48. Heald, "Mary F. Thomas"; Tuchman, *Science Has No Sex;* Elizabeth Blackwell, *Pioneer Work in Opening the Medical Profession to Women* (London: Longmans Green, 1895); Morantz-Sanchez, *Sympathy and Science.*

49. The impact of the war on the promotion of women physicians in the United States deserves further research; in particular, comparative studies of women in Europe and the United States might offer insights into the generative power of the war for encouraging women in medicine. This war, like the two world wars, generally promoted the acceptance of women in the workplace, as the shortage of men generated by military assignment and war casualties created openings that women filled. Virginia Nicholson, *Singled Out: How Two Million Women Survived without Men after the First World War* (New York: Viking, 2007); Emily Yellin, *Our Mothers' War: American Women at Home and at the Front during World War II* (New York: Free Press, 2004).

50. Henry Wadsworth Longfellow, "Santa Filomena," *Atlantic Monthly* 1, no. 1 (1857): 22–24. Longfellow is making a play on words here that would have been familiar to his readers. Philomela and Procne were turned into birds in Greek myth (although whether Philomela or Procne became the nightingale differs between versions). St. Philomena was discovered in the early nineteenth century and believed to be a martyr of the early church. Longfellow merges the two stories, and his readers would have easily understood the reference to Florence Nightingale.

51. Florence Nightingale, *Notes on Nursing: What It Is, and What It Is Not* (London: Harrison, 1859); Florence Nightingale, *Notes on Hospitals* (London: J. W. Parker, 1859).

52. Quoted in Kristie R. Ross, "'Women Are Needed Here': Northern Protestant Women as Nurses during the Civil War, 1861–1865" (PhD diss., Columbia University, 1993), 68.

53. Charles Dickens, *The Life and Adventures of Martin Chuzzlewit* (1843; repr., London: Penguin, 1995).

54. Ann Douglas Wood claims that female nurses sought to usurp the medical professionals' role in the hospital. "The War within a War: Women Nurses in the Union Army," *Civil War History* 18 (1972): 197–212. But Jane Schultz argues differently in her essay "The Inhospitable Hospital: Gender and Professionalism in Civil War Medicine," *Signs* 17 (1992): 363–92. Schultz recognizes that the power struggles between nurse and doctor were about improving patient care, not "replacing male decision makers with females" (375).

55. This may have been particularly true of southern, slave-holding women who managed large establishments before the war. See, for example, Jean V. Berlin, ed., *A Confederate Nurse: The Diary of Ada W. Bacot, 1860–1863* (Columbia: University of South Carolina Press, 1994), 168.

56. The most thoughtful and thorough work on nurses in the Civil War is Jane Schultz's *Women at the Front.* See also Nina Bennett Smith, "The Women Who Went to War: The Union Army Nurse in the Civil War" (PhD diss., Northwestern University, 1981); Libra Rose Hilde, "Worth a Dozen Men: Women, Nursing, and Medical Care during the Ameri-

can Civil War" (PhD diss., Harvard University, 2003). There are an abundance of diaries from Civil War hospital nurses on both sides that provide much detail on attitudes and day-to-day activities. See McDevitt, *Women and the American Civil War*, for a list of published diaries.

57. Schultz, *Women at the Front*, 20–21.

58. David Gollaher, *Voice for the Mad: The Life of Dorothea Dix* (New York: Free Press, 1995); and Thomas J. Brown, *Dorothea Dix: New England Reformer* (Cambridge, MA: Harvard University Press, 1998).

59. Schultz, *Women at the Front*, 15; Ross, "Women Are Needed Here," 83.

60. Katharine Prescott Wormeley, *The Cruel Side of War with the Army of the Potomac: Letters from the Headquarters of the United States Sanitary Commission during the Peninsular Campaign in Virginia in 1862* (1888; repr., Boston: Roberts Brothers, 1898); Nancy Scripture Garrison, *With Courage and Delicacy: Civil War on the Peninsula: Women and the U.S. Sanitary Commission* (Mason City, IA: Savas, 1999).

61. Berlin, *Confederate Nurse;* Phoebe Yates Pember, *A Southern Woman's Story* (1879; repr., Columbia: University of South Carolina Press, 2002); Richard Barksdale Harwell, ed., *Kate: The Journal of a Confederate Nurse* (1959; repr., Baton Rouge: Louisiana State University Press, 1987).

62. Henrietta Stratton Jaquette, ed., *Letters of a Civil War Nurse: Cornelia Hancock, 1863–1865* (Lincoln: University of Nebraska Press, 1998); Schultz, *Women at the Front*, 18.

63. John R. Brumgardt, ed. *Civil War Nurse: The Diary and Letters of Hannah Ropes* (Knoxville: University of Tennessee Press, 1980), 53.

64. Susan Cheever, *Louisa May Alcott* (New York: Simon and Schuster, 2010), 158–59; Louisa May Alcott, *Hospital Sketches*, ed. Alice Fahs (1863; rev. ed., Boston: Bedford St. Martins, 2004), 28–29.

65. Pember, *Southern Woman's Story*, 48–49.

66. On the nursing sisters, see Mary Denis Maher, *To Bind up the Wounds: Catholic Sister Nurses in the U.S. Civil War* (New York: Greenwood Press, 1989); Barbara Mann Wall, "Called to a Mission of Charity: The Sisters of St. Joseph in the Civil War," *Nursing History Review* 6 (1998): 85–113.

67. This was Phoebe Pember's constant struggle. See Pember, *Southern Woman's Story*.

68. Brumgardt, *Civil War Nurse*, 80–90.

69. Theophilis Parsons, ed., *Memoir of Emily Elizabeth Parsons* (Boston: Little, Brown, 1880); Humphreys, *Intensely Human*, chap. 5.

70. Jane Schultz, "Seldom Thanked, Never Praised, and Scarcely Recognized: Gender and Racism in Civil War Hospitals," *Civil War History* 48 (2002): 220–36.

71. Susie King Taylor, *Reminiscences of My Life in Camp* (1902; repr,. New York: M. Wiener, 1988).

72. Susan Reverby, *Ordered to Care: The Dilemma of American Nursing, 1850–1945* (Cambridge: Cambridge University Press, 1987).

73. Stephen B. Oates, *A Woman of Valor: Clara Barton and the Civil War* (New York: Free Press, 1994).

74. Blackwell, *Pioneer Work*, 176, 234–36; Jeanie Attie, *Patriotic Toil: Northern Women and the American Civil War* (Ithaca, NY: Cornell University Press, 1998), 39–57; Judith Ann

Giesberg, *Civil War Sisterhood: The U.S. Sanitary Commission and Women's Politics in Transition* (Boston: Northeastern University Press, 2000), 22–45. William Quentin Maxwell, *Lincoln's Fifth Wheel: The Political History of the U.S. Sanitary Commission* (New York: Longmans, Green, 1956); and George M. Fredrickson, *The Inner Civil War: Northern Intellectuals and the Crisis of the Union* (New York: Harper and Row, 1965) are the classic sources on the men of the USSC, their motivations, and their political activity. For the many functions of the USSC, see USSC Microfilm Papers. The activities of the Sanitary Commission are described in more detail in chapter 4 of this volume.

75. Suellen Hoy, *Chasing Dirt: The American Pursuit of Cleanliness* (New York: Oxford University Press, 1995).

76. Giesberg, *Civil War Sisterhood.*

77. Attie, *Patriotic Toil.*

78. Drew Gilpin Faust, *Mothers of Invention: Women of the Slaveholding South in the American Civil War* (New York: Vintage Books, 1997), 24.

79. Harwell, *Kate,* 83.

80. Drew Gilpin Faust, "Altars of Sacrifice: Confederate Women and the Narratives of War," *Journal of American History* 76 (1990): 1200–1228.

81. A. Clair Siddall, "Bloodletting in American Obstetric Practice, 1800–1945," *Bulletin of the History of Medicine* 54 (1980): 101–10.

82. Judith Walzer Leavitt, "'Science' Enters the Birthing Room: Obstetrics in America since the 18th Century," *Journal of American History* 70 (1983): 281–304.

83. Sally G. McMillen, *Motherhood in the Old South: Pregnancy, Childbirth, and Infant Rearing* (Baton Rouge: Louisiana State University Press, 1990). Prissy's mother Dilcey, in *Gone with the Wind,* was needed when Melanie went into labor, and much is made of Prissy claiming to have midwife skills but not, in the end, having any.

84. Michael A. Flannery, "Another House Divided: Union Medical Service and Sectarians during the Civil War," *Journal of the History of Medicine and Allied Sciences* 54 (1999): 478–510.

85. J. Matthew Gallman, *The North Fights the Civil War: The Home Front* (Chicago: Ivan R. Dee, 1994), 92–108; Attie, *Patriotic Toil.*

86. Giesberg, *Army at Home.*

87. See, for example, Walker, "Incidents"; and Harwell, *Kate.*

88. Faust, "Altars of Sacrifice"; and Edwards, *Scarlett Doesn't Live Here Anymore,* 93–99; Bynum, *Unruly Women.*

89. Thomas P. Lowry, *The Story the Soldiers Wouldn't Tell: Sex in the Civil War* (Mechanicsburg, PA: Stackpole Books, 1994), 99–108. See also James Boyd Jones Jr., "A Tale of Two Cities: The Hidden Battle against Venereal Disease in Civil War Nashville and Memphis," *Civil War History* 31 (1985): 270–76.

90. Glymph, *Out of the House of Bondage.*

91. Jim Downs, "The Other Side of Freedom: Destitution, Disease, and Dependency among Freedwomen and their Children during and after the Civil War," in *Battle Scars: Gender and Sexuality in the American Civil War,* ed. Catherine Clinton and Nina Silber (New York: Oxford University Press, 2006), 78–103; and Downs, *Sick from Freedom.*

92. James E. Yeatman, *A Report on the Condition of the Freedmen of the Mississippi,*

Presented to the Western Sanitary Commission, December 17th, 1863 (Saint Louis: Western Sanitary Commission, 1864), 12.

93. Gaines M. Foster, "The Limitations of Federal Health Care for Freedmen, 1862–1868," *Journal of Southern History* 48 (1982): 349–72.

94. Glymph, *Out of the House of Bondage.*

95. Dorothy Sheahan, "The Social Origins of American Nursing and Its Movement in the University: A Microscopic Approach" (PhD diss., New York University, 1980).

Chapter 3 · Infectious Disease in the Civil War

1. Paul Steiner, *Disease in the Civil War: Natural Biological Warfare in 1861–1865* (Springfield, IL: Charles C. Thomas, 1968).

2. Charles Rosenberg, *The Cholera Years: The United States in 1832, 1849, and 1866* (Chicago: University of Chicago Press, 1962), 193.

3. Steiner, *Disease in the Civil War.* In Margaret Mitchell's fictional account, Scarlett's first husband, Charles Hamilton, dies of measles not long after he arrives in camp.

4. William G. Williams and James Andrews, "An Account of the Yellow Fever Which Prevailed at Rodney, Mississippi, during the Autumn of 1843," *New-Orleans Medical Journal* 1 (1844): 35–43, quotation on 40.

5. Margaret Humphreys, *Yellow Fever and the South* (New Brunswick, NJ: Rutgers University Press, 1992); Rosenberg, *Cholera Years;* Erwin Ackerknecht, "Anticontagionism between 1821 and 1867," *Bulletin of the History of Medicine* 22 (1948): 562–93; and Martin Pernick, "Politics, Parties, and Pestilence: Epidemic Yellow Fever in Philadelphia and the Rise of the First Party System," *William and Mary Quarterly* 29 (1972): 559–86.

6. These generalizations do not do justice to the rich literature on disease theory that was produced during the mid-nineteenth century, especially in regard to cholera and yellow fever. Much of the debate occurred in England and France and filtered into the U.S. medical literature in abstracted form. If I have simplified these notions, it is in an attempt to summarize the common thought of American physicians, gleaned as nearly as possible from their published articles, the information in the USSC papers, and official government reports on disease during the war. In addition to works cited in subsequent notes, see Margaret Pelling, *Cholera, Fever, and English Medicine, 1825–1865* (Oxford: Oxford University Press, 1978); and John M. Eyler, *Victorian Social Medicine: The Ideas and Methods of William Farr* (Baltimore: Johns Hopkins University Press, 1979); Christopher Hamlin, "Predisposing Causes and Public Health in Early Nineteenth-Century Medical Thought," *Social History of Medicine* 5 (1992): 43–70; and William Coleman, *Yellow Fever in the North: The Methods of Early Epidemiology* (Madison: University of Wisconsin Press, 1987). On the history of disease in the United States, see Gerald Grob, *A Deadly Truth: A History of Disease in America* (Cambridge, MA: Harvard University Press, 2002).

7. Although he may not agree with all my conclusions, my thinking on the question of etiology and infectious diseases has been much clarified by K. Codell Carter, *The Rise of Causal Concepts of Disease* (Aldershot, UK: Ashgate, 2003).

8. Ibid.

9. Elisha Harris, *Control and Prevention of Infectious Diseases,* in *Military, Medical, and*

Surgical Essays Prepared for the United States Sanitary Commission, ed. William A. Hammond (Philadelphia: Lippincott, 1864), 45–90.

10. Christopher Lawrence and Richard Dixey, "Practicing on Principle: Joseph Lister and the Germ Theories of Disease," in *Medical Theory, Surgical Practice: Studies in the History of Surgery*, ed. Christopher Lawrence (London: Routledge, 1992), 153–215.

11. Robert Angus Smith reviewed the history of disinfectants in the introductory chapter of his book *Disinfectants and Disinfection* (Edinburgh: Edmonston and Douglas, 1869).

12. B. F. Craig, "Report on Disinfectants and Their Use in Connection with Cholera" (appendix), in *Report on Epidemic Cholera in the Army of the United States, during the year 1866, Circular No. 5, Washington, May 4, 1867*, by War Department, Surgeon General's Office (Washington, DC: Government Printing Office, 1867), 63–65, quotation on 63.

13. Caroline Alexander, *The Bounty: The True Story of the Mutiny on the Bounty* (New York: Viking, 2003), 84.

14. On theories of water purity in this period, see Christopher Hamlin, *A Science of Impurity: Water Analysis in Nineteenth Century Britain* (Berkeley: University of California Press, 1990).

15. William Alexander Hammond, *A Treatise on Hygiene, with Special References to the Military Service* (Philadelphia: Lippincott, 1863), 217.

16. J. J. Woodward to Joseph Barnes, May 1, 1867, in War Department, *Report on Epidemic Cholera, 1866*, xvii.

17. James D. Birkett, "The 1861 de Normandy Desalting Unit at Key West," *Desalination and Water Reuse Quarterly* 7 (1997): 53–57.

18. Hammond, *Treatise on Hygiene*, 218 and 221.

19. Alfred Stillé, *Dysentery*, in Hammond, *Military, Medical, and Surgical Essays*, 329–82.

20. Harris, *Control and Prevention*, 80.

21. J[ohn] K[earsley] Mitchell, *On the Cryptogamous Origin of Malarious and Epidemic Fevers* (Philadelphia: Lea and Blanchard, 1849).

22. Michael Worboys, *Spreading Germs: Disease Theories and Medical Practice in Britain, 1865–1900* (Cambridge: Cambridge University Press, 2000).

23. Florence Nightingale, *Notes on Hospitals*, 3rd ed. (London: Longman, Green, Longman, Roberts, and Green, 1863), 10; Charles Rosenberg, "Florence Nightingale on Contagion: The Hospital as Moral Universe," in *Healing and History: Essays for George Rosen*, ed. Charles Rosenberg (New York: Science History, 1979), 116–36.

24. Nightingale, *Notes on Hospitals*, 66.

25. Carol C. Green, *Chimborazo: The Confederacy's Largest Hospital* (Knoxville: University of Tennessee Press, 2004).

26. Phoebe Yates Pember, *A Southern Woman's Story* (1879; repr., Columbia: University of South Carolina Press, 2002), 20–21.

27. *Regulations for the Medical Department of the C. S. Army* (Richmond: Ritchie and Dunnavant, 1862), 8. Available at Regulations for the Medical Department of the C.S. Army: Electronic Edition, http://docsouth.unc.edu/imls/regulations/regulations.html (accessed Oct. 10, 2007).

28. Joseph Janvier Woodward, *Outlines of the Chief Camp Diseases of the United States*

Armies as Observed during the Present War: A Practical Contribution to Military Medicine (Philadelphia: Lippincott, 1863; repr., New York: Hafner, 1964), 120.

29. J. L. Cabell, trans. and ed., "Remarks concerning the Hygiene of Military Hospitals, Extracted from a Paper Read before the Imperial Academy of Medicine at Paris, by M. H. Baron Larrey. Paris, 1862," *CSMSJ* 1 (1864): 26–28, 47–48, quotation on 47.

30. Nightingale, *Notes on Hospitals,* 6.

31. Rudolf Virchow, *Die Cellularpathologie, in ihrer begründung auf physiologische und pathologische Gewebelehre* (Berlin: A. Hirshfeld, 1858).

32. Lawrence and Dixey, "Practicing on Principle"; John Harley Warner, "'The Nature-Trusting Heresy': American Physicians and the Concept of the Healing Power of Nature in the 1850's and 1860's," *Perspectives in American History* 11 (1977–78): 291–324. Contemporary surgeons could learn of these complex theories of inflammation, pus, and healing in such books as Edward Warren, *An Epitome of Practical Surgery for Field and Hospital* (1863; repr., San Francisco: Norman, 1989), 13–52.

33. Lawrence and Dixey, "Practicing on Principle."

34. Samuel Preston Moore, *A Manual of Military Surgery, Prepared for the Use of the Confederate States Army* (1863; repr., San Francisco: Norman, 1989), 13–32, quotation on 32 (emphasis in original). See also sections on gangrene, traumatic erysipelas, and pyaemia in the *MSHW,* 12:823–66.

35. Moore, *Manual of Military Surgery,* 32.

36. M[iddleton] Goldsmith, *A Report on Hospital Gangrene, Erysipelas, and Pyaemia* (Louisville: Bradley and Gilbert, 1863).

37. John Hooker Packard, *A Manual of Minor Surgery* (Philadelphia: Lippincott, 1863; repr., San Francisco: Norman, 1990), 249 and 255.

38. See, for example, Moritz Shuppert, *A Treatise on Gun-Shot Wounds Written for and Dedicated to the Surgeons of the Confederate States Army* (New Orleans: Bulletin Book and Job Office, 1861; repr., San Francisco: Norman, 1990); Charles Stuart Tripler and George Curtis Blackman, *Hand-Book for the Military Surgeon* (1861; repr., San Francisco: Norman, 1989); Stephen Smith, *Hand-Book of Surgical Operations* (New York: Baillière Brothers, 1862; repr., San Francisco: Norman, 1990).

39. Frank Hastings Hamilton, *A Practical Treatise on Military Surgery* (New York: Baillière Brothers, 1861; repr., San Francisco: Norman, 1989), 193.

40. See, for example, Richard Barksdale Harwell, ed., *Kate: The Journal of a Confederate Nurse* (1866; repr., Baton Rouge: Louisiana State University Press, 1959, 1987), 171, page number in the 1987 edition; and Gerald Schwartz, ed., *A Woman Doctor's Civil War: Esther Hill Hawks Diary* (Columbia: University of South Carolina Press, 1994), 51.

41. Elizabeth A. Fenn, *Pox Americana: The Great Smallpox Epidemic of 1775–82* (New York: Hill and Wang, 2001); Elizabeth A. Fenn, "Biological Warfare in Eighteenth-Century North America: Beyond Jeffrey Amherst," *Journal of American History* 86 (2000): 1552–80; John Blake, "The Inoculation Controversy in Boston, 1721–1722," *New England Quarterly* 25 (1952): 489–506; and Charles Rosenberg, "The Cause of Cholera: Aspects of Etiological Thought in Nineteenth Century America," *Bulletin of the History of Medicine* 34 (1960): 331–54.

42. Edward Jenner, *An Inquiry into the Various Causes and Effects of the Variolae Vaccinae, a Disease Discovered in Some of the Western Counties of England, particularly Gloucestershire, and Known by the Name of Cow Pox* (1798; repr., Birmingham, UK: Classics of Medicine Library, 1978).

43. Roberts Bartholow, "Camp Measles," in *Contributions Relating to the Causation and Prevention of Disease and to Camp Diseases; Together with a Report of the Diseases, etc., among the Prisoners at Andersonville, GA,* ed. Austin Flint (New York: Hurd and Houghton, 1867), 218–35, quotation on 219.

44. Andrew Cliff, Peter Haggett, and Matthew Smallman-Raynor, *Measles: An Historical Geography of a Major Human Viral Disease from Global Expansion to Local Retreat, 1840–1990* (Oxford: Blackwell, 1993), 101–7.

45. J. H. Salisbury, "Remarks on Fungi," *Boston Medical and Surgical Journal* 65 (1862): 509–15, 532–36.

46. Bartholow, "Camp Measles."

47. Mitchell, *On the Cryptogamous Origin.*

48. John T. Metcalf, "Miasmatic Fevers," in Hammond, *Military, Medical and Surgical Essays,* 207–34.

49. Andrew McIlwaine Bell, *Mosquito Soldiers: Malaria, Yellow Fever, and the Course of the American Civil War* (Baton Rouge: Louisiana State University Press, 2010).

50. Rosenberg, *Cholera Years;* and Humphreys, *Yellow Fever and the South.*

51. Steiner, *Disease in the Civil War,* 98–155.

52. Roberts Bartholow, "Effects of a Malarious Atmosphere as Regards Physical Endurance . . . ," in Flint, *Contributions Relating to Disease,* 118–36.

53. Elisha Harris, "Yellow Fever on the Atlantic Coast and at the South during the War," in Flint, *Contributions Relating to Disease,* 236–68.

54. Ibid. Harris does not refer to the Charleston outbreak, but Confederate surgeon R. A. Kinloch mentions it in a September 1864 letter to surgeon A. W. Thomson, telling him that certain surgeons under his command had been detailed to hospitals in Charleston because yellow fever had broken out there and these men were acclimated. R. A. Kinloch to A. W. Thomson, Sept. 26, 1864, Second NC Hospital, Columbia, Order Book &tc. 1863–1865, S192115, SCSA.

55. William T. Wragg, "Report on the Yellow Fever at Wilmington, N.C., in the Autumn of 1862," *CSMSJ* 1 (1864): 17–20, 33–36.

56. Harris, "Yellow Fever," 266–67.

57. Ira Russell, "Cerebro Spinal Meningitis, as It Appeared among the Colored Soldiers at Benton Barracks, MO, during the Winter of 1863," *St. Louis Medical and Surgical Journal,* n.s., 1 (1864): 121–28.

58. Ira Russell, "Cerebro-Spinal Meningitis as It Appeared among the Troops Stationed at Benton Barracks, MO," *Boston Medical and Surgical Journal* 70 (1864): 309–13, quotation on 313. On concepts of sthenia and asthenia, along with the use of bloodletting for one and stimulants for the other, see John Harley Warner, *The Therapeutic Perspective: Medical Practice, Knowledge, and Identity in America, 1820–1885* (Cambridge, MA: Harvard University Press, 1986). For a description of the outbreaks of meningitis that preceded

the war, see Alfred Stillé, *Epidemic Meningitis, or Cerebro-Spinal Meningitis* (Philadelphia: Lindsay and Blakiston, 1867).

59. Ira Russell, "Pneumonia as It Appeared among the Colored Troops at Benton Barracks, MO, during the Winter of 1864," in Flint, *Contributions Relating to Disease*, 322–23.

60. Margaret Humphreys, "A Stranger in Our Camps: Typhus in American History," *Bulletin of the History of Medicine* 80 (2006): 269–90.

61. Joseph Jones, "Investigations upon the Diseases of the Federal Prisoners Confined in Camp Sumpter, Andersonville, Ga.," in Flint, *Contributions Relating to Disease*, 469–655, quotation on 600.

62. W. H. Van Buren, *Rules for Preserving the Health of the Soldier* (Washington, DC: U.S. Sanitary Commission, 1861).

63. Paula Canavese, "Factors Determining Health Status: An Analysis of Diarrhea in the Union Army, 1861–1920" (PhD diss., University of Chicago, 2005).

64. See case reports in the *MSHW*, vols. 5 and 6, on continued fevers.

65. William Budd, *On the Causes and Mode of Propagation of the Common Continued Fevers of Great Britain and Ireland (1839)*, ed. Dale Smith (Baltimore: Johns Hopkins University Press, 1984).

66. Harris, *Control and Prevention*, 66.

67. Woodward, *Outlines of the Chief Camp Diseases*, 54.

68. [M. M. Marsh], "Report on the Sanitary State of the Troops at Charleston," *Sanitary Commission Bulletin* 1 (1863): 78–84, quotation on 82.

69. P. H. Bailhache, quoted in Sanford B. Hunt, "Camp Diarrhoea and Dysentery," in Flint, *Contributions Relating to Disease*, 291–318, quotation on 315.

70. Benjamin Woodward, quoted in Hunt, "Camp Diarrhoea," 311.

71. Alfred Jay Bollet, *Civil War Medicine: Challenges and Triumphs* (Tucson: Galen Press, 2002), 365–74; and Humphreys, "Stranger to Our Camps."

72. Nancy Tomes, "American Attitudes toward the Germ Theory of Disease: Phyllis Allen Richmond Revisited," *Journal of the History of Medicine* 52 (1997): 17–50, quotation on 29. Tomes discussed Richmond's older paper: Phyllis Allen Richmond, "American Attitudes toward the Germ Theory of Disease (1860–1880)," *Journal of the History of Medicine and Allied Sciences* 9 (1954): 58–84.

Chapter 4 · Connecting Home to Hospital and Camp

1. Charles J. Stillé, *History of the United States Sanitary Commission Being the General Report of Its Work during the War of the Rebellion* (Philadelphia: Lippincott, 1866), 38.

2. Major primary sources for the history of the USSC are its papers, its published document collection, and histories of the organization written in the 1860s. See USSC Microfilm Papers; *Documents of the U.S. Sanitary Commission*, 2 vols. (New York: U.S. Sanitary Commission, 1866); U.S. Sanitary Commission, *The Sanitary Commission of the United States Army: A Succinct Narrative of Its Works and Purposes* (New York: U.S. Sanitary Commission, 1864). The best-known secondary sources on the Sanitary Commission are George M. Fredrickson, *The Inner Civil War: Northern Intellectuals and the Crisis of the*

Union (New York: Harper and Row, 1965); and William Quentin Maxwell, *Lincoln's Fifth Wheel: The Political History of the U.S. Sanitary Commission* (New York: Longmans, Green, 1956). See also Suellen Hoy, *Chasing Dirt: The American Pursuit of Cleanliness* (New York: Oxford University Press, 1995), 29–62.

3. Frederickson, *Inner Civil War; Documents of the U.S. Sanitary Commission,* docs. 1–4.

4. Mary C. Gillett, *The Army Medical Department, 1818–1865* (Washington, DC: Center of Military History, U.S. Army, 1987), 151–76.

5. Doris Kearns Goodwin, *Team of Rivals: The Political Genius of Abraham Lincoln* (New York: Simon and Schuster, 2005), 403.

6. Frederick Law Olmsted to James Murray Forbes, Esq., Dec. 15, 1861, in *The Papers of Frederick Law Olmsted,* vol. 4, *Defending the Union: The Civil War and the U.S. Sanitary Commission, 1861–1863,* ed. Jane Turner Censer (Baltimore: Johns Hopkins University Press, 1986), 240.

7. Censer, *Papers of Olmsted,* introduction, 17; Stillé, *History of the Sanitary Commission,* 130–31.

8. Frederick Law Olmsted to John Olmsted, April 19, 1862, in Censer, *Papers of Olmsted,* 310.

9. Historian Frank Freemon has Stanton opposing Hammond's appointment from the outset, and he sees McClellan's influence as the reason Lincoln chose Hammond. If even Olmsted did not know the reasons for the sudden breakthrough, it seems unlikely that historians will be able to settle the question at this late date. See Frank R. Freemon, "Lincoln Finds a Surgeon General: William A. Hammond and the Transformation of the Union Army Medical Bureau," *Civil War History* 33 (1987): 5–21.

10. "An Act to Reorganize and Increase the Efficiency of the Medical Department of the Army," *Statutes at Large,* 37th Cong., 2nd sess., chap. 50, 378–79, quotation on 379. Facsimile reproduction in *A Century of Lawmaking for a New Nation: U.S. Congressional Documents and Debates, 1774–1875,* http://memory.loc.gov/cgi-bin/ampage (accessed June 18, 2008).

11. Bonnie Ellen Blustein, "'To Increase the Efficiency of the Medical Department': A New Approach to U.S. Civil War Medicine," *Civil War History* 33 (1987): 22–41; Bonnie Ellen Blustein, *Preserve Your Love for Science: Life of William A. Hammond, American Neurologist* (Cambridge: Cambridge University Press, 1991).

12. "To the Presidents and Officers of the Various Life Insurance Companies of the United States," in *Documents of the U.S. Sanitary Commission,* doc. 5, vol. 1.

13. "Statement of the Object and Methods of the Sanitary Commission," 1863, in *Documents of the U.S. Sanitary Commission,* doc. 69, 9, vol. 2.

14. Katharine P. Wormeley, *The United States Sanitary Commission: A Sketch of Its Purpose and Its Work* (Boston: Little, Brown, 1863), 1.

15. Women were less able to preserve the domestic traditions surrounding dying and death in the maelstrom of war. See Drew Gilpin Faust, *This Republic of Suffering: Death and the American Civil War* (New York: Knopf, 2008).

16. Henry W. Bellows, "Sanitary Commission," a printed letter dated Oct. 22, 1862, to be circulated to USSC-affiliated women's aid societies, printed as doc. 54, vol. 1, in *Documents of the U.S. Sanitary Commission.*

17. See Virginia Mescher, "Lint and Charpie: It's Not Your Dryer Lint," at Ragged Sol-

dier Sutlery and Vintage Volumes, www.raggedsoldier.com/lint.pdf (accessed June 20, 2012). This article describes the lint scraping process, its use, and the development of machine-made lint during the war.

18. Stillé, *History of the Sanitary Commission*, 166–96; Mary A. Livermore, *My Story of the War: A Woman's Narrative of Four Years Personal Experience* (Hartford, CT: A. D. Worthington, 1889), 112–13.

19. Sarah Jane Full Hill of St. Louis described making bedticks for the army in a diary entry for Dec. 3, 1861, reproduced in Robert E. Denney, *Distaff Civil War* (Victoria, BC: Trafford Press, 2001), 81. The USSC *Bulletin* frequently listed bedticks and other items donated to hospitals.

20. Livermore, *My Story*, 181–82.

21. Ibid., 182.

22. H. Lyman to E. Harris, Oct. 11, 1865, USSC Microfilm Papers, reel 3, frame 1027–28.

23. Stillé, *History of the Sanitary Commission*, 505–6.

24. General Orders, No. 226, July 8, 1864, in *OR*, series 3, vol. 2, p. 481.

25. Bell Irwin Wiley, *The Life of Billy Yank: The Common Soldier of the Union* (Indianapolis: Bobbs-Merrill, 1951), 224–46.

26. Ibid., 230.

27. John Strong Newberry, *The U.S. Sanitary Commission in the Valley of the Mississippi, during the War of the Rebellion, 1861–1866* (Cleveland: Fairbanks, Benedict, 1871), 222.

28. Ibid., 224.

29. J. S. Newberry, "Report on the Operations of the U.S. Sanitary Commission in the Valley of the Mississippi Made September 1st, 1863," in *Documents of the U.S. Sanitary Commission*, doc. 75, 6, vol. 2.

30. Stillé, *History of the Sanitary Commission*, 342–66. See also E. B. Wolcott to Edward Salomon, Nov. 5, 1863, USSC Microfilm Papers, reel 4, frames 894–98.

31. Henry Bellows, "Sanitary Commission."

32. Livermore, *My Story*, 156.

33. Henry W. Bellows, letter of Sept. 24, 1862, no address, 1, in *Documents of the U.S. Sanitary Commission*, doc. 48, vol. 1.

34. *What Becomes of the Money Raised for the Sanitary Commission?* broadside in Washington Office Scrapbook, ca. June 1864, vol. 1067, box 939, USSC Records, NYPL.

35. Wormeley, *United States Sanitary Commission*, 162.

36. J. S. Newberry, "A Visit to Fort Donelson, Tenn. for the Relief of the Wounded of Feb. 15, 1862," in *Documents of the U.S. Sanitary Commission*, doc. 42, 7, vol. 1.

37. J. S. Newberry, quoted in Wormeley, *United States Sanitary Commission*, 166.

38. Andrew McIlwaine Bell, *Mosquito Soldiers: Malaria, Yellow Fever, and the Course of the American Civil War* (Baton Rouge: Louisiana State University Press, 2010); Kathryn S. Meier, "'No Place for the Sick': Nature's War on Civil War Soldier Mental and Physical Health in the 1862 Peninsula and Shenandoah Valley Campaigns," *Journal of the Civil War Era* 1 (2011): 176–206.

39. U.S. Sanitary Commission, *Hospital Transports: A Memoir of the Embarkation of the Sick and Wounded from the Peninsula of Virginia in the Summer of 1862* (Boston: Ticknor and Fields, 1863), 106.

40. Ibid., 100.

41. Ibid.; Wormeley, *United States Sanitary Commission;* Nancy Scripture Garrison, *With Courage and Delicacy: Civil War on the Peninsula: Women and the U.S. Sanitary Commission* (Mason City, IA: Savas, 1999).

42. John S. Haller, *Farmcarts to Fords: A History of the Military Ambulance, 1790–1925* (Carbondale: Southern Illinois University Press, 1992), 33–39.

43. James M. McPherson, *Crossroads of Freedom: Antietam: The Battle That Changed the Course of the Civil War* (New York: Oxford University Press, 2002).

44. W. M. Chamberlain, "Report of W. M. Chamberlain, M.D., Inspector," [late Sept. 1862], in *Documents of the U.S. Sanitary Commission,* doc. 48, 1–8, vol. 1.

45. H. A. Dubois to Louis Steiner, Sept. 19, 1862, in June–Dec. 1862 folder, Lewis Henry Steiner letters, NYPL. See also the folder of telegrams for 1862 in this box. Dubois's writing is hasty and obscure; the town name might also be Brinkettville, but none by that name survives on the modern map, and there is a Burkittsville, MD, near the Antietam battlefield.

46. C. R. Agnew, "Letter from Dr. Agnew," Sept. 22, 1862, in *Documents of the U.S. Sanitary Commission,* doc. 48, 9–14, vol. 1, quotations on 11 and 12 (emphasis in original).

47. Ibid., 12.

48. For more on the battle of Perryville, see James McPherson, *Battle Cry of Freedom: The Civil War Era* (New York: Oxford University Press, 1988), 518–22.

49. Newberry, *Sanitary Commission in the Valley of the Mississippi,* 52–66.

50. A. N. Read, "Report of Dr. A. N. Read, Oct. 23, 1862," in J. C. Newberry, "Operations of the Sanitary Commission at Perryville, Ky.," Oct. 24, 1862, in *Documents of the U.S. Sanitary Commission,* doc. 55, 7–18, vol. 1, quotations on 8, 10, 11.

51. Ibid., 9, 11, quotations on 9.

52. Frederick Law Olmsted, "Preliminary Report of the Operations of the Sanitary Commission with the Army of the Potomac, during the Campaign of June and July, 1863," in *Documents of the U.S. Sanitary Commission,* doc. 68, 1, vol. 2. Aquia is also spelled "Acquia" in some Civil War sources; "Aquia" is the modern spelling.

53. Ibid., 4. For more on the Battle of Gettysburg, see Gary Gallagher, ed., *Three Days at Gettysburg: Essays on Confederate and Union Leadership* (Kent, OH: Kent State University Press, 1999).

54. Olmsted, "Preliminary Report," 3–4.

55. Henrietta Stratton Jaqette, ed., *Letters of a Civil War Nurse: Cornelia Hancock, 1863–1865* (Lincoln: University of Nebraska Press, 1998), 6.

56. Olmsted, "Preliminary Report."

57. Livermore, *My Life,* 533–34. See also Adele Deleeuw, *Civil War Nurse: Mary Ann Bickerdyke* (New York: Simon and Schuster, 1973).

58. Margaret Humphreys, *Intensely Human: The Health of the Black Soldier in the American Civil War* (Baltimore: Johns Hopkins University Press, 2008), 119–41.

59. Wilson G. Smillie, *Public Health: Its Promise for the Future: A Chronicle of the Development of Public Health in the United States, 1607–1914* (New York: Macmillan, 1955), 280.

60. These pamphlets were combined into a single volume as William A. Hammond,

ed., *Military, Medical, and Surgical Essays Prepared for the United States Sanitary Commission* (Philadelphia: Lippincott, 1864).

61. Frederick Law Olmsted, "A Report to the Secretary of War of the Operations of the Sanitary Commission and upon the Sanitary Condition of the Volunteer Army, Its Medical Staff, Hospitals, and Hospital Supplies, December 1861," in *Documents of the U.S. Sanitary Commission,* doc. 40, 6, vol. 1; "Statement of the Object and Methods," ibid., doc. 69, 18–19, vol. 2.

62. W. H. Van Buren, *Rules for Preserving the Health of the Soldier* (Washington, DC: U.S. Sanitary Commission, 1861), quotations on 4, 6–7.

63. Stillé, *History of the Sanitary Commission,* 53.

64. "Sanitary Commission," in *Documents of the U.S. Sanitary Commission,* doc. 9 [summer 1861], vol. 1; "Camp Inspection Return," ibid., doc. 19a.

65. Thousands of USSC inspection reports can be found in USSC Microfilm Papers, reels 15–28.

66. Unnamed officer quoted in Wormeley, *United States Sanitary Commission,* 289.

67. Frederick Law Olmsted, "General Instructions to Sanitary Inspectors" [summer 1861], in *Documents of the U.S. Sanitary Commission,* doc. 24.2 (the second doc. 24 in this source) [summer 1861], vol. 1, p. 2.

68. Frederick N. Knapp, "Two Reports concerning the Aid and Comfort Given by the Sanitary Commission to Sick Soldiers Passing through Washington," Sept. 23, 1861, in *Documents of the U.S. Sanitary Commission,* doc. 35, vol. 1, quotation on 1–2.

69. Stillé, *History of the Sanitary Commission.*

70. "Statement of the Object and Methods," 4–5.

71. Ibid., 56–57.

72. USSC, *Sanitary Commission,* 256–57.

Chapter 5 · The Sanitary Commission and Its Critics

1. "Report of a Preliminary survey of the Camps of a Portion of the Volunteer Forces near Washington," July 9, 1861, in *Documents of the U.S. Sanitary Commission,* 2 vols. (New York: USSC, 1866), doc. 17, vol. 1, quotations on 4 and 8.

2. Frederick Law Olmsted, "A Report to the Secretary of War of the Operations of the Sanitary Commission and upon the Sanitary Condition of the Volunteer Army, Its Medical Staff, Hospitals, and Hospital Supplies, December, 1861," in *Documents of the United States Sanitary Commission,* doc. 40, 20, vol. 1.

3. U.S. Sanitary Commission, *Hospital Transports: A Memoir of the Embarkation of the Sick and Wounded from the Peninsula of Virginia in the Summer of 1862* (Boston: Ticknor and Fields, 1863), 80.

4. Mary Livermore, *My Story of the War* (Hartford, CT: A. D. Worthington, 1889), 53.

5. A. N. Read to Dear Doctor [Elisha Harris,] Dec. 26, 1865, USSC Microfilm Papers, reel 7, frame 347.

6. USSC, *Hospital Transports,* 120.

7. Walter I. Trattner, *From Poor Law to Welfare State: A History of Social Welfare in Amer-*

ica (1974; 6th ed., New York: Free Press, 1999); Charles E. Rosenberg, *The Care of Strangers: The Rise of America's Hospital System* (New York: Basic Books, 1987).

8. "General Instructions to Sanitary Inspectors," n.d. [ca. summer 1861], in *Documents of the United States Sanitary Commission*, doc. 24, 5, vol. 1.

9. Henry W. Bellows, Sept. 24, 1862, untitled circular letter to women's aid societies, in *Documents of the United States Sanitary Commission*, doc. 48, 2, vol. 1.

10. Livermore, *My Story*, 139, 141.

11. "Statement of the Object and Methods of the Sanitary Commission," 1863, in *Documents of the United States Sanitary Commission*, doc. 69, 25, vol. 2; Alfred J. Bloor, "Letters from the Army of the Potomac Written during the Month of May, 1864, to Several of the Supply Correspondents of the U.S. Sanitary Commission," in *Documents of the United States Sanitary Commission*, doc. 80, 32, 64, vol. 1.

12. Henry W. Bellows to Stephen G. Perkins, Esq., Aug. 15, 1862, in *Documents of the United States Sanitary Commission*, doc. 49, 1–2, vol. 1.

13. Robert Bremner, *The Public Good: Philanthropy and Welfare in the Civil War Era* (New York: Knopf, 1980), 144–78.

14. Ibid., 54.

15. Livermore, *My Story*, 156.

16. Jeanie Attie, *Patriotic Toil: Northern Women and the American Civil War* (Ithaca, NY: Cornell University Press, 1998); Judith Ann Giesberg, *Civil War Sisterhood: The U.S. Sanitary Commission and Women's Politics in Transition* (Boston: Northeastern University Press, 2000).

17. "Statement of the Object and Methods," 7.

18. Robert H. Bremner, *American Philanthropy* (Chicago: University of Chicago Press, 1960), 72–84.

19. A. N. Read, Report of Oct. 23, in "Operations of the Sanitary Commission at Perryville, Ky.," in *Documents of the United States Sanitary Commission*, doc. 55, 15, vol. 1; Granville Moody (colonel commanding the Seventy-Fourth Regiment of the Ohio Volunteer Infantry) to A. N. Read, Feb. 5, 1863, USSC Microfilm Papers, reel 11, frame 171.

20. William Hannaman, Report of the Indiana State Sanitary Commission, in "Message of the Governor of Indiana, to the General Assembly, Jan. 6, 1865," USSC Microfilm Papers, reel 13, frame 949; Jane Turner Censer, ed., *The Papers of Frederick Law Olmsted*, vol. 4, *Defending the Union: The Civil War and the U.S. Sanitary Commission, 1861–1863* (Baltimore: Johns Hopkins University Press, 1986), introduction, 14.

21. "Statement of the Object and Methods," 6.

22. Robert Patrick Bender, "Old Boss Devil: Sectionalism, Charity, and the Rivalry between the Western Sanitary Commission and the United States Sanitary Commission during the Civil War" (PhD diss., University of Arkansas at Fayetteville, 2001).

23. See, for example, "The Western Sanitary Commission," an advertisement in the *New York Times*, March 2, 1862 (listed in the weeks after the battles of Donelson and Shiloh).

24. Bender, "Old Boss Devil," 22; the mixed tenses ("seems" and "emphasized") occur in the original.

25. Ibid.

26. Ibid.

27. Livermore, *My Story*, 410–37; William Y. Thompson, "Sanitary Fairs and the Civil War," *Civil War History* 4 (1958): 51–67; Bender, "Old Boss Devil"; *The Western Sanitary Commission; A Sketch of Its Origin, History, Labors for the Sick and Wounded of the Western Armies, and Aid Given to Freedmen and Union Refugees, with Incidents of Hospital Life* (St. Louis: R. P. Studley, 1864); Attie, *Patriotic Toil*, 210; Giesberg, *Civil War Sisterhood*, 106; Alvin Robert Kantor and Marjorie Sered Kantor, *Sanitary Fairs: A Philatelic and Historical Study of Civil War Benevolences* (Glencoe, IL: SF, 1992).

28. Censer, *Papers of Olmsted*, introduction, 17.

29. "The Sanitary Commission," *New York World*, Nov. 8, 1861, 4. Given the style and themes of the article, it is likely that Frederick Law Olmsted had a major role in its production.

30. "The Medical Bureau," *New York World*, Nov. 16, 1861, 4. "Medical Bureau" here refers to the Medical Department of the U.S. Army.

31. See, for example, "New-York Sanitary Association: Bellows on the Health of the Army," *New York Times*, Nov. 16, 1861, 2.

32. "The Medical Bureau of the Army," *New York Times*, Nov. 25, 1861, 4. The correct name is "Medical Department of the U.S. Army."

33. *New York World*, Nov. 27, 1861, 4.

34. "Camp Inspection and the Sanitary Commission," *New York Times*, Dec. 4, 1861, 5.

35. TRUTH, "The Sanitary Commission—What It Has and What It Has Not Done," *New York Times*, Dec. 4, 1861.

36. TRUTH, "The Sanitary Commission and the Hospitals in Washington," *New York Times*, Dec. 6, 1861. A similar letter signed "Justitia" appeared in the *National Republican* ca. Dec. 1861–Jan. 1862. Undated newspaper clipping, which may have also been by Powell, is found in Washington Office Scrapbook, vol. 1067, box 939, United States Sanitary Commission Records, NYPL.

37. Fred. Law Olmsted, "The Sanitary Commission—Note from the Secretary," *New York Times*, Dec. 7, 1861, 3.

38. "Report of the Sanitary Commission," *New York Times*, Dec. 17, 1861, 4.

39. Frederick Law Olmsted to Henry Whitney Bellows, Dec. 20, 1861, in Censer, *Papers of Olmsted*, 242.

40. "What Has Been Done for the Health of the Army," *New York Times*, Jan. 9, 1862, 4. The report is on p. 3 of the same issue.

41. "Medical Humbug," *New York Times*, Feb. 2, 1862, 1; "General News," *New York Times*, April 23, 1862, 4.

42. Frederick Law Olmsted to Henry W. Bellows, Jan. 11, 1862, in Censer, *Papers of Olmsted*, 244–45.

43. Del [Elizabeth M. Powell Delescluze] to Adolph Ochs, Feb. 16, 1901, in Correspondence, 1901 folder, box 2, New York Times Company Records, George Jones and Henry J. Raymond Papers, NYPL. Powell married Mr. Delescluze in 1867, according to this correspondence. This and subsequent letters to Ochs were written concerning an essay that Delescluze was writing on the biography of the young Henry Raymond. An undated, unaddressed letter in the folder provides the fact that Raymond wrote her 150 letters. She

returned these to Raymond in 1868, except for a few she kept back for their sentimental value. She then sent these to Ochs in 1901, but they are not in this archive. A letter dated Aug. 25, 1901 says that she is near death.

44. Del to Ochs, Feb. 16, 1901.

45. George Templeton Strong to Frederick Law Olmsted, May 18, 1862, in Censer, *Papers of Olmsted*, 245.

46. William Quentin Maxwell, *Lincoln's Fifth Wheel: The Political History of the U.S. Sanitary Commission* (New York: Longmans, Green, 1956), 139.

47. Henry J. Raymond, *History of the Administration of President Lincoln* (New York: J. C. Derby and N. C. Miller, 1864).

48. Francis Brown, *Raymond of the Times* (New York: Norton, 1951), described Raymond's estrangement from his wife, his reputation for philandering, and his death. This volume does not mention Powell by name.

49. Maxwell, *Lincoln's Fifth Wheel*, 140–43.

50. Ibid., 171, 195.

51. J. S. Newberry, "Operations of the Sanitary Commission at Perryville, Ky.," Oct. 24, 1862, in *Documents of the United States Sanitary Commission*, doc. 55, 3, vol. 1.

52. Maxwell, *Lincoln's Fifth Wheel*, 199–200. In a letter to Henry Bowditch in fall 1862, Montgomery Meigs, quartermaster general of the United States, asked him to read the rhetoric of the USSC with skepticism, telling him that its leaders built their case for public sympathy and donations by downplaying the army's care and trumpeting its own. He particularly defended the keeping of transport and supply in the hands of the quartermaster's office. Montgomery Meigs to Henry I. Bowditch, Oct. 20, 1862, John Samuel Blatchford Letter Book, 1865–1866, b MS Am 1084 (1246) MOLLUS, Houghton Library, Harvard University, Cambridge, MA.

53. Bonnie Blustein, *Preserve Your Love for Science: Life of William A. Hammond, American Neurologist* (Cambridge: Cambridge University Press, 1991), 84; John Harley Warner, *The Therapeutic Perspective: Medical Practice, Knowledge, and Identity in America, 1820–1885* (Cambridge, MA: Harvard University Press, 1986), 221–24; Michael A. Flannery, *Civil War Pharmacy: A History of Drugs, Drug Supply and Provision, and Therapeutics for the Union and Confederacy* (New York: Haworth Press, 2004), 150–55.

54. Maxwell, *Lincoln's Fifth Wheel*, 234–47; Blustein, *Preserve Your Love for Science*, 86–93.

55. Joseph K. Barnes to E. M. Stanton, Oct. 20, 1864, in *OR*, series 3, 4:790–93; *MSHW*.

56. "The Work," in USSC Microfilm Papers, reel 2, frame 546. Many of the documents and letters of the first eight reels of the USSC Microfilm Papers concern this project.

57. Austin Flint, ed., *Contributions Relating to the Causation and Prevention of Disease and to Camp Diseases; Together with a Report of the Diseases, etc., among the Prisoners at Andersonville, GA* (New York: Hurd and Houghton, 1867); Benjamin Apthorp Gould, *Investigations in the Military and Anthropological Statistics of American Soldiers* (New York: Hurd and Houghton, 1869); Frank Hastings Hamilton, ed., *Surgical Memoirs of the War of the Rebellion* (New York: Hurd and Houghton, 1871).

58. Elisha Harris to Cornelius Agnew, June 23, 1865, USSC Microfilm Papers, reel 2, frame 79.

59. Harris to Agnew, July 22, 1865, reel 1, frame 932; Harris to Standing Committee of the U.S. Sanitary Commission, July 22, 1865, reel 1, frame 936, all in USSC Microfilm Papers. On the outbreak of scurvy in Texas among black troops, see Margaret Humphreys, *Intensely Human: The Health of the Black Soldier in the Civil War* (Baltimore: Johns Hopkins University Press, 2008), 119–41.

60. Trattner, *From Poor Law to Welfare State*, 79.

61. Bremner, *American Philanthropy;* and Bremner, *Public Good.*

62. Allan Nevins, preface to Maxwell, *Lincoln's Fifth Wheel*, viii.

63. Maxwell, *Lincoln's Fifth Wheel.*

64. Attie, *Patriotic Toil*, 210; Giesberg, *Civil War Sisterhood.*

65. George M. Frederickson, *The Inner Civil War: Northern Intellectuals and the Crisis of the Union* (New York: Harper and Row, 1965), quotations on 99, 111, 112.

66. J. Matthew Gallman, *Mastering Wartime: A Social History of Philadelphia during the Civil War* (Cambridge: Cambridge University Press, 1990), 146–69.

67. Multiple letters to Lewis Steiner concerning agents at Brandy Station and their rowdy behavior, Jan. 1864 folder, Lewis Henry Steiner letters, NYPL.

68. Livermore, *My Story*, 164.

69. Robert H. Wiebe, *The Search for Order, 1870–1920* (New York: Hill and Wang, 1967).

70. Lorien Foote, *The Gentlemen and the Roughs: Manhood, Honor, and Violence in the Union Army* (New York: New York University Press, 2010).

Chapter 6 · *The Union's General Hospital*

1. Robert W. Rydell, John E. Findling, and Kimberly D. Pelle, *Fair America: World's Fairs in the United States* (Washington, DC: Smithsonian Institution Press, 2000); and Linda P. Gross and Theresa R. Snyder, *Philadelphia's 1876 Exhibition* (Charleston, SC: Arcadia, 2005).

2. Bert Hansen, *Picturing Medical Progress from Pasteur to Polio: A History of Mass Media Images and Popular Attitudes in America* (New Brunswick, NJ: Rutgers University Press, 2009), 32. See also Julie K. Brown, *Health and Medicine on Display: International Expositions in the United States, 1876–1904* (Cambridge, MA: MIT Press, 2009), 11–41. The display included papier-mâché dummies rigged up in traction to show the treatment of fractures, photographs of amputations, a case of medical and surgical curiosities, displays of surgical and medical supplies, cases of artificial limbs, and even a medical mess room complete with silverware. In addition, hospital tents were set up near the post hospital building. James D. McCabe, *The Illustrated History of the Centennial Exhibition* (Cincinnati: Jones Brothers, 1876), 649–51.

3. "The Satterlee Hospital," *Philadelphia Inquirer*, June 2, 1863, 1. For simplicity's sake the hospital is referred to as Satterlee throughout this chapter.

4. "The General Hospitals," *MSHW*, 6:896–966. The report on Satterlee is on 926–30; Mower is on 932.

5. Nathaniel West, *Sketch of the General Hospital, U.S. Army, at West Philadelphia* (Philadelphia: Ringwalt and Brown, 1862); Nathaniel West, *History of Satterlee U.S.A. Gen. Hos-*

pital, at West Philadelphia, PA., from October 8, 1862, to October 8, 1863 (Philadelphia: Hospital Press, 1863); *West Philadelphia Hospital Register 1–3* (1863–65).

6. West, *Sketch*, 4.

7. "The U.S. Army General Hospital, West Philadelphia," *Hospital Register* 1, no. 13 (May 19, 1863): 49–50.

8. "Monthly Report of the Post Office," *Hospital Register* 3, no. 2 (Sept. 3, 1864): 6.

9. William Alexander Hammond, *A Treatise on Hygiene: With Special Reference to the Military Service* (Philadelphia: Lippincott, 1863), 299–300. Kenneth J. Carpenter, in *The History of Scurvy and Vitamin C* (Cambridge: Cambridge University Press, 1986), 230–33, describes the amount of vitamin C available in raw meat and fish, the use of this food by Eskimos, and the discussion and use of this food by polar explorers.

10. Douglas W. Wamsley, *Polar Hayes: The Life and Contributions of Isaac Israel Hayes, M.D.* (Philadelphia: American Philosophical Society, 2009), 303.

11. Thomas Low, Diary 1861–1863, Thomas Low Papers, Duke SC.

12. Wamsley, *Polar Hayes*, 311–16, quotation on 316.

13. Hammond, *Treatise on Hygiene*, 364.

14. Ibid., 367–68.

15. Dr. Samuel Gross described the drop in class size during the war at Jefferson Medical School, the other large school in Philadelphia, as "under three hundred." *Autobiography of Samuel D. Gross, M.D.* (Philadelphia: George Barrie, 1887), 1:128. When Gross arrived in Philadelphia to take a post at Jefferson in 1857, there were a total of 1,139 students attending the four medical schools in Philadelphia (1:131).

16. Wamsley, *Polar Hayes*, 311–40; Frank H. Taylor, *Philadelphia in the Civil War, 1861–1865* (Philadelphia: Published by the City, 1913), 232.

17. Autopsy reports on cases of chronic dysentery from Satterlee Hospital, *MSHW*, 3:109–23. See also the autopsy record books in entry 621, Record Group 94, NARA.

18. Wamsley, *Polar Hayes*, 317.

19. Comments of unnamed surgeon at Satterlee Hospital, in "Remittent Fever with Cerebral Symptoms," *MSHW*, 5:115.

20. Article on pneumonia, *MSHW*, 6:806.

21. Story of Hiram Straight, 154th New York Infantry Regiment, as told in Mark H. Dunkelman, *Brothers One and All: Esprit de Corps in a Civil War Regiment* (Baton Rouge: Louisiana State University Press, 2004), 116–17, 134.

22. Many cases from Satterlee are reported in the *MSHW*. Quinine was often used for remittent fever, as one would expect. This diagnosis included many cases of malaria as well as typhoid fever. See, for example, a case report from Satterlee Hospital, "Remittent Following Typhoid," *MSHW*, 5:116; "Continued Fever Quickly Changing to Remittent and Intermittent," *MSHW*, 5:116–17; "Malarial Congestions," *MSHW*, 5:117–18; "Malarial Rheumatism," *MSHW*, 5:118; "Malarial Neuralgia, Debility, and Oedema," *MSHW*, 5:118–19.

23. C. R. H., "Cal. O'Melas," *Hospital Register* 2 (March 19, 1864): 119, originally published in Benjamin Colby, *A Guide to Health, Being an Exposition of the Principles of the Thomsonian System of Practice, and Their Mode of Application in the Cure of Every Form of Disease* (Milford, NH: John Burns, 1846); "Homeopathic Soup," *Hospital Register* 2 (May

14, 1864): 149; see slightly different versions in *Harper's Weekly,* Dec. 26, 1863; and *Punch,* Nov. 28, 1863.

24. Mary Denis Maher, *To Bind up the Wounds: Catholic Sister Nurses in the U.S. Civil War* (1989; paperback ed., Baton Rouge: Louisiana State University Press, 1999), 2, 15. Elizabeth Seton was canonized as a Roman Catholic saint in 1975.

25. Ibid., 26–43.

26. Sarah Trainer Smith, ed., "Notes on Satterlee Military Hospital, West Philadelphia, Penna., from 1862 until Its Close in 1865. From the Journal Kept at the Hospital by a Sister of Charity," *Records of the American Catholic Historical Society of Philadelphia* 8 (1897): 408; Maher, *To Bind up the Wounds,* 69. While there were other nursing orders in the United States called the Sisters of Charity, all of the nuns working at Satterlee were part of the order based in Emmitsburg.

27. I. I. Hayes, "Satterlee Hospital, West Philadelphia," Oct. 31, 1862, *MSHW,* 6:926–30, quotation on 926. The editor of this volume labeled this report "Satterlee Hospital," but when Hayes wrote it in fall 1862, the hospital had not yet been renamed.

28. Taylor, *Philadelphia in the Civil War,* 233.

29. Smith, "Notes on Satterlee," 407.

30. Ibid., 407–8.

31. Ibid., 422.

32. Ibid., 441, 404.

33. Nancy Lusignan Schultz, *Fire and Roses: The Burning of the Charleston Convent, 1834* (New York: Simon and Schuster, 2001).

34. Maher, *To Bind up the Wounds,* 13.

35. Mack Ewing to Nan Ewing, Feb. 12, 19, 1865, Civil War Letters of Mack and Nan Ewing, Henry McKendree Ewing Collection, Archives of Michigan, Michigan Library and Historical Center, Lansing, MI (hereafter Ewing Letters).

36. Maher, *To Bind up the Wounds;* Constant C. Hanks to Mother, April 14, 1862, Constant C. Hanks Papers, 1861–65, Hunter, NY, Duke SC. The grave of Hanks behind a Methodist Episcopal Church is listed at Hunter Village Cemetery, www.rootsweb.ancestry .com/~nygrecn2/hunter_village_cemetery.htm (accessed Oct. 14, 2009).

37. Unnamed Sister of Charity, quoted in Wamsley, *Polar Hayes,* 328; J. E. MacLane, letter to *Philadelphia Evening Star,* unknown date (ca. 1897), reprinted in George Barton, *Angels of the Battlefield: A History of the Labors of the Catholic Sisterhoods in the Late Civil War* (Philadelphia: Catholic Art, 1897), 114. There was a James E. McLane in the 142nd Pennsylvania Infantry Regiment, which participated in the Battle of Gettysburg. It is possible that the letter writer was one of the thousands sent to Satterlee after that battle.

38. Douglas Wamsley, in his biography of Hayes, *Polar Hayes,* described his search for the Satterlee Hospital record books in the National Archives. Such hospital records, if they exist, are no longer cataloged in an accessible way at NARA.

39. West, *History of Satterlee,* 34. Smith, "Notes on Satterlee," 408, gives the "greater than 1500" number for the second Bull Run battle—but also claims that 6,000 came after Gettysburg, which is 2,000 more than other sources claim. And the hospital could hold only 4,500 when crammed to its fullest.

40. Smith, "Notes on Satterlee," 401.

41. Richard A. F. Penrose, "Remarks on the Treatment of Chronic Dysentery," *MSHW*, 3:44–47, quotations on 44–45. Penrose also commented that many of the men were skillful malingerers.

42. Case reports in the section of postmortem reports of the continued fevers, *MSHW*, 5:399–400.

43. *MSHW*, 3:12. Satterlee's outcome data compared favorably with those of most other hospitals listed. This particular listing includes statistics only for diarrhea and dysentery cases.

44. See, for example, Horace N. Snow to his parents, Nov. 10, 1862, Snow Family Papers, Duke SC; Edward W. Benham to dear Jennie, Sept. 18, 1864, Edward W. Benham Papers, 1864–1964, Duke SC; W. S. Shockley to Eliza, Nov. 23, 1862; Dec. 26, 1863; and Jan. 3, 1864, W. S. Shockley Papers, 1861–1864, Duke SC.

45. E. T. Stetson to dear Mother McKim, May 28, Aug. 29, 1863, Stetson Family Letters, 1861–1865, Duke SC.

46. C. H. T., "A Sergeant of the Mass. Volunteers," *Hospital Register* 2 (Sept. 12, 1863): 10–11, quotation on 11; Cyrus Beamenderfer to Daniel Musser, Sept. 11, 1864, Daniel Musser Collection, MG 95, Pennsylvania State Archives, Harrisburg, PA. On malingering, and tricks to scope it out, see W. W. Keen et al., "On Malingering, Especially in Regards to Simulation of Diseases of the Nervous System," *American Journal of the Medical Sciences* 48 (1864): 367–94.

47. William B. Fullerton to sister Lydia, Feb. 17, March 13, 1863; and to sister Sue, July 4, 1863, William B. Fullerton Civil War Letters, 1862–1865, MS183, Special Collections and Archives, University Library, University of California–Santa Cruz.

48. *Hospital Register* 2 (Aug. 22, 1863): 165–66. The same story was told in Smith, "Notes on Satterlee," 433–34, with the coda that the unfortunate man received baptism at the end after resisting the sister's remonstrations for days.

49. Cyrus W. Beamenderfer to Daniel Musser, May 20, 1864 (from Fredericksburg, VA), June 9, 1864 (from Wolfe St. Hospital, Alexandria, VA), June 15, 1864 (from Alexandria), July 7, 1864 (from Satterlee), Musser Collection. Pa-Roots, 84th Infantry, www .pa-roots.com/pacw/infantry/84th/84thcoa.html includes two C. Beamenderfers who enlisted for three years on the same day in December 1861 (accessed Oct. 19, 2009). So even if Cyrus did not receive a medical discharge, his term of service would have been up a month after the last letter to his friend Daniel Musser dated Oct. 31, 1864. "Beamenderfer" is also spelled "Beamendorfer" in some references. For another Satterlee case that required prolonged care and repeat surgeries, finally ending with amputation of an infected femur at the hip joint, see George A. Otis, "Memorandum of a Case of Re-amputation at the Hip, with Remarks on the Operation," *American Journal of Medical Sciences* 61 (1871): 141–49.

50. Anonymous diary entry, reprinted in Smith, "Notes on Satterlee," 410.

51. "Smallpox," *MSHW*, 6:625–48.

52. Smith, "Notes on Satterlee," 405.

53. *MSHW*, 6:623. All told, there were 12, 236 cases of smallpox among Union troops and 4,717 deaths from it. There is no independent evidence to bolster the sisters' memories of the number of cases and deaths at Satterlee.

54. Dennis L. Stevens, "Necrotizing Fasciitis, Gas Gangrene, Myositis, and Myonecrosis," in *Infectious Diseases,* 2nd ed., ed. Jonathan Cohen and William G. Powderly (Edinburgh: Mosby, 2004), 145–55, esp. 145.

55. *MSHW,* 12:824–25. These numbers do not include Union men who died of gangrene in prisoner-of-war camps.

56. John H. Packard, "On Hospital Gangrene and Its Efficient Treatment," *American Journal of the Medical Sciences* 49 (1865): 114–19, quotations on 115. See also Thomas G. Morton, "A Brief Description of the System of Forced Ventilation Which Has Been Lately Introduced into the Pennsylvania Hospital, with some Surgical Statistics and Remarks on Hospital Construction," *American Journal of Medical Sciences* 73 (1877): 424–31, which mentions the hospital gangrene outbreak in Satterlee and Mower Hospitals.

57. Report of W. W. Keen, in *MSHW,* 12:826–29, quotation on 827. Since this report is not dated, and the relevant volume of the *MSHW* appeared in the 1880s, it is possible that Keen wrote this after learning of Lister's work (1867) and theories of microbial contamination that were not dominant during the Civil War. Keen was referring to a few cases of gangrene that he treated in the winter of 1862–63; there is nothing in the nurses' memoir to support his reported regimen for preventing contagion. But given the hospital's low mortality rates, such as we know them, it is possible that these steps did become common and have an effect.

58. Untitled article on cumulative hospital statistics, *Cartridge Box* 1 (March 19, 1864), no page number.

59. "Deserters from Satterlee Hospital," *Philadelphia Inquirer,* July 6, 1864, 3.

60. E. S., "Reported Deserter," *Hospital Register* 2 (Nov. 21, 1863): 49.

61. Henry Smith, "Concerning the Invalid Corps," *Hospital Register* 2 (Sept. 5, 1863): 8. See also Rusticus, "The Invalid Corps," *Cartridge Box* 1 (March 12, 1864), no pagination.

62. J. H. Moody to Susie [Fullerton?], n.d., Fullerton Letters. Evidence from that collection puts Moody in the Fifth New Hampshire Infantry; this regiment had a John H. Moody as well as a James H. Moody. See Civil War Soldier Sailor System, www.itd.nps.gov/cwss/ (accessed Oct. 2, 2009).

63. "The General Hospitals," *MSHW,* 6:966.

64. C. A. Waite, "Notice," *Crutch* 1 (Jan. 9, 1864), no page numbers.

65. "Mullavell's Bowling Saloon!" *Crutch* 1 (March 5, 1864), no page numbers.

66. "Justice to our Soldiers," *Hospital Register* 3 (Aug. 27, 1864): 2. The *Philadelphia Evening Standard* claims (from Aug. 23, 1864) are quoted in this article.

67. Shakespeare, *The Tragedy of Hamlet, Prince of Denmark,* act 3, scene 2.

68. "The Canvas-Back Chair," *Hospital Register* 2 (Sept. 12, 1863): 10.

69. Ward Z, "A Voice from the Guard House," *Hospital Register* 1 (Aug. 8, 1863): 152.

70. D. E. W., "The Temperance Movement in This Hospital," *Hospital Register* 2 (April 16, 1864): 134.

71. "The Philadelphia Natatorium, on Broad St.," *Hospital Register* 1 (July 11, 1863): 119; "Base Ball," *Hospital Register* 3 (June 3, 1865): 150; Mack Ewing to Nan Ewing, Feb. 19, 1865, Ewing Letters.

72. "Venereal Diseases," *MSHW,* 6:891–96, quotations on 891. See also Thomas P. Lowry's books, which draw extensively on court-martial records: *The Story the Soldiers*

Wouldn't Tell: Sex in the Civil War (Mechanicsburg, PA: Stackpole Books, 1994); and *Sexual Misbehavior in the Civil War: A Compendium* (Bloomington, IN: Xlibris, 2006). Cyrus W. Beamenderfer to Daniel Musser, Oct. 31, 1864, Musser Collection.

73. C. H. T., "A Sergeant of Mass. Volunteers," *Hospital Register* 2 (Sept. 12, 1863): 10–11, quotations on 10; Robert H. Wiebe, *The Search for Order: 1877–1920* (New York: Hill and Wang, 1967).

74. "The Soldier's Mail," *Hospital Register* 2 (July 23, 1864): 190.

75. Ward H., "Lay of a Used Up Patient," *Hospital Register* 1 (April 4, 1863): 32.

76. Wamsley, *Polar Hayes.* Other hospital papers included the *Crutch* (Annapolis, MD), the *Cartridge Box* (York, PA), the *Convalescent* (Jefferson Barracks Hospital, St. Louis), the *Hammond Gazette* (Point Lookout, MD), and the *Haversack* (Annapolis, MD).

77. "Our Hospital," *Hospital Register* 3 (Jan. 14, 1865): 80.

78. "Anniversary of the Reading-Room," *Hospital Register* 2 (Feb. 20, 1864): 102.

79. "Parlor Dramatics—Ingrate," *Hospital Register* 2 (April 30, 1864): 142–43; "Satterlee Literary and Debating Association," *Hospital Register* 2 (April 30, 1864): 143; untitled notice, *Hospital Register* 2 (May 21, 1864): 154. The question about capital punishment was also discussed by the debating society at Hayes's boarding school; perhaps he was the source of the debating topics. Wamsley, *Polar Hayes,* 27.

80. John Harley Warner, "'Exploring the Inner Labyrinths of Creation': Popular Microscopy in Nineteenth-Century America," *Journal of the History of Medicine and Allied Sciences* 37 (1982): 7–33.

81. Untitled article, *Hospital Register* 2 (May 28, 1864): 159.

82. Untitled article, *Hospital Register* 2 (Dec. 17, 1864): 64.

83. "Our School for Soldiers," *Hospital Register* 3 (Jan. 21, 1865): 84.

84. "Re-opening of a Military Hospital," *Hospital Register* 3 (Jan. 14, 1865): 90.

85. Wamsley, *Polar Hayes,* 330.

86. W. W. Keen, "Egg-Shells at Army Hospitals," *Journal of the American Medical Association* 71 (1918): 592.

87. "The General Hospital," *MSHW,* 6:959.

88. References to fruits are frequent in the *Hospital Register,* the *Crutch,* and the *Cartridge Box.* See especially a young Tennessean's wonder at receiving his first orange. "An Orange," *Crutch,* April 9, 1864.

89. "I'm Too Sic to Go Bak to the War," *Hospital Register* 1 (April 11, 1863): 36.

90. "The Delusions of Soldiering," *Hospital Register* 1 (Aug. 1, 1863): 141–42, quotations on 141. James McPherson reviews the reasons soldiers fought in the war in *For Cause and Comrades: Why Men Fought in the Civil War* (New York: Oxford University Press, 1998).

91. Allen Galyean, June 26, 1864, letter, quoted in Alan D. Gaff, *On Many a Bloody Field: Four Years in the Iron Brigade* (Bloomington: Indiana University Press, 1996), 366.

92. Untitled article, *Cartridge Box,* March 19, 1864, unpaginated.

93. "The Satterlee Hospital," *Hospital Register* 2 (July 30, 1864): 194.

94. L. S. I., "From a Soldier's Letter," *Hospital Register* 2 (Dec. 26, 1863): 70.

95. "Ventilation of the West Philadelphia Hospital," *Hospital Register* 1 (March 7, 1863): 16.

96. "Presentation," *Hospital Register* 2 (July 30, 1864): 195; "Presentation," *Hospital Register* 2 (Aug. 13, 1864): 202.

97. "An Acceptable Testimonial," *Philadelphia Inquirer,* July 27, 1865, 8; Auction advertisement, *Philadelphia Inquirer,* Sept. 21, 1865, 7; Taylor, *Philadelphia in the Civil War,* 234.

98. See Philly History.org, www.phillyhistory.org for information on Clark Park (accessed Oct. 20, 2009). The monument to Satterlee Hospital was erected in 1916.

99. James Downs, "Diagnosing Reconstruction: Sickness, Dependency, and the Medical Division of the Freedmen's Bureau, 1861–1870" (PhD diss., Columbia University, 2005).

100. "The General Hospitals," *MSHW,* 6:950–51.

101. Nitin K. Ahuja, "Fordism in the Hospital: Albert Kahn and the Design of Old Main, 1917–25," *Journal of the History of Medicine and Allied Sciences* 67 (2012): 398–427.

Chapter 7 · *Medicine for a New Nation*

1. On the medical history of the Confederacy, see Horace H. Cunningham, *Doctors in Gray: The Confederate Medical Service* (Baton Rouge: Louisiana State University Press, 1958); Glenna R. Schroeder-Lein, *Confederate Hospitals on the Move: Samuel H. Stout and the Army of Tennessee* (Columbia: University of South Carolina Press, 1994); and Nancy Schurr, "Inside the Confederate Hospital: Community and Conflict during the Civil War" (PhD diss., University of Tennessee–Knoxville, 2004). Cunningham's volume provides the most useful overview but is a frustrating book to use, as his quotations are not specifically tied to the bibliography of sources at the end of the book.

2. An original handwritten and signed copy of the Confederate Constitution is held by the Museum of the Confederacy, Richmond, Virginia. It is available online, with comparisons to the U.S. Constitution, at The Constitution of the Confederate States of America, www.usconstitution.net/csa.html (accessed March 2, 2010). In 1858 a statue of George Washington flanked by such Virginia notables as Patrick Henry and Thomas Jefferson was unveiled on the grounds of the Virginia state capitol in Richmond. Virginia was the "mother of presidents," the state that birthed seven of the first twelve presidents. As the southern states seceded, their politicians emphasized the return to the true republic that Lincoln had destroyed and spoke of a new battle for independence, of reclaiming their country. These arguments are well described in classic texts on the Confederacy, including Charles P. Roland, *The Confederacy* (Chicago: University of Chicago Press, 1960); and Emory M. Thomas, *The Confederate Nation, 1861–1865* (New York: Harper and Row, 1979). For a more recent treatment, see Drew Faust, *The Creation of Confederate Nationalism: Ideology and Identity in the Civil War South* (Baton Rouge: Louisiana State University Press, 1988), 14–15.

3. John Harley Warner, *Against the Spirit of System: The French Impulse in Nineteenth-Century American Medicine* (Princeton, NJ: Princeton University Press, 1998); Erwin Ackerknecht, *Medicine at the Paris Hospital, 1794–1848* (Baltimore: Johns Hopkins Press, 1967); and Jacalyn Duffin, *To See with a Better Eye: A Life of R. T. H. Laennec* (Princeton, NJ: Princeton University Press, 1998).

4. Michael Sappol, *A Traffic of Dead Bodies: Anatomy and Embedded Social Identity in*

Nineteenth-Century America (Princeton, NJ: Princeton University Press, 2002); and Robert L. Blakely and Judith M. Harrington, eds., *Bones in the Basement: Postmortem Racism in Nineteenth Century Medical Training* (Washington, DC: Smithsonian Institution Press, 1998).

5. Margaret Humphreys, "Public Health in the Old South," in *Science and Medicine in the Old South,* ed. Ronald L. Numbers and Todd L. Savitt (Baton Rouge: Louisiana State University Press, 1989), 226–55.

6. William G. Rothstein, *American Physicians in the Nineteenth Century: From Sects to Science* (Baltimore: Johns Hopkins University Press, 1972); Joseph F. Kett, *The Formation of the American Medical Profession: The Role of Institutions, 1780–1860* (New Haven, CT: Yale University Press, 1968); and Thomas Neville Bonner, *Becoming a Physician: Medical Education in Britain, France, Germany, and the United States, 1750–1945* (New York: Oxford University Press, 1995).

7. Chad Morgan, *Planter's Progress: Modernizing Confederate Georgia* (Gainesville: University Press of Florida, 2005).

8. Cunningham, *Doctors in Gray;* Ira Rutkow, "Biographical Introduction" to Samuel Preston Moore, *Regulations for the Amy of the Confederate States* (1862; repr., San Francisco: Norman, 1992), v–xiv; Carol C. Green, *Chimborazo: The Confederacy's Largest Hospital* (Knoxville: University of Tennessee Press, 2004), 1–5.

9. John Harley Warner, *The Therapeutic Perspective: Medical Practice, Knowledge, and Identity in America, 1820–1885* (Cambridge, MA: Harvard University Press, 1986).

10. John Harley Warner, "The Idea of Southern Medical Distinctiveness: Medical Knowledge and Practice in the Old South"; and "A Southern Medical Reform: The Meaning of the Antebellum Argument for Southern Medical Education," both in Numbers and Savitt, *Science and Medicine,* 179–225.

11. Margaret Humphreys, *Yellow Fever and the South* (New Brunswick, NJ: Rutgers University Press, 1992); and Humphreys, *Malaria: Poverty, Race, and Public Health in the United States* (Baltimore: Johns Hopkins University Press, 2001).

12. Cunningham, *Doctors in Gray,* 10.

13. James H. Cassedy, "Medical Men and the Ecology of the Old South," in Numbers and Savitt, *Science and Medicine,* 166–78.

14. Gary W. Gallagher reviews the literature on these questions in "Disaffection, Persistence, and Nation: Some Directions in Recent Scholarship on the Confederacy," *Civil War History* 55 (2009): 329–53; and Gallagher, *The Confederate War* (Cambridge, MA: Harvard University Press, 1997). See particularly Faust, *Creation of Confederate Nationalism.* As Faust notes, "Scholars have continued to fear that accepting the reality of Confederate nationalism would somehow imply its legitimacy" (3). The fact that medical nationalists within the Confederate States of America (CSA) promoted medical reforms that modern-day scholars might applaud, such as increased clinical training, drug research, and familiarity with the autopsied body, should not be confused with approval of the overarching goals of the Confederacy, secession and the preservation of slavery.

15. Steven M. Stowe, *Doctoring the South: Southern Physicians and Everyday Medicine in the Nineteenth Century* (Chapel Hill: University of North Carolina Press, 2004), describes the essential and persistent localism of southern medical practice.

16. Stephen C. Kenny, "'A Dictate of Both Interest and Mercy'? Slave Hospitals in the Antebellum South," *Journal of the History of Medicine and Allied Sciences* 65 (2010): 1–47.

17. Southern medical schools did "benefit" from access to bodies for autopsies; slaves had little control over their bodies after death. Northerners imported preserved black corpses in barrels for medical school use. See Blakely and Harrington, *Bones in the Basement;* and Sappol, *Traffic of Dead Bodies.*

18. Schurr, "Inside the Confederate Hospital," 3, 10; and Cunningham, *Doctors in Gray,* 118.

19. Joseph Jones, "The Medical History of the Confederate States Army and Navy," *Southern Historical Society Papers* 20 (1892): 109–40, specifically 112.

20. John J. Chisolm, *A Manual of Military Surgery for the Use of Surgeons in the Confederate Army; with an Appendix of the Rules and Regulations of the Medical Department of the Confederate Army* (Richmond: West and Johnston, 1861; repr., San Francisco: Norman, 1989).

21. S. P. Moore, circular, Nov. 28, 1862, "Instructions to Asst. Surgeons Applying for Promotion in the Medical Department," chap. 6, vol. 664, Letters, orders, and circular relating to various hospitals in AL, GA, MS, and TN, 1862–1863 (received from R. T. Durrett, Louisville, KY, March 14, 1887), Record Group 109, NARA.

22. Cunningham, *Doctors in Gray,* 21 and 37.

23. Joseph Jones, "Medical History," 114.

24. Schroeder-Lein, *Confederate Hospitals on the Move;* and Cunningham, *Doctors in Gray.*

25. Confederate States of America War Department, *Regulations for the Medical Department of the C. S. Army* (Richmond: Ritchie and Dunnavant, 1862).

26. D[uncan] U. B[uie] to Miss Kate [McGeachy], July 8, 1862, Catherine McGeachy Buie Papers, 1819–1899, Duke SC. Other letters in this collection describe the disease climate, the soldiers' general comfort, and the fact that they were able to raid Union ships that ran aground on the treacherous shoals of North Carolina. Fort Caswell was a small mainland fort near the present Bald Head Island.

27. A. P. Mason, Special Orders 146, July 9, 1862, Record Book of Special Orders, Circulars, and Letters, Isaac Scott Tanner Collection, MS-19610930.5, MOC.

28. Hunter Holmes McGuire, order, April 20, 1863, in ibid. McGuire served as brigade surgeon under General Stonewall Jackson and cared for him during the general's last illness. See John W. Schildt, *Hunter Holmes McGuire, Doctor in Gray* (Chewsville, MD: Privately published, 1986).

29. Rebecca Barbour Calcutt, *Richmond's Wartime Hospitals* (Gretna, LA: Pelican, 2005); and Green, *Chimborazo.*

30. Peter W. Houck, *A Prototype of a Confederate Hospital Center in Lynchburg, Virginia* (Lynchburg, VA: Warwick House, 1986).

31. Ibid., 8, 13.

32. S. Cooper, Adjutant and Inspector General's Office, Richmond, VA, Nov. 25, 1862, General Orders no. 95, Second NC Hospital, Columbia, inside front cover, Order Book &tc., 1863–1865, S192115, SCSA.

33. Thomas More Downey, "A Call to Duty: Confederate Hospitals in South Carolina" (MA thesis, University of South Carolina, 1992).

34. Jerrold Northrop Moore, *Confederate Commissary General: Lucius Bellinger Northrop and the Subsistence Bureau of the Southern Army* (Shippensburg, PA: White Mane: 1996), 53.

35. Calcutt, *Richmond's Wartime Hospitals,* 43.

36. Drew Gilpin Faust, *Mothers of Invention: Women of the Slaveholding South in the American Civil War* (New York: Vintage Books, 1997).

37. Jack D. Welsh, *Two Confederate Hospitals and Their Patients: Atlanta to Opelika* (Macon, GA: Mercer University Press, 2005); and Jane E. Schultz, *Women at the Front: Hospital Workers in Civil War America* (Chapel Hill: University of North Carolina Press, 2004).

38. S. P. Moore, Circular to Medical Directors of Hospitals, July 1, 1863, General Hospital No. 2, 3rd Ward, 3rd Division, Lynchburg, VA, Supplies and Orders book Oct. 1862 to April 1865, MOC.

39. Faust, *Mothers of Invention,* 97–99; Cunningham, *Doctors in Gray,* 70–98.

40. Mark A. Weitz documented some 8,746 deserters from Confederate military hospitals; this number is almost certainly a significant underestimate. See his *More Damning than Slaughter: Desertion in the Confederate Army* (Lincoln: University of Nebraska Press, 2005), 261.

41. Michel Foucault, *Discipline and Punish: The Birth of the Prison* (New York: Vintage Books, 1995).

42. Moore, Circular, July 1, 1863, MOC.

43. Benjamin Blackford to D[aniel] S. Green at General Hospital, Lynchburg, VA, May 21 1862, General Hospital, Front Royal, VA, Letter and Order Book, Sept. 1861–Jan. 1865, MOC. Blackford wrote from a hospital in Liberty, VA, about twenty-eight miles west of Lynchburg on the Virginia and Tennessee railroad. The town is now named Bedford, perhaps because there is another Liberty, VA, east of Interstate 95. Blackford's hospital had begun at Front Royal, VA, but had moved further south to be less susceptible to Union attack.

44. Blackford to S. P. Moore, May 29, 1862, ibid.; Weitz, *More Damning than Slaughter.*

45. See *Guide for Inspection of Hospitals, and for Inspector's Report* (Richmond: Surgeon General's Office, n.d.), at Documenting the American South, http://docsouth.unc.edu/imls/guide/guide.html (accessed July 19, 2010). Medical directors and inspectors are listed in *CSMSJ* 1:152, 176.

46. N. S. Crowell, Circular, May 31, 1864, Second NC Hospital, Columbia, Order Book &tc., 1863–1865, S192115, SCSA.

47. Robert Libby, Surgeon in Charge of Louisiana Hospital no. 1, Charleston, SC, to Surgeon O. Becker, Wayside Hospital, March16, 1864, Charleston, SC, Order and Letter Book, S19115, SCSA.

48. John Herbert Roper, ed., *Repairing the "March of Mars": The Civil War Diaries of John Samuel Apperson, Hospital Steward in the Stonewall Brigade, 1861–1865* (Macon, GA: Mercer University Press, 2001).

49. Schurr, "Inside the Confederate Hospital," 6.

50. Samuel Preston Moore, Circular no. 8, Richmond, May 9, 1864, to Medical Directors in the Field and Hospital, Second NC Hospital, Columbia, SC, Order Book &tc., 1863–1865, S192115, SCSA.

51. Thomas H. Williams to S. P. Moore, Oct. 17, 1862, chap. 6, vol. 461, Letters sent

by Surgeon Thomas H. Williams, Inspector of Various Hospitals in Virginia, 1862–1863, War Department Collection of Confederate Records, Record Group 109, NARA. This letter explains why a certain lieutenant was being cared for at the hospital instead of in private quarters at his own expense (the typical path); in this case the man had been elected but not yet commissioned.

52. Jodi L. Koste, "'Medical School for a Nation': The Medical College of Virginia, 1860–1865," in *Years of Change and Suffering: Modern Perspectives on Civil War Medicine,* ed. James M. Schmidt and Guy R. Hasegawa (Roseville, MN: Edinborough Press, 2009), 13–35.

53. Edward S. Dunster, "The Comparative Mortality in Armies from Wounds and Disease," in *Contributions Relating to the Causation and Prevention of Disease, and to Camp Diseases; Together with a Report of the Diseases, etc., among the Prisoners at Andersonville, GA,* ed. Austin Flint, (New York: Hurd and Houghton, 1867), 169–92, 181–83.

54. Daniel Scott Smith, "Seasoning, Disease Environment, and Conditions of Exposure: New York Union Army Regiments and Soldiers," in *Health and Labor Force Participation over the Life Cycle: Evidence from the Past,* ed. Dora Costa (Chicago: University of Chicago Press, 2003), 89–112, 97.

55. James M. Fry, "Provost-Marshal-General's Report, March 17, 1866," *OR,* series 3, 5:666; Joseph T. Glatthaar, *General Lee's Army: From Victory to Collapse* (New York: Free Press, 2008), 438.

56. Joseph Jones, "Observations upon the Losses of the Confederate Armies from Battle, Wounds, and Disease, during the American Civil War of 1861–65," *Richmond and Louisville Medical Journal* 8 (1869): 339–58, 451–80; 9 (1870): 257–75, 635–57.

57. Margaret Humphreys, *Intensely Human: The Health of the Black Soldier in the American Civil War* (Baltimore: Johns Hopkins University Press, 2008).

58. William A. Davis to J. B. McCaw, March 14, 1862, chap. 6, vol. 707, Letters Received and Sent, Chimborazo Hospital, Richmond, VA, Oct. 1861 to Dec. 1862, War Department Collection of Confederate Records, Record Group 109, NARA.

59. L[afayette] Guild, Medical Director A. N. V. to Surgeon Isaac Tanner, Dec. 20, 1864, Correspondence, MS-19610930.1, MOC.

60. Downey, "Call to Duty," 120–21.

61. W. C. N. Randolph, circular no. 5, April 1864, conveying Moore's order striking a long list of drugs and equipment from the Confederate supply table, General Hospital No. 2, 3rd Ward, 3rd Division, Lynchburg, VA, Supplies and Orders book Oct. 1862 to April 1865, MOC; and S. P. Moore's order, recorded March 15, 1864, in Wayside Hospital, Charleston, SC, Order and Letter Book, S192115, SCSA. Moore sent these orders to medical directors, who in turn distributed them to the hospitals under their direction. In some cases the orders are recorded as coming from Moore, and in others from the intermediate in the medical bureaucracy. The clerk copying the orders in the "letters received book" often truncated the language to shorten his task.

62. S. P. Moore, circular, Aug. 15, 1863, General Hospital No. 2, 3rd Ward, 3rd Division, Lynchburg, VA, Supplies and Orders Book, Oct. 1862 to April 1865, MOC.

63. Schroeder-Lein, *Confederate Hospitals on the Move.*

64. J. W. Powell Notebook, "Confederate States Hospital Department, Notes of Sur-

gical Cases," printed by Ritchie and Dunnavant, Richmond, VA, 1862. In the front is a printed letter from S. P. Moore, Aug. 1, 1862, that says in part: "This book is issued to Medical Officers, solely for the purpose designated: and they will be expected to record herein, with faithful zeal and accuracy, all cases of surgical interest occurring within their professional control."

65. Blank notebook, "Confederate States Hospital Department, Notes of Surgical Cases," printed by Ritchie and Dunnavant, Richmond, VA, 1862. This bound volume contains pages labeled for notes of surgical cases, with lines for name, rank, regiment, company, nature and date of injury, operation and method, site, date, and disposition. There is a section for remarks on cases of special interest. This book is entirely blank. Tucked inside the front cover are loose sheets that order physicians to fill the volume with appropriate case notes. James Wright Tracy Papers, Duke SC.

66. Printed form, Surgeon General's Office, Richmond, April 25, 1863, signed S. P. Moore, Second NC Hospital, Columbia, Order Book &tc., 1863–1865, S192115, SCSA.

67. S. P. Moore to Surgeons and Assistant Surgeons, C.S.A., Sept. 16, 1863, General Hospital No. 2, 3rd Ward, 3rd Division, Lynchburg, VA, Supplies and Orders Book Oct. 1862 to April 1865, MOC. Further questions can be found in letters of Sept. 21 and Oct. 3, 1863, and later.

68. *CSMSJ* 1 (1864): 112; and *CSMSJ* 2 (1865): 1–2.

69. "Salutatory," *CSMSJ* 1 (1864): 13.

70. Cunningham, *Doctors in Gray,* 29.

71. The AMA published *Transactions of the American Medical Association* from 1848 to 1882 (vols. 1–33); it was then continued as the *Journal of the American Medical Association.*

72. Confederate Association of Army and Navy Surgeons Papers, folder 2, "List of Members," and folder 3, "Minutes, August 1863–March 1865," MS-19060831, MOC.

73. "Association of Army and Navy Surgeons," *CSMSJ* 1 (1864): 13–16, 14.

74. "Transactions of the Association of Army and Navy Surgeons," *CSMSJ* 1 (1864): 44–46.

75. Ibid., 90–92.

76. Ayers and Wade, publishers, "Prospectus," undated and unpaginated but interleaved between Dec. 1864 and Jan. 1865 issues, *CSMSJ.*

77. James O. Breeden, *Joseph Jones, M. D.: Scientist of the Old South* (Lexington: University of Kentucky Press, 1975), 117–66.

78. Joseph Jones, "Outline of the Results of an Examination of the Statistics and Records of Certain General Hospitals," Nov. 10, 1865, reel 4, frames 1132–1203, USSC Microfilm Papers, describes his research efforts for the Confederacy. Joseph Jones to Elisha Harris, Nov. 13, 1865, reel 5, frame 8, ibid., mentions his ill health during this research. Jones was short of money after the war and wrote essays for the USSC in return for compensation.

79. "Editorial and Miscellaneous," *CSMSJ* 2 (1865): 11.

80. Robert A. Kinloch, a Confederate medical director and inspector of hospitals, conveyed Moore's orders to a hospital in Columbia, SC, in December 1864. In various messages he advised that "endorsements on returned reports must specify the regulations, general order, circular and clause under which comment is furnished" (Kinloch circular, Dec. 12, 1864); advised physicians on telegram and postage costs; and told

them that new paperwork was required in order to purchase supplies. It was as if Moore hoped to hold back the tide by increasing the medical bureaucracy. Second North Carolina Hospital, Columbia, Order Book &tc., 1863–1865, S192115, SCSA. William T. Sherman's troops spent Christmas in Savannah and then started marching north through South Carolina in early January 1865. Columbia burned on Feb. 17, 1865.

81. S. P. Moore to J. B. McCaw, Nov. 15, 1862, chap. 6, vol. 707, Letters Received and Sent, Chimborazo Hospital, Richmond, VA, Oct. 1861 to Dec. 1862, War Department Collection of Confederate Records, Record Group 109, NARA.

82. Francis Peyre Porcher, *Resources of the Southern Fields and Forests* (Charleston, SC: Evans and Cogswell, 1863), available online at Documenting the American South, http://docsouth.unc.edu/imls/porcher/porcher.html (accessed July 19, 2010). W. T. Grant likewise strove to educate his fellow physicians about indigenous plants in his "Indigenous Medical Plants," *CSMSJ* 1 (1864): 84–86. Michael A. Flannery discusses the indigenous remedy search in detail in his *Civil War Pharmacy: A History of Drugs, Drug Supply and Provision, and Therapeutics for the Union and Confederacy* (New York: Haworth Press, 2004); he particularly addresses the quinine shortage in "Hapless or Helpmate? The Effectiveness of the Union's Blockade from a Medical Perspective," *North and South* 8, no. 3 (2005): 72–80. See also Guy R. Hasegawa, "Southern Resources, Southern Medicines," in Schmidt and Hasegawa, *Years of Change and Suffering*, 107–25.

83. "On the External Application of Oil of Turpentine as a Substitute for Quinine in Intermittent Fever, with Reports of Cases," *CSMSJ* 1 (1864): 7–8; Joseph Jacobs, "Some of the Drug Conditions during the War between the States, 1861–5," *Southern Historical Society Papers* 33 (1905), at Home of the American Civil War, www.civilwarhome.com/drugsshsp.htm (accessed online Aug. 3, 2010), mentions that *Pinckney pubens* was found to contain colchinine by a Dr. Fair; Mary Chesnut refers to a "Dr. Fair" in Columbia, South Carolina, in her diary. C. Vann Woodward, ed., *Mary Chesnut's Civil War* (New Haven, CT: Yale University Press, 1981), 368. One modern website describing *P. pubens* reports that its traditional name was fever-bark tree and also that the plant was relatively uncommon: "Its native habitat is wet acidic soils on the margins of bays, swamps, and streams, often in the light shade of scattered pine trees. Cultivated plants can grow on drier sites, but may need special attention during droughts. Root rot can be a problem, especially on poorly-drained clay soils." It is found in a small band from southern Georgia into north Florida. Tree Trail, www.treetrail.net/pinckneya.html (accessed June 22, 2012).

84. Surgeon General's Circular, "Special Report on the Use of Turpentine," Sept. 20, 1863, Second NC Hospital, Columbia, Order Book &tc. 1863–1865, SCSA. Other record books in this collection indicate how common intermittent fevers were—not surprising given the medical history of the Carolina low country.

85. "External Application of Turpentine as a Substitute for the Internal Use of Quinine in the Treatment of Intermittent Fever," *CSMSJ* 1 (1864): 119–20. See also the preliminary report on this topic: "On the External Application of Oil of Turpentine as a Substitute for Quinine in Intermittent Fever, with Reports of Cases," *CSMSJ* 1 (1864): 7–8.

86. W. A. Carrington to Surgeon H. Barton, Feb. 29, 1864, chap. 6, vol. 364, Letters Sent and Received, Medical Director's Office, Richmond, VA, 1864–65, Record Group 109, War Department Collection of Confederate Records, NARA.

87. W. A. Carrington to Surgeon McCaw, March 7, 1864, ibid.

88. S. P. Moore to J. B. McCaw, June 10, 1863, chap. 6, vol. 708, Letters Received and Sent by the Surgeon of Chimborazo Hospital, Jan. to Dec. 1863, Record Group 109, War Department Collection of Confederate Records, NARA.

89. W. A. Carrington, Circular no. 5, March 11, 1865 (maxillary fractures); and W. A. Carrington, Circular no. 2, ca. March 1865 (orthopedic hospital), in General Hospital No. 2, 3rd Ward, 3rd division, Lynchburg, VA, Supplies and Orders Book, Oct. 1862 to April 1865, MOC. On Bean and southern dentistry, see Colin F. Baxter, "Dr. James Baxter Bean, Civil War Dentist: An East Tennessean's Victorian Tragedy," *Journal of East Tennessee History* 67 (1995): 3–57; and Glenna R. Schroeder-Lein, "Dentistry," in *The Encyclopedia of Civil War Medicine,* by Schroeder-Lein (Armonk, NY: M.E. Sharpe, 2008), 84–85.

90. "History of the Southern Medical Association," *Southern Medical Journal* 54 (1961): 87. See also the history section of the website of the Southern Medical Association, www .sma.org (accessed Aug. 24, 2011). The group began publishing their proceedings in 1908 as the *Southern Medical Journal.*

91. Glenna R. Schroeder-Lein, "'While the Participants Are Yet Alive': The Association of Medical Officers of the Army and Navy of the Confederacy," in *Inside the Confederate Nation: Essays in Honor of Emory M. Thomas,* ed. Lesley J. Gordon and John C. Inscoe (Baton Rouge: Louisiana State University Press, 2005), 335–48.

92. Moore's office, containing all of his correspondence and paperwork, burned in the Richmond evacuation fire of 1865. There are scattered letters sent to hospitals in various archives and a small collection in the National Library of Medicine, but little archival material that offers clues to his thinking outside of official pronouncements.

93. R. A. Kinloch, "Annual Address by the President: A Plea for Education as the Means for Unifying the Profession and Strengthening the Association," *Transactions of the South Carolina Medical Association,* 1884, 23–48, quotation on 30.

94. There is an abundant literature on Sims. One recent article with citations to this literature is L. Lewis Wall, "Did J. Marion Sims Deliberately Addict His First Fistula Patients to Opium?" *Journal of the History of Medicine and Allied Sciences* 62 (2007): 336–56.

95. Kinloch, "Annual Address by the President," 30–31.

Chapter 8 · Confederate Medicine

1. Jerold Northrop Moore, *Confederate Commissary General: Lucius Bellinger Northrop and the Subsistence Bureau of the Southern Army* (Shippensburg, PA: White Mane, 1996).

2. Invoice for medicines, instruments, hospital stores, bedding, etc., issued to (blank for name of surgeon) by (blank for name of surgeon), medical purveyor, (blank for place issued), James Lawrence Cabell Papers, 1862–1865, Cabell Family Papers, 1755–1909, Duke SC. Cabell managed a hospital in Charlottesville, where he later became president of the university. Surgeons had to keep track of all this equipment and, when they left a post, make sure it was transferred to the next man in charge. For an example of the Medical Department Forms, see Ira Rutkow, *Regulations for the Amy of the Confederate States* (1862; repr., San Francisco: Norman, 1992), 244–85.

3. Joseph Jones, "Observations upon the Losses of the Confederate Armies from

Battle, Wounds, and Disease during the American Civil War of 1861–1865," *Richmond and Louisville Medical Journal* 8 (1869): 339–58, 451–80; 9 (1870): 257–75, 635–57, 466; Thomas L. Livermore, *Numbers and Losses in the Civil War in America: 1861–65* (1900; repr., with introduction by Edward E. Barthell, Bloomington: Indiana University Press, 1957), 1–50; Gary W. Gallagher, *The Confederate War* (Cambridge, MA: Harvard University Press, 1997), 28–29.

4. Jones, "Observations," 470. Gallagher reviewed recent scholarship on Confederate Army numbers in "Disaffection, Persistence, and Nation: Some Directions in Recent Scholarship on the Confederacy," *Civil War History* 55 (2009): 329–53, where he puts the total at 850,000–900,000. Initially the age range of white male conscription was 18–35; later acts extended this range, ultimately to 17–45. Livermore, *Numbers*, 11–18.

5. See title pages of *MSHW*, vols. 3 and 5.

6. *MSHW*, 1:641. I have used "wounds" in the text for simplicity. The actual category was "Class V: Wounds, Accidents, and Injuries," and it included burns, contusions, fractures, and drowning. Gunshot wounds were by far the most-represented injury of the lot, making up 58% of the total cases in the Union figures and 90% of the mortality. The second- and third-most-common diagnoses within this group were sprains and contusions, which, not surprisingly, had a much lower case mortality rate.

7. Jack D. Welsh, *Two Confederate Hospitals and Their Patients: Atlanta to Opelika* (Macon, GA: Mercer University Press, 2005), 99, 128.

8. Robertson Hospital Register Collection, Virginia Commonwealth University, Richmond, VA, http://dig.library.vcu.edu/cdm4/index_rhr.php?CISOROOT=/rhr (accessed July 22, 2010).

9. This absence would make the most difference in cases of falciparum malaria (as opposed to vivax) but it is difficult enough to identify malaria at all from the unfamiliar Civil War nosology, without drawing further distinctions. Falciparum is in general much more likely to cause death than vivax, but even vivax had a 5 percent case mortality rate among soldiers in World War I. See Margaret Humphreys, *Malaria: Poverty, Race, and Public Health in the United States* (Baltimore: Johns Hopkins University Press, 2001).

10. *MSHW*, 5:30.

11. Welsh, *Two Confederate Hospitals*, 112.

12. *MSHW*, 1:637.

13. Quotation from *MSHW*, 5:33; other information from 5:29–33, 69.

14. Jones, "Observations."

15. Daniel Scott Smith, "Seasoning, Disease Environment, and Conditions of Exposure: New York Union Army Regiments and Soldiers," in *Health and Labor Force Participation over the Life Cycle: Evidence from the Past*, ed. Dora Costa (Chicago: University of Chicago Press, 2003), 89–112, 97.

16. Samuel Preston Moore, "Address of the President of the Association of Medical Officers of the Confederate States Army and Navy," *Southern Practitioner* 31 (1909): 491–98, quotation on 493. Moore gave this address in 1874; it was printed then in the Richmond newspapers and reprinted (posthumously) in 1909.

17. Joseph Jones, "On the Prevalence and Fatality of Pneumonia and of Typhoid Fever in the Confederate Army during the War of 1861–1865," in *Contributions Relating to the*

Causation and Prevention of Disease; Together with a Report of the Diseases, etc., among the Prisoners at Andersonville, GA, ed. Austin Flint (New York: Hurd and Houghton, 1867), 335–59, quotation on 352.

18. Jones, "Observations," 636.

19. Joseph T. Glatthaar, *Soldiering in the Army of Northern Virginia* (Chapel Hill: University of North Carolina Press, 2011), 18–20.

20. Ibid.

21. Jones, "Observations," 635.

22. Ibid., 637.

23. Historical accounts that discuss some aspects of these shortages include James Breeden, "Medical Shortages and Confederate Medicine," *Southern Medical Journal* 86 (1993): 1040–47; Horace H. Cunningham, *Doctors in Gray: The Confederate Medical Service* (Baton Rouge: Louisiana State University Press, 1958); Nancy Schurr, "Inside the Confederate Hospital: Community and Conflict during the Civil War" (Ph.D. diss., University of Tennessee–Knoxville, 2004); and Glenna R. Schroeder-Lein, *Confederate Hospitals on the Move: Samuel H. Stout and the Army of Tennessee* (Columbia: University of South Carolina Press, 1994).

24. John Christopher Schwab, *The Confederate States of America, 1861–1865: A Financial and Industrial History of the South during the Civil War* (New York: Charles Scribner's, 1901); Moore, *Confederate Commissary General.*

25. Moore, *Confederate Commissary General,* 264.

26. Robert C. Black III, *The Railroads of the Confederacy* (1952; repr., Chapel Hill: University of North Carolina Press, 1998); John E. Clark, *Railroads in the Civil War: The Impact of Management on Victory and Defeat* (Baton Rouge: Louisiana State University Press, 2001); Moore, *Confederate Commissary General.*

27. Pat Brown to Kate McGeachy, Nov. 12, 1862, in Catherine McGeachy Buie Papers, 1819–1899, Duke SC.

28. Moore, *Confederate Commissary General.*

29. Richard E. Beringer, Herman Hattaway, Arthur Jones, and William N. Still Jr., *The Elements of Confederate Defeat* (Athens: University of Georgia Press, 1988), 17.

30. Ludwell H. Johnson, "Trading with the Union: The Evolution of Confederate Policy," *Virginia Magazine of History and Biography* 78, no. 3 (1970): 306–25; and his "Contraband Trade during the Last Year of the Civil War," *Mississippi Valley Historical Review* 49 (1963): 635–52.

31. Schroeder-Lein, *Confederate Hospitals on the Move,* 97–101.

32. See the *West Philadelphia Hospital Register,* issues for June and July 1863.

33. For a description of the hospitals in Louisville, see letters by J. S. Newberry and F. H. Bushnell in the *Louisville Journal,* Feb. 16, 1863; J. S. Newberry described the medical response to the battle of Perryville in *The U.S. Sanitary Commission in the Valley of the Mississippi, during the War of the Rebellion, 1861–1865* (Cleveland: Fairbanks, Benedict, 1871), 52–60.

34. Schroeder-Lein, *Confederate Hospitals on the Move,* 101.

35. S. S. Satchwell (general military hospital no. 2, Wilson NC) to Surgeon Graves, July 24, 1863, William A Holt Papers, 1862–1865, 2516-z, UNC SHC.

36. Benjamin Blackford to W. A. Carrington, June 28, 1864, General Hospital, Front Royal, VA, Letter and Order Book, Sept. 1861–Jan. 1865, MOC.

37. Schroeder-Lein, *Confederate Hospitals on the Move*, 91, 142.

38. James King Wilkerson to Mrs. Wilkerson and family, April 4, 1865, James Wilkerson Papers, 1820–1899, Duke SC.

39. Thomas Williams to S. P. Moore, July 2, 1862, chap. 101, vol. 366, Letters sent by Surgeon Thomas H. Williams, Inspector of Various Hospitals in Va., 1862, War Department Collection of Confederate Records, Record Group 109, NARA.

40. S. P. Moore, Circular, April 1, 1863, chap. 6, vol. 664, Letters, orders, and cir. relating to various hospitals in Ala., Ga., Miss., and Tenn., 1862–63. (received from R. T. Durrett, Louisville, Ky., March 14, 1887), War Department Collection of Confederate Records, Record Group 109, NARA.

41. Lafayette Guild, Circular, March 23, 1863, Record Book of Special Orders, Circulars, and Letters, MS-19610930.5, Isaac Scott Tanner Collection, MOC.

42. Moore, *Confederate Commissary General*.

43. Schroeder-Lein, *Confederate Hospitals on the Move*.

44. Ibid., 113–14.

45. Pat Brown to Kate McGeachy, Nov. 12, 1862, in Buie Papers; W. S. Shockley to his wife Eliza, March 7, 1862, W. S. Shockley Papers, Duke SC.

46. W. C. N. Randolph, general order no. 11, Oct. 9, 1862, General Hospital No. 2, 3rd Ward, 3rd Division, Lynchburg, VA, Supplies and Orders Book, Oct. 1862–April 1865, MOC.

47. W. S. Shockley to wife Eliza, Oct. 15, Nov. 28, 1862, Shockley Papers.

48. Cunningham, *Doctors in Gray*, 80.

49. Moore, *Confederate Commissary General*, 188.

50. Peter W. Houck, *A Prototype of a Confederate Hospital Center in Lynchburg, Virginia* (Lynchburg, VA: Warwick House, 1986), 97.

51. James C. Franklin to his wife Sookey, June 22, 1862, James C. Franklin Correspondence, 1862–64, Duke SC.

52. J. B. Farthing to G. F. Adams, May 18, 1864, Alfred Adams Papers, 1862–64, Duke SC. By "hospital rats" he meant malingerers, men faking illness. T. G. Richardson to Doctor [Stout], March 10, 1864, quoted in Schroeder-Lein, *Confederate Hospitals on the Move*, 120; John H. Kinyoun to wife Bettie, Aug. 8, Oct. 9, 1864, John H. Kinyoun Papers, 1851–1898, Duke SC.

53. [Duncan Buie] to Kate McGeachy, April 20, 1862, Buie Papers. Duncan Buie and his men in North Carolina coastal forts fared well during the war. They stayed local, were able to plant gardens, and salvaged food off the Union ships that ran afoul of the "Graveyard of the Atlantic," the treacherous shoals off Cape Hatteras. It was with great regret that, when retreating in January 1865, Duncan's men had to leave fifty pounds of coffee and one hundred pounds of English butter salvaged from a recent wreck (ibid., Jan. 23, 1865). Buie even gained twenty pounds during the war, not the usual experience for Confederate soldiers (ibid., Feb. 4, 1865).

54. John. H. Kinyoun to wife Bettie, Oct. 13 1861, Kinyoun Papers; Thomas W. Scott to J. B. McCaw, Dec. 16 1862, chap. 6, vol. 707, Letters Received and Sent, Chimborazo

Hospital, Richmond, Virginia, Oct. 1861–Dec. 1862, War Department Collection of Confederate Records, Record Group 109, NARA.

55. S. P. Moore, Circular no. 15, Aug. 10, 1864, Second NC Hospital, Columbia, Order Book &tc., 1863–1865, S192115, SCSA; Blackford to S. P. Moore, June 11, 1864, in Order and Letter Book, General Hospital, Front Royal, Jan. 1864–March 1865 (hospital moves to Liberty, Va. by May, 1862), MOC.

56. Moore, *Confederate Commissary General,* 171, 212.

57. William Blair, *Virginia's Private War: Feeding Body and Soul in the Confederacy, 1861–1865* (New York: Oxford University Press, 1998).

58. Joseph T. Glatthaar, *General Lee's Army: From Victory to Collapse* (New York: Free Press, 2008), 463.

59. Schurr, "Inside the Confederate Hospital," 43, 115.

60. Brigadier General R. E. Colston, general order 16, Aug. 8, 1864, General Hospital No. 2, 3rd Ward, 3rd Division, Lynchburg, VA, Supplies and Orders book, Oct. 1862–April 1865, 262–63, MOC.

61. Emily V. Mason, "Memories of a Hospital Matron," *Atlantic Monthly,* 1902, 113.

62. Ibid., 475–77.

63. Phoebe Yates Pember, *A Southern Woman's Story* (Charleston: University of South Carolina Press, 2002), 45.

64. Ibid., 45–47.

65. S. Cooper, Department of Inspector General, general orders 60, July 21, 1864, Second NC Hospital, Columbia, SC, Order Book &tc., 1863–1865, S192115, SCSA; Glatthaar, *General Lee's Army,* 446.

66. Gallagher, *Confederate War,* 34–35.

67. Emory M. Thomas, *The Confederate Nation: 1861 to 1865* (New York: Harper & Row, 1979), 196, 260.

68. T. H. Williams, General Orders, Dec. 13–14, 1861, in Order and Letter Book, General Hospital, Front Royal 1861 (hospital moves to Liberty, Va., by May 1862), MOC; Benjamin Blackford to Williams, Dec. 17, 1861, ibid.

69. Circular, copied from S. P. Moore master circular, issued by N. S. Crowell, Medical Director of Hospitals, Department of S.C., Geo. and Fla., Charleston, SC, Aug. 15, 1863, in Second N. C. Hospital, Columbia, Order Book &tc, 1863–1865, S192115, SCSA.

70. Benjamin Blackford to W. A. Carrington, Oct. 24, 1864, in Order and Letter Book, General Hospital, Front Royal 1861 (hospital moves to Liberty, Va., by May, 1862), MOC. See also other letters in that collection.

71. W. C. N. Randolph, general orders, April 1, 1863, General Hospital No. 2, 3rd Ward, 3rd Division, Lynchburg, VA, Supplies and Orders book Oct. 1862–April 1865, MOC. This order says that all nurses fit for duty are to return to their regiments, leaving only one nurse per ward to protect property and superintend cleaning.

72. William A. Carrington to Capt [John A.] Coke, E[xecutive] O[fficer] Rich[mond], April 30, 1864, chap. 6, vol. 364, letters sent and received, medical director's office, Richmond, VA, 1864–65, War Department Collection of Confederate Records, Record Group 109, NARA.

73. W. C. N. Randolph, General Order No. 11, Oct. 9, 1862, General Hospital No. 2, 3rd Ward, 3rd Division, Lynchburg, VA, Supplies and Orders book, Oct. 1862–April 1865, MOC.

74. S. P. Moore, July 14, 1864, General Hospital No. 2, 3rd Ward, 3rd Division, Lynchburg, VA, Supplies and Orders Book, Oct. 1862–April 1865, MOC.

75. CSA War Department Circular to citizens of Halifax, Lunenberg, Charlotte, Pittsylvania, and Campbell Counties, June 13, 1862, frame 750, roll 1, "The Negro in the Military Service of the United States, 1639–1886," Adjutant General's Office, NARA M858; Carol C. Green, *Chimborazo: The Confederacy's Largest Hospital* (Knoxville: University of Tennessee Press, 2004), 41–63; and Schurr, "Inside the Confederate Hospital," 144–70.

76. Benjamin Blackford to S. P. Moore, Oct. 31, 1862, in Order and Letter Book, General Hospital, Front Royal 1861 (hospital moves to Liberty, VA, by May, 1862), MOC.

77. Benjamin Blackford to S. P. Moore, Nov. 18, 1864, ibid.

78. J. L. Cabell to S. P. Moore, June 20, 1862, chap. 6, vol. 369, Letters received by Surgeon Thomas H. Williams, Inspector of Various Hospitals in Virginia, 1862, War Department Collection of Confederate Records, Record Group 109, NARA. For a similar plea from the Chimborazo director, see J. B. McCaw to S. P. Moore, May 17, 1862, chap. 6, vol. 707, Letters Received and Sent, Chimborazo Hospital, Richmond, VA, Oct. 1861–Dec. 1862, War Department Collection of Confederate Records, Record Group 109, NARA.

79. Robert Libby, June 26, 1864; and Robert Libby to Major General Jones, June 28, 1864, Wayside Hospital, Charleston, SC, Order and Letter Book, S192115, SCSA.

80. Cunningham, *Doctors in Gray,* 37; George Worthington Adams, *Doctors in Blue: The Medical History of the Union Army in the Civil War* (Baton Rouge: Louisiana State University Press, 1952), 9. These numbers are approximations.

81. William Carrington, Medical Director, to Surgeon General, May 24, 1864, chap. 6, vol. 364, letters sent and received, Medical Director's Office, Richmond, VA, 1864–1865, War Department Collection of Confederate Records, Record Group 109, NARA.

82. S. P. Moore to medical directors of hospitals, July 1, 1863, General Hospital No. 2, 3rd Ward, 3rd Division, Lynchburg VA, Supplies and Orders Book, Oct. 1862–April 1865, MOC.

83. John Terrell, "A Confederate Surgeon's Story," *Confederate Veteran* 39 (1931): 457–59, 458.

84. Richard O. Curry to P. E. Hines, Oct. 14, 1864, chap. 6, vol. 35, Letters. Orders, and Circulars Issued and Received, Military Prison Hospital, Salisbury, NC, 1864–1865, War Department Collection of Confederate Records, Record Group 109, NARA; P. E. Hines to Richard O. Curry, Oct. 19, 1864, ibid.

85. Richard O. Curry to Samuel Moore and P. E. Hines, Nov. 21, 1864, ibid.

86. Cunningham, *Doctors in Gray,* 40.

87. Carnot Posey to Thomas Williams, Oct. 16, 1861, chap. 6, vol. 369, Letters received by Surgeon Thomas H. Williams, Inspector of Various Hospitals in Virginia, 1862, War Department Collection of Confederate Records, War Department Collection of Confederate Records, Record Group 109, NARA.

88. G. W. Nelson to J. B. McCaw, July 13, 1862, chap. 6, vol. 707, Letters Received and

Sent, Chimborazo Hospital, Richmond, VA, Oct. 1861–Dec. 1862, War Department Collection of Confederate Records, Record Group 109, NARA.

89. Benjamin Blackford to S. P. Moore, May 29, 1862, in Order and Letter Book, General Hospital, Front Royal 1861 (hospital moves to Liberty, VA, by May, 1862), MOC; S. P. Moore, Circular no. 22, Dec. 3, 1864, chap. 6, vol. 664. Letters, orders, and cir. relating to various hospitals in Ala., Ga., Miss., and Tenn., 1862–63 (received from R. T. Durrett, Louisville, KY, March 14, 1887), War Department Collection of Confederate Records, Record Group 109, NARA.

90. Thomas H. Williams to W. C. N. Randolph, Nov. 17, 1862; and W. C. N. Randolph to Surgeon [Henry] Chalmers, May 12, 1864, General Hospital No. 2, 3rd Ward, 3rd Division, Lynchburg, VA, Supplies and Orders book, Oct. 1862–April 1865, MOC.

91. N. S. Crowell to Thomson, May 30, 1864, in Second N.C. Hospital, Columbia, Order Book &tc, 1863–1865, S192115, SCSA.

92. E. A. Flewellen to Corps Directors, Oct. 30, 1863, chap. 6, vol. 749, Letters Sent to Medical Director, Army of Tennessee, 1863–1864, Letters. Orders, and Circulars Issued and Received, Military Prison Hospital, Salisbury, NC 1864–1865, War Department Collection of Confederate Records, Record Group 109, NARA.

93. Alfred Raoul to sir, Jan. 22, 1865, Second NC Hospital, Columbia, SC, Order Book &tc., 1863–1865, S192115, SCSA.

94. S. Cooper, Aug. 6, 1864, General Orders no. 63, Wayside Hospital, Charleston, SC, Order and Letter Book, S192115, SCSA; Carrington, Aug. 11, 1864, circular no. 10, General Hospital No. 2, 3rd Ward, 3rd Division, Lynchburg, VA, Supplies and Orders book Oct. 1862–April 1865, 230–32, MOC.

95. Rebecca Barbour Calcutt, *Richmond's Wartime Hospitals* (Gretna, LA: Pelican, 2005), 93.

96. W. C. N. Randolph, general order no. 43, Aug. 2, 1864, General Hospital No. 2, 3rd Ward, 3rd Division, Lynchburg, VA, Supplies and Orders book Oct. 1862–April 1865, 226, MOC.

97. Thomas H. Williams to Colonel A. C. Myers, Dec. 12, 1862, chap. 6, vol. 461, Letters sent by Surgeon Thomas H. Williams, Inspector of Various Hospitals in Virginia, 1862–1863, War Department Collection of Confederate Records, Record Group 109, NARA.

98. Glatthaar, *General Lee's Army*, 215. On the shoe myth, see Encyclopedia Virginia, www.encyclopediavirginia.org/Shoes_at_Gettysburg (accessed Aug. 6, 2010). On the persistence of this myth in popular culture, including the claim that Lee's men were targeting a shoe factory in Gettysburg, see Ted Kerin, "What Did You Pay for Those Shoes?" The Ultimate Lieutenant Columbo site, www.columbo-site.freeuk.com/ (accessed Oct. 22, 2010).

99. F. G. Nicholls, General Orders, Feb. 26, 1864, General Hospital No. 2, 3rd Ward, 3rd Division, Lynchburg, VA, Supplies and Orders book Oct. 1862–April 1865, 197–98, MOC.

100. Blackford to S. P. Moore, Oct. 22, 1862, General Hospital, Front Royal, VA, Letter and Order Book, Sept. 1861–Jan. 1865, MOC.

101. Blackford to Cunningham, Oct. 16, 1863, ibid.

102. Inside front cover, Second NC Hospital, Columbia, Order Book &tc., 1863–1865, S192115, SCSA.

103. Medical Director to A. B. Lawrence, Jan. 30, 1865, chap. 6, vol. 364, letters sent and received, medical director's office, Richmond, VA, 1864–65, War Department Collection of Confederate Records, Record Group 109, NARA.

104. Historian Horace H. Cunningham emphasized the success of blockade runners in *Doctors in Gray*, 135–37.

105. Lafayette Guild, circular letter, April 25, 1863, Record Book of Special Orders, Circulars, and Letters, MS-19610930.5, Isaac Scott Tanner Collection, MOC; N. S. Crowell, Medical Director of Hospitals, Department of S.C., Geo. and Fla., Charleston, SC, Circular, Feb. 18, 1864, in Second N. C. Hospital, Columbia, Order Book &tc, 1863–1865, #S192115, SCSA.

106. Michael Flannery, *Civil War Pharmacy: A History of Drugs, Drug Supply and Provision, and Therapeutics for the Union and Confederacy* (New York: Haworth Press, 2004), 192–209.

107. Michael Flannery, "Hapless or Helpmate? The Effectiveness of the Union's Blockade of the Confederacy from a Medical Perspective," *North and South* 8, no. 3 (May 2005): 72–80; Norman H. Franke, "Pharmaceutical Conditions and Drug Supply in the Confederacy," *Georgia Historical Quarterly* 37 (1953): 287–98.

108. Flannery, *Civil War Pharmacy*, 247; and Norman H. Franke, "Official and Industrial Aspects of Pharmacy in the Confederacy," *Georgia Historical Quarterly* 37 (1953): 175–87.

109. Wayside Hospital, Charleston, SC, Order and Letter Book, S192115, SCSA.

110. Flannery, *Civil War Pharmacy*, 192–202; on the doll, named Lucy Ann, see Steve Szkotak, "Museum of the Confederacy Going on the Road," SF Gate (*San Francisco Chronicle*), http://articles.sfgate.com/2008-09-14/news/17159165_1_museum-s-president-confedcrate-museum-rawls (accessed Aug. 8, 2010), or visit the Museum of the Confederacy, Richmond, VA.

111. S. P. Moore circular no. 10, Nov. 28, 1864, Record Book of Special Orders, Circulars, and Letters, MS-19610930.5, Isaac Scott Tanner Collection, MOC; Kinloch circular, July 13, 1864, Second NC Hospital, Columbia, Order Book &tc., 1863–1865, S192115, SCSA; and Paul W Gates, *Agriculture and the Civil War* (New York: Knopf, 1965), 96–99.

112. Cunningham, *Doctors in Gray*, 158.

113. Thomson to N. S. Crowell, May 14, 1864, Second NC Hospital, Columbia, Order Book &tc., 1863–1865, S192115, SCSA.

114. N. S. Crowell to Thomson, May 27, 1864, ibid.

115. R. Libby to S. P. Moore, June 22, 1864, Wayside Hospital, Charleston, SC, Order and Letter Book, S192115, SCSA; S. H. Stout, Medical Director, Oct. 11, 1864, Columbus, GA, in *OR*, series 4, vol. 3, 1864.

116. Green, *Chimborazo*, 142–51; and Calcutt, *Richmond's Wartime Hospitals*, 87–99.

117. P. E. Hines to J. M. Otey, letters from March 20 to 23, 1865, and April 15, 1865, John M. Otey Papers, 1864–1865, Duke SC.

118. Margaret Mitchell, *Gone with the Wind* (New York: Scribner, 1936).

119. Glatthaar, *General Lee's Army*, 395.

Chapter 9 · Mitigating the Horrors of War

1. Abraham Lincoln, Second Inaugural Address, April 5, 1865, at Bartleby.com: Great Books on Line, www.bartleby.com/124/pres32.html (accessed July 20, 2012).

2. Benjamin G. Cloyd, *Haunted by Atrocity: Civil War Prisoners in American Memory* (Baton Rouge: Louisiana State University Press, 2010).

3. Steven Pinker, *The Better Angels of Our Nature: Why Violence Has Declined* (New York: Viking, 2011).

4. Dorence Atwater, "Introductory Remarks," in *A List of the Union Soldiers Buried at Andersonville,* by Atwater and Clara Barton (New York: Tribune Association, 1866), iii.

5. Clara Barton, "Report of an Expedition to Andersonville, etc.," ibid., vii.

6. Augustus C. Hamlin, *Martyria; or, Andersonville Prison* (Boston: Lee and Shepard, 1866), 29.

7. Harry S. Stout, *Upon the Altar of the Nation: A Moral History of the American Civil War* (New York: Viking, 2006); George C. Rable, *God's Almost Chosen Peoples: A Religious History of the American Civil War* (Chapel Hill: University of North Carolina Press, 2010). See also Mark Grimsley, *The Hard Hand of War: Union Military Policy toward Southern Civilians, 1861–1865* (Cambridge: Cambridge University Press, 1995); and George S. Burkhardt, *Confederate Rage, Yankee Wrath: No Quarter in the Civil War* (Carbondale: Southern Illinois University Press, 2007). For one philosopher's discussion of the attempt to grasp morality in war, see Michael Ignatieff's works, especially *The Warrior's Honor: Ethnic War and the Modern Conscience* (New York: Henry Holt, 1997).

8. Charles J. Stillé, *History of the United States Sanitary Commission Being the General Report of Its Work during the War of the Rebellion* (Philadelphia: Lippincott, 1866), 510.

9. Ibid., 517.

10. *Report of the Joint Select Committee Appointed to Investigate the Condition and Treatment of Prisoners of War,* presented to the Confederate Congress, March 3, 1865. Reproduced in Documenting the American South, www.docsouth.unc.edu/imls/report/report .html, 3 (accessed Sept. 16, 2008).

11. [Francis Lieber,] General Orders No. 100, issued by E. D. Townsend, assistant Adjutant General, "Instructions for the Government of Armies of the United States in the Field, in *OR,* series 2, 5:671–82, quotation on 675. Lieber immediately qualified this by saying a commander could order that no quarter be given if "in great straits when his own salvation makes it impossible to cumber himself with prisoners" (675). The document is also reprinted in full, with commentary, in Richard Shelly Hartigan, *Lieber's Code and the Law of War* (Chicago: Precedent, 1983).

12. Andrew Ward, *River Run Red: The Fort Pillow Massacre in the American Civil War* (New York: Viking, 2005).

13. [Lieber,] "Instructions for the Government of Armies," 676.

14. U.S. Sanitary Commission, *Narrative of Privations and Sufferings of United States Officers and Soldiers While Prisoners of War in the Hands of the Rebel Authorities. Being the Report of a Commission of Inquiry, Appointed by the United States Sanitary Commission. With an Appendix Containing the Testimony* (Philadelphia: USSC, King & Baird, 1864), 28, 29.

15. Charles W. Sanders Jr., *While in the Hands of the Enemy: Military Prisons of the Civil War* (Baton Rouge: Louisiana State University Press, 2005), 1.

16. Indeed the reinterpretation of Andersonville ("it wasn't so bad") can be found on the same page as Holocaust denials. See, e.g., Revisionist History, www.revisionisthistory .org/revisionist4.html. Another website that includes historical works from the southern perspective, suitable for "in-home schooling," is Southern Heritage Site, www.dixie general.com/webdoc1.htm. On Elmira, see Michael Horigan, *Elmira: Death Camp of the North* (Mechanicsburg, PA: Stackpole Books, 2002).

17. Sanders, *While in the Hands of the Enemy*, 5.

18. George Levy, *To Die in Chicago: Confederate Prisoners at Camp Douglas, 1862–1865* (Evanston, IL: Evanston, 1994).

19. Henry W. Bellows to William Hoffman, June 30, 1862, in *OR*, series 2, 4:106. "Police," used as a verb here, means to attend to sanitary cleanup and enforce sanitary behavior. In this context it would have meant that excrement and other trash was evident on the grounds and that the place needed cleaning up. On Camp Douglas, see Levy, *To Die in Chicago*.

20. W. Hoffman to M. C. Meigs, July 10, 1862, in *OR*, series 2, 4:166; ibid., July 17, 1862, 238.

21. W. Hoffman to H. Bellows, July 12, 1862, in *OR*, series 2, 4:178–79; W. Hoffman to E. M. Stanton, May 21, 1863, in *OR*, series 2, 5:686–87.

22. Wm. H. Van Buren to E. M. Stanton, May 10, 1863, in *OR*, series 2, 5:587–89.

23. Ibid.

24. W. Hoffman to E. M. Stanton, May 21, 1863, in *OR*, series 2, 5:686–87.

25. Frederick Law Olmsted to Robert Morton Lewis, July 30, 1863, in *The Papers of Frederick Law Olmsted*, vol. 4, *Defending the Union: The Civil War and the U.S. Sanitary Commission, 1861–1863*, ed. Jane Turner Censer (Baltimore: Johns Hopkins University Press, 1986), 684.

26. HWB[ellows] to the editor of the *Christian Inquirer*, "The Union Prisoners in Richmond," Oct. 1, 1863, in USSC, Washington Scrapbook of Newspaper Cuttings, vol. 1025, USSC Records, NYPL; Frederick Law Olmsted, "Preliminary Report of the Operations of the Sanitary Commission with the Army of the Potomac, during the Campaign of June and July, 1863," *Documents of the U.S. Sanitary Commission*, 2 vols. (New York: USSC, 1866), doc. no. 68, 2:4.

27. John Hancock Douglas, "Report on the Operations of the Sanitary Commission during and after the Battles at Gettysburg, July 1st, 2d, and 3d, 1863," doc. no. 71, in *Documents of the U.S. Sanitary Commission*, doc. no. 68, 2:15.

28. Benjamin Butler to Robert Ould, Dec. 25, 1863, in *OR*, series 2, 6:756; Robert Ould to S. A. Meredith, Aug. 1, 1863, ibid., 166–67; S. A. Meredith to E. A. Hitchcock, Sept. 14, 1863, ibid., 285.

29. Sanders, *While in the Hands of the Enemy*, 173–76.

30. Orlando B, Willcox, *Forgotten Valor: The Memoirs, Journals, and Civil War Letters of Orlando B. Willcox*, ed. Robert Garth Scott (Kent, OH: Kent State University Press, 1999), 307.

31. W. F. Swalm to J. H. Douglas, Nov. 13, 1863, in *OR*, series 2, 6:575–81, quotations on 575 and 576.

32. Ibid., quotations on 577 and 579.

33. Ibid.

34. Frederick N. Knapp to William Hoffman, Nov. 26, 1863, in *OR*, series 2, 6:575.

35. Hoffman to Knapp, Nov. 27, 1863, in *OR*, series 2, 6:586.

36. Gilman Marston to Hoffman, Dec. 4, 1863, in *OR*, series 2, 6:645–46.

37. A. M. Clark, "Report of Inspection of the Hammond U.S. General Hospital at Point Lookout, Md., December 17, 1863," in *OR*, series 2, 6:740–41; and "Report of Inspection of Camp and Field Hospitals for Prisoners of War at Point Lookout, Md., December 17 and 18, 1863," ibid., 741–45. See Hoffman's concluding letter to Marston, Dec. 24, 1863, ibid., 753–54. A letter from General Benjamin Butler, then in charge of prisoner-of-war exchanges, likewise hailed the camp as well-ordered. Butler to E. M. Stanton, Dec. 27, 1863, ibid., 763–64.

38. Hoffman to Knapp, Dec. 15, 1863, in *OR*, series 2, 6:705–6, quotation on 706.

39. USSC, *Narrative of Privations*, 151; "Our Prisoners at Richmond," *New York Times*, Dec. 13, 1863, in Washington Office Scrapbook, No. 1067, box 939, United States Sanitary Commission Records, NYPL. This scrapbook contains many clippings on this topic, some reporting that the materials got through and others saying that materials had been confiscated or stolen. See also *Crutch,* the hospital newspaper at Annapolis, USA General Hospital, Division no. 1, especially articles in spring 1864, when prisoners from Richmond arrived at the hospital. Newspaper available at Archives of Maryland Online, http://aomol.net/megafile/msa/speccol/sc2900/sc2908/crutch/html/index.html (accessed Aug. 23, 2011).

40. John C. Waugh, *Reelecting Lincoln: The Battle for the 1864 Presidency* (New York: Crown, 1997).

41. James M. McPherson, *Battle Cry of Freedom: The Civil War Era* (New York: Oxford University Press, 1988); George S. *Burkhardt, Confederate Rage, Yankee Wrath: No Quarter in the Civil War* (Carbondale: Southern Illinois University Press, 2007); and U.S. Congress, *Reports of the Committee on the Conduct of the War: Fort Pillow Massacre, Returned Prisoners (1864),* 38th Cong., 1st sess., reports no. 63 and 68. A facsimile reprint of the pamphlet (Whitefish, MT: Kessinger, n.d. [ca. 2006]) contained the two reports.

42. USSC, *Narrative of Privations*, 13, 14. Committee members were Valentine Mott, Ellersbee Wallace, Gouvenor Wilkins, Edward Delafeld, Clark Hare, and Treadwell Walden.

43. USSC, *Narrative of Privations,* 21–22.

44. Ibid., 25.

45. Ibid., quotation on 58.

46. Ibid., 70.

47. Ibid., 58–59.

48. Ibid., 73–74.

49. Ibid., 70–85. While the protective power of vaccination was well known, the committee recognized that imprisoned rebels became infected with smallpox faster than the Union could vaccinate them. The rebel army should have vaccinated them upon enrollment, and the smallpox outbreaks were hardly the Union's fault.

50. Ibid., 90.

51. J. Ogden Murray, *The Immortal Six Hundred: A Story of Cruelty to Confederate Prison-*

ers of War (Roanoke, VA: Stone, 1911), 191. See also Brian Temple, *The Union Prison at Fort Delaware: A Perfect Hell on Earth* (Jefferson, NC: McFarland, 2003).

52. Sanders, *While in the Hands of the Enemy*, 249.

53. Ibid., 98 (emphasis in original).

54. Ibid., 99.

55. Ibid., 28.

56. *Report of the Joint Select Committee*, 17.

57. See published review, "Treatment of Our Prisoners at the South," *National Intelligencer*, Oct. 7, 1864, 3 for these date estimations.

58. See Letters for March–Oct. 1864, MS N-1829, Henry W. Bellows Papers, Massachusetts Historical Society, Boston. There may be documents in the USSC papers at the New York Public Library, but a large portion of that collection is closed. Still, William Maxwell, who had full access to this collection in the 1950s, mentions only the published document.

59. Henry W. Bellows to my dear wife, Oct. 26, 1864, in Bellows Papers.

60. See letters from Oct. to Dec. 1864, Bellows Papers.

61. Henry W. Bellows to my dear wife, Oct. 26, 1864 (second letter for that day), Bellows Papers.

62. Henry W. Bellows to Russell Bellows, Nov. 2, 1864, Bellows Papers.

63. Ibid.; and other letters in the October and November 1864 folders, Bellows Papers. Emphasizing the prisoner-of-war tragedy may have backfired to some extent for the Republican partisans. Lincoln's government had refused to continue the open exchange of prisoners for a number of reasons, including protest against the ways southerners treated black prisoners and Grant's reluctance to return able-bodied southerners to the armies he was trying to deplete. So even though the prisoner-of-war news painted the Confederacy as evil, it also raised antagonism against the Lincoln administration for not doing all it could to bring the Union prisoners home. See Sanders, *While in the Hands of the Enemy*, for a more detailed discussion of the role of such rhetoric in the election of 1864.

64. See "Treatment of our Prisoners at the South," a review in a Democratic-leaning newspaper that did not challenge the physical description of the released prisoners but did take issue with the question of whether the Confederate government had deliberately starved the men or had instead been dealing with conditions beyond human control.

65. Stillé, *History of the Sanitary Commission*, 307n.

66. The three volumes published were Austin Flint, ed., *Contributions Relating to the Causation and Prevention of Disease and to Camp Diseases; Together with a Report of the Diseases, etc., among the Prisoners at Andersonville, GA* (New York: Hurd and Houghton, 1867); Benjamin Apthorp Gould, *Investigations in the Military and Anthropological Statistics of American Soldiers* (New York: Hurd and Houghton, 1869); Frank Hastings Hamilton, ed., *Surgical Memoirs of the War of the Rebellion* (New York: Hurd and Houghton, 1871). Copies of the prospectus, handwritten and printed, are in USSC Microfilm Papers, reel 12, frames 362–445.

67. A. M. Clark to Elisha Harris, Dec. 8, 1865, in USSC Microfilm Papers, reel 5, frame 505.

68. A. M. Clark, "Report of Inspection of Camp and Field Hospital, Indianapolis, Ind., Oct. 22, 1863," in *OR*, ser 2, 6:424–26, quotation on 425.

69. Clark to Harris, Dec. 15, 1865, in USSC Microfilm Papers, reel 6, frame 831.

Clark's letter to Harris on Dec. 11 says he is getting started and is pleased that his outline was acceptable (reel 5, frame 961). Frame 511 on reel 5 is an image of the envelope the outline came in, with the notation that it had been borrowed by Dr. Van Buren.

70. A. M. Clark, April 4, 1866, USSC Microfilm Papers, reel 9, frame 877.

71. Clark to Harris, May 15 1866, USSC Microfilm Papers, reel 10, frame 81.

72. Clark's reports can be found easily in the digitized *Official Records*. A search in Indexcat, National Library of Medicine, turned up no publications by A. M. Clark, although he may have been one of the "A. Clark" authors from the mid-nineteenth century. There are no publications listed in Indexcat of an A. Clark on the health of prisoners of war.

73. Joseph Jones, "Investigations upon the Diseases of the Federal Prisoners Confined in Camp Sumpter, Andersonville, Ga.," in Flint, *Contributions Relating to Disease*, 469–655.

74. William Hesseltine, *Civil War Prisons: A Study in War Psychology* (Columbus: Ohio State University Press, 1930), 172.

75. Eugene Sanger's correspondence is quoted in Jesse Waggoner, "The Role of the Physician: Eugene Sanger and a Standard of Care at the Elmira Prison Camp," *Journal of the History of Medicine and Allied Sciences* 63 (2007): 1–22, quotations on 18. This process of creating the "other" and then feeling free to hate that category of person has become a common theme among historians seeking to understand prejudicial and cruel behavior.

76. Martin Gumpert, *Dunant: The Story of the Red Cross* (New York: Oxford University Press, 1938).

77. William E. Barton, *The Life of Clara Barton: Founder of the American Red Cross* (Boston: Houghton Mifflin, 1922), 2:2; Caroline Moorehead, *Dunant's Dream: War, Switzerland, and the History of the Red Cross* (New York: Carroll and Graf, 1998), 37–39.

78. J. Henry Dunant to Henry W. Bellows, Dec. 18, 1865, Archives Register, Letters Received, Association for the Relief of the Misery of Battle Fields Archives, box 941, United States Sanitary Commission Records, NYPL; Minute Book, ibid. Previous USSC affiliates who came to the meeting included C. A. Agnew, John Blatchford, George Templeton Strong, F. N. Knapp, J. F. Jenkins, and Henry W. Bellows.

79. Minute Book, Association for the Relief of the Misery of Battle Fields, ibid.; American Branch of the International Association for Relief of Misery of Battlefields, *The Work of Humanity in War: Plan and Results of the Geneva Congress and International Treaty, Securing to the Sick and Wounded in War the Benefits of Neutrality and Sanitary Care* (New York: Anson D. F. Randolph, 1870).

80. Clara Barton, *The Red Cross: A History of This Remarkable International Movement in the Interest of Humanity* (Washington, DC: American National Red Cross, 1898), 35–55. See also Stephen B. Oates, *A Woman of Valor: Clara Barton and the Civil War* (New York: Free Press, 1994).

Chapter 10 · A Public Health Legacy

1. Howard Kramer, "Effect of the Civil War on the Public Health Movement," *Mississippi Valley Historical Review* 35 (1948): 449–62, quotation on 453.

2. John Griscom, *The Sanitary Condition of the Laboring Population of New York, with Suggestions for Its Improvement* (New York: Harper, 1845); Lemuel Shattuck, *Report of the*

Sanitary Commission of Massachusetts (1850; facsimile repr., Cambridge, MA: Harvard University Press, 1948).

3. Charles E. Rosenberg, *The Cholera Years: The United States in 1832, 1849, and 1866* (Chicago: University of Chicago Press, 1962); Barbara G. Rosenkrantz, *Public Health and the State: Changing Views in Massachusetts, 1842–1936* (Cambridge, MA: Harvard University Press, 1972); Margaret Humphreys, *Yellow Fever and the South* (New Brunswick, NJ: Rutgers University Press, 1992); John Duffy, *The Sanitarians: A History of American Public Health* (Urbana: University of Illinois Press, 1990).

4. The first two volumes of the *MSHW* are composed of such statistics.

5. Elisha Harris, "Yellow Fever on the Atlantic Coast and at the South during the War," in *Contributions Relating to the Causation and Prevention of Disease and to Camp Diseases; Together with a Report of the Diseases, etc., among the Prisoners at Andersonville, GA*, ed. Austin Flint (New York: Hurd and Houghton, 1867), 236–68, quotation on 242.

6. William T. Wragg, "Report on the Yellow Fever at Wilmington, N.C., in the Autumn of 1862," *CSMSJ* 1 (1864): 17–20, 33–36.

7. Harris, "Yellow Fever," 266–67.

8. Stanford E. Chaillé, "The Yellow Fever, Sanitary Condition, and Vital Statistics of New Orleans during Its Military Occupation: The Four Years 1862–1865," *New Orleans Journal of Medicine* 23 (1870): 563–98, quotations on 563, 567–68.

9. Ibid.

10. Harvey E. Brown, *Quarantine on the Southern and Gulf Coasts of the United States* (New York: William Wood, 1873).

11. Ibid., quotations on 2, 85, 87.

12. Humphreys, *Yellow Fever and the South;* Margaret Humphreys, "Hunting the Yellow Fever Germ: The Principle and Practice of Etiological Proof in Late Nineteenth-Century America," *Bulletin of the History of Medicine* 59 (1985): 361–82.

13. Jim Downs, *Sick from Freedom: African-American Illness and Suffering during the Civil War and Reconstruction* (New York: Oxford University Press, 2012), 113–16.

14. Rosenberg, *Cholera Years,* 193; John Duffy, *A History of Public Health in New York City, 1866–1966* (New York: Russell Sage Foundation, 1974), 1–31.

15. Henry Wentworth Acland, in *Memoir on the Cholera at Oxford, in the Year 1854* (London: John Churchill, 1856), mentioned the possible risk of cholera excretions and recommended destruction with acids and caustic alkalis. He in turn cited William P. Alison, "On the Communicability of Cholera by Dejections," *Edinburgh Medical Journal* 1 (1855–56): 481–92, who credited Budd as the source of the plan to disinfect rice-water discharges. For Budd's persistent pleas for this action sixteen years later, see William Budd, *Asiatic Cholera in Bristol in 1866* (Bristol: Thomas Kerslake, 1871), quotation on iii–iv. See also Richard J. Evans, "Epidemics and Revolutions: Cholera in Nineteenth-Century Europe," *Past and Present* 120 (1988): 123–46; and Anne Hardy, "Cholera, Quarantine, and the English Preventive System, 1850–1895," *Medical History* 37 (1993): 250–69.

16. U.S. Army, Office of the Surgeon General, *Report on Epidemic Cholera and Yellow Fever in the Army of the United States during the Year 1867. War Dept. Surgeon General's Office, Washington, June 10, 1868. Circular no. 1* (Washington, DC: Government Printing Office, 1868), quotation on vi.

17. Joseph K. Barnes, Circular no. 3, dated April 20, 1867, reprinted ibid., 17, 24.

18. Ibid., 38.

19. House of Representatives, *The Cholera Epidemic of 1873 in the United States*, Ex. Doc., 43rd Cong., 2nd sess. (Washington, DC: Government Printing Office, 1875), introduction by John M. Woodworth, quotations on 8.

20. F. A. P. Barnard, "The Germ Theory of Disease and Its Relation to Hygiene," *Public Health Reports and Papers* 1 (1873): 70–87, 72. David Barnes describes how this compromise between sanitation and germ theories of disease further evolved (as the "Sanitary Bacteriological Synthesis") in French public health in the late nineteenth century. See his *The Great Stink of Paris and the Nineteenth-Century Struggle against Filth and Germs* (Baltimore: Johns Hopkins University Press, 2006).

21. Barnard, "Germ Theory," 87.

22. *MSHW*, 6:627–28.

23. Elisha Harris, "Vaccination in the Army—Observations on the Normal and Morbid Results of Vaccination and Revaccination during the War and on Spurious Vaccination," in Flint, *Contributions Relating to Disease*, 137–65, quotation on 138.

24. Ira Russell, quoted in *MSHW*, 6:636.

25. Owen M. Long, "Spurious Vaccination," USSC Microfilm Papers, reel 8, frames 116–17.

26. See Harris, "Vaccination"; and *MSHW*, 6:636–38, for examples of these less-than-optimal vaccination practices.

27. Harris, "Vaccination," 165.

28. *MSHW*, 6:638. This page includes an extensive footnote with references to southern medical journal articles on the failures of vaccination during the war.

29. *MSHW*, 6:624.

30. *MSHW*, 6:629.

31. Joseph Jones, "A Report of the Diseases, etc. among the Prisoners at Andersonville, Ga.," in Flint, *Contributions Relating to Disease*, 608–18.

32. Edward B. Janes, "Report on the Practical Lessons of the Recent Prevalence of Small-Pox, with Reference to Its Prevention in the Future," *Public Health Papers and Reports* 1 (1873): 173–83.

33. Ibid., quotation on 175.

34. Judith Walzer Leavitt, *The Healthiest City: Milwaukee and the Politics of Health Reform* (Princeton, NJ: Princeton University Press, 1982); the quotation, from a letter printed in the *Milwaukee Sentinel*, Jan. 26, 1869, is quoted on 82.

35. Martin Kaufman, in "The American Anti-vaccinationists and Their Arguments," *Bulletin of the History of Medicine* 41 (1967): 463–78, argued that many antivaccination proponents came from the sectarian medical community.

36. Elisha Harris, "A Report on Laws, Sanitary Provisions, and Methods for Securing the Benefits of General Vaccination throughout the Country," *Public Health Papers and Reports* 2 (1875): 140–53, quotations on 142 and 141.

37. Horace H. Cunningham, *Doctors in Gray: The Confederate Medical Service* (Baton Rouge: Louisiana State University Press, 1958), 197–98.

38. Henry A. Martin, "Report on Animal Vaccination," *Transactions of the American*

Medical Association 28 (1877): 187–248; Arthur Allen, *Vaccine: The Controversial Story of America's Greatest Lifesaver* (New York: Norton, 2007), 61–62. See also James Colgrove, *State of Immunity: The Politics of Vaccination in Twentieth-Century America* (Berkeley: University of California Press, 2006), which takes up the smallpox vaccination story with the 1890s epidemics and the fierce backlash that erupted against vaccination.

39. Jacob Steere-Williams, "The Perfect Food and the Filth Disease: Milk Borne Typhoid and Epidemiological Practice in Late Victorian Britain," *Journal of the History of Medicine and Allied Sciences* 65 (2010): 514–45. His references include the large literature on this subject. On the American case, see particularly Werner Troesken, *Water, Race, and Disease* (Cambridge, MA: MIT Press, 2004). Anne Hardy is currently completing a history of salmonella that will further enrich this historiography.

Chapter 11 · Medicine in Postwar America

1. Morton Keller, *Affairs of State: Public Life in Late Nineteenth Century America* (Cambridge, MA: Harvard University Press, 1977), 2.

2. J. David Hacker, "A Census-Based Account of the Civil War Dead," *Civil War History* 57 (2011): 307–48, quotation on 311, CSA percentage on 313.

3. Eric T. Dean, "'The Awful Shock and Rage of Battle': Rethinking the Meaning and Consequences of Combat in the Civil War," in Kent Gramm, ed., *Battle: The Nature and Consequences of Civil War Combat* (Tuscaloosa: University of Alabama Press, 2008), 92–110. Jeffrey McClurken, *Take Care of the Living: Reconstructing Confederate Veteran Families in Virginia* (Charlottesville: University of Virginia Press, 2009).

4. *MSHW*, 1:637. Peter Donnelly was my grandfather's grandfather and came to Anoka, Minnesota in the 1850s from Ireland. He named my great-grandfather Hugh (b. 1869) after his recently deceased brother.

5. *MSHW*, 1:637. Because tuberculosis is a slower disease, from onset to termination, there would have been more time for diagnosis and discharge than in the case of the more rapid typhoid fever. Another daughter of Peter Donnelly died at 19, thirty years after Hugh Donnelly's death.

6. Laurann Figg and Jane Farrel-Beck, "Amputation in the Civil War: Physical and Social Dimensions," *Journal of the History of Medicine and Allied Sciences* 48 (1993): 454–75.

7. Patrick J. McCawley, *Artificial Limbs for Confederate Soldiers* (Columbia: South Carolina Department of Archives and History, 1992). This figure has at least two sources of inaccuracy. One is that some years are missing from the records of the program. The other is that a single name may be repeated in the list, either because of real duplication or the change of residence of the soldier with that name. Also, most limbs were distributed after 1880, meaning a successful candidate for one had to have survived the fifteen years following the war.

8. Ansley Herring Wegner, *Phantom Pain: North Carolina's Artificial-Limbs Program for Confederate Veterans* (Raleigh: North Carolina Office of Archives and History, 2004), 20–35.

9. Lisa Marie Herschbach, "Fragmentation and Reunion: Medicine, Memory, and Body in the American Civil War" (PhD diss., Harvard University, 1997); Lisa A. Long,

Rehabilitating Bodies: Health, History, and the American Civil War (Philadelphia: University of Pennsylvania Press, 2004).

10. Bert Hansen, *Picturing Medical Progress from Pasteur to Polio: A History of Mass Media Images and Popular Attitudes in America* (New Brunswick, NJ: Rutgers University Press, 2009), 32. See also Julie K. Brown, *Health and Medicine on Display: International Expositions in the United States, 1876–1904* (Cambridge, MA: MIT Press, 2009), 11–41.

11. The Literature, Arts, and Medicine Database, http://litmed.med.nyu.edu, contains information on all four paintings as well as references on their interpretation.

12. In his *New York Times* blog *Disunion,* Pat Leonard lauded the "immense impact" of Letterman's work and offered a lengthy list of sources about the "father of modern battle-field medicine." See New York Times, http://opinionator.blogs.nytimes.com/2012/07/05/the-end-of-the-gutbuster (accessed July 10, 2012).

13. Thomas A. McParlin, "Report of the Medical Director Army of the Potomac from January 1, 1865 to the Close of the War," *MSHW,* 2:202–6, statistics on 205, quotation on 206.

14. George M. Fredrickson, *The Inner Civil War: Northern Intellectuals and the Crisis of the Union* (New York: Harper and Row, 1965), 98–112.

15. Joseph Janvier Woodward, *Outlines of the Chief Camp Diseases of the United States Armies as Observed during the Present War: A Practical Contribution to Military Medicine* (Philadelphia: Lippincott, 1863; repr., New York: Hafner, 1964), 209, page reference in the 1964 edition.

16. The poisonous power of tartar emetic is evident in a case posted in the *New England Journal of Medicine* in July 2012. It reports that when a laborer from Central America became ill, his wife gave him a powder labeled tártaro emétic. The powder included trivalent antimony, which poisons the liver, kidneys, and heart, while causing violent and repeated vomiting. Wendy Macías Konstantopoulos, Michele Burns Ewald, and Daniel E. Pratt, "Case 22-2012: A 34-Year-Old Man with Intractable Vomiting after Ingestion of an Unknown Substance," *New England Journal of Medicine* 367 (2012): 259–68. On the antibiotic properties of antimony, see Jacalyn Duffin and Pierre René, "'Anti-moine Anti-biotique': The Public Fortunes of the Secret Properties of Antimony Potassium Tartrate," *Journal of the History of Medicine and Allied Sciences* 46 (1991): 440–56. See also John Haller, "On the Use and Abuse of Tartar Emetic in 19th-Century Materia Medica," *Bulletin of the History of Medicine* 49 (1975): 235–57.

17. Woodward, *Outlines of the Chief Camp Diseases,* 222.

18. Margaret Humphreys, *Intensely Human: The Health of the Black Soldier in the American Civil War* (Baltimore: Johns Hopkins University Press, 2008), 74.

19. Mark M. Ravitch, *A Century of Surgery: The History of the American Surgical Association,* 2 vols. (Philadelphia: Lippincott, 1981), 1:19–43.

20. Constitution of the American Surgical Association, adopted May 31, 1882, reprinted ibid., 1:10.

21. On the history of specialization, see George Weisz, *Divide and Conquer: A Comparative History of Medical Specialization* (New York: Oxford University Press, 2005). Weisz ignores the effect of the Civil War on medical and surgical specialization in the United States, a story that deserves more than my brief mention here.

22. "Fellows of the American Surgical Association, April, 1885," *Transactions of the American Surgical Association* 3 (1885): v–xiv; James Ewing Mears, "Reminiscences of the Early Days of the American Surgical Association," *Transactions of the American Surgical Association* 26 (1908): 15–33.

23. John Harley Warner, *The Therapeutic Perspective: Medical Practice, Knowledge, and Identity in America, 1820–1885* (Cambridge, MA: Harvard University Press, 1986). These two sentences in the text admittedly simplify his argument, owing to space limitations.

24. Keller, *Affairs of State*.

25. Margaret Humphreys, *Yellow Fever and the South* (New Brunswick, N.J.: Rutgers University Press, 1992); Peter Bruton, "The National Board of Health" (PhD diss., University of Maryland, 1974).

26. Peter C. English, *Shock, Physiological Surgery, and George Washington Crile: Medical Innovation in the Progressive Era* (Westport, CT: Greenwood Press, 1980), 22–26.

27. Michael Flannery, "Another House Divided: Union Medical Service and Sectarianism during the Civil War," *Journal of the History of Medicine and Allied Sciences* 54 (1999): 478–510. The "gouty bigots" quotation is from an anonymous article in the *Eclectic Medical Journal of Philadelphia*, 1861, quoted by Flannery on 489. Bonnie Blustein describes the tension between medical elites and less well educated physicians during the war in her discussion of the bill that established the Union medical hierarchy and organization of hospitals in 1862. See her "'To Increase the Efficiency of the Medical Department': A New Approach to U.S. Civil War Medicine," *Civil War History* 33 (1987): 22–41.

28. Flannery, "Another House Divided."

29. George Rosen, *The Specialization of Medicine with Particular Reference to Ophthalmology* (New York: Froben Press, 1944).

30. This phrase was commonly voiced during my own medical training in Harvard teaching hospitals as the core principle involved in learning minor surgical procedures.

31. Michael Sappol, *A Traffic of Dead Bodies: Anatomy and Embodied Social Identity in Nineteenth-Century America* (Princeton, NJ: Princeton University Press, 2002).

32. John Harley Warner, "Ideals of Science and Their Discontents in Late Nineteenth-Century American Medicine," *Isis* 82 (1991): 454–78, reviews this controversy and its historiography in detail.

33. Mears, "Reminiscences," 26.

34. The Flexner Report rated medical schools by a set of quality measurements. The literature concerning the report's meaning and influence, as well as the medical school reform movement that preceded it, is voluminous. Lynn E. Miller and Richard M. Weiss review this literature in "Medical Education Reform Efforts and Failures of U.S. Medical Schools, 1870–1930," *Journal of the History of Medicine and Allied Sciences* 63 (2008): 348–87. See particularly Kenneth Ludmerer, *Learning to Heal: The Development of American Medical Education* (New York: Basic Books, 1985); and Thomas Neville Bonner, *Iconoclast: Abraham Flexner and a Life in Learning* (Baltimore: Johns Hopkins University Press, 2002).

35. Henry D. Noyes, "Account of the Origin and of the First Meeting of the American Ophthalmological Society," *Transactions of the American Ophthalmological Society* 2 (1875): 11–16. War-service information was sought via the website Civil War Soldiers and Sailors System, www.itd.nps.gov/cwss/soldiers.cfm (accessed June 26, 2011).

36. For the list of associations with dates, see James G. Burrow, *AMA: Voice of American Medicine* (Baltimore: Johns Hopkins Press, 1963), 7.

37. Silas Weir Mitchell, George Read Morehouse, and William Williams Kean, *Gunshot Wounds and Other Injuries of Nerves* (1864; repr., San Francisco: Norman, 1989).

38. Charles F. Wooley, *The Irritable Heart of Soldiers and the Origins of Anglo-American Cardiology: The US Civil War (1861) to World War I (1918)* (Burlington, VT: Ashgate, 2002).

39. Charles Edward Amory Winslow, *The Conquest of Epidemic Disease: A Chapter in the History of Ideas* (Madison: University of Wisconsin Press, 1943).

40. Nancy J. Tomes, "American Attitudes toward the Germ Theory of Disease: Phyllis Allen Richmond Revisited," *Journal of the History of Medicine* 52 (1997): 17–50.

41. Shauna Devine devotes considerable attention to the understanding and treatment of hospital gangrene in her forthcoming book (publication information not yet available).

42. W[alter] Kempster, "Hospital Gangrene, as It Occurred at Patterson Park, U.S. General Hospital, Baltimore," *American Journal of the Medical Sciences*, n.s., 51 (1866): 351–56, quotations on 356. His military service is listed on the website Civil War Soldiers and Sailors System, www.itd.nps.gov/cwss/soldiers.cfm (accessed June 26, 2011). For his biography, see J[udith] W[alzer] Leavitt, "Kempster, Walter," in *Dictionary of American Medical Biography*, ed. Martin Kaufman, Stuart Galishoff, and Todd L. Savitt, 2 vols. (Westport, CT: Greenwood Press, 1984), 1:410–11. See prior list of disinfectants in chapter 3.

43. W[alter]. Kempster, "Carbolic Acid as a Remedial Agent," *American Journal of Medical Sciences*, n.s., 56 (1868): 31–39, quotations on 31, 34, 37.

44. On the introduction of Lister's ideas to American medicine, see Thomas P. Gariepy, "The Introduction and Acceptance of Listerian Antisepsis in the United States," *Journal of the History of Medicine and Allied Sciences* 49 (1994): 167–206.

45. George Derby, "Carbolic Acid in Surgery," *Boston Medical and Surgical Journal* 77 (1867): 271–73.

46. Ibid., 271–72.

47. [George] Derby, "[Letter from] House Near Battle Ground, 4 Miles from Newbern, N.C., March 18, 1862, " *Boston Medical and Surgical Journal* 66 (1862): 197–98.

48. See, for example, George Derby, "The Disinfection of Rooms," *Boston Medical and Surgical Journal* 87 (1873): 403.

49. George Derby, "The Lessons of the War to the Medical Profession," *Publications of the Massachusetts Medical Society* 2 (1866–68): 217–28, 221, 224; biographical material is from Barbara Gutmann Rosenkrantz, Public Health and the State: Changing Views in Massachusetts, 1842–1936 (Cambridge, MA: Harvard University Press, 1972), 50–51.

50. Gariepy, "Introduction and Acceptance of Listerian Antisepsis," 179.

Afterword

1. David J. Bodenhamer, John Corrigan, and Trevor M. Harris, eds., *The Spatial Humanities: GIS and the Future of Humanities Scholarship* (Bloomington: Indiana University Press, 2010); Elizabeth Fenn, *Pox Americana: The Great Smallpox Epidemic of 1775–82* (New York: Hill and Wang, 2001).

University of Michigan Medical School, 62
University of Nashville, 220
University of Pennsylvania Medical School, 154, 158, 187
U.S. Army: ambulance corps, 120; bilking of, 106; and Centennial International Exhibition, 152; and cholera, 278–79; and escaped slaves, 8; and federal and state control, 137; and measles, 92; medical leadership of, 21; and medical reform, 9, 107; mortality rates in, 8; and scurvy, 221; size of, 105–6; and smallpox vaccination, 92, 282, 285; supplies and food distribution by, 125; supply trains of, 113; surgeons in, 105; in Tenn., 114; and transport, 120–21; and USSC, 9, 105, 134; and vaccination, 283; in Va., 113; and Walker, 55–58; and women physicians, 54–55
U.S. Army medical department, 21, 72, 105, 106, 130, 140, 141, 142; and Hammond, 145, 146; law reorganizing, 144; and medical history of Civil War, 147–48; and Moore, 186; and USSC, 4, 108, 131, 145, 150; and vital statistics, 272–73
U.S. Public Health Service, 288
U.S. War Department, 105
Utica, N.Y., State Lunatic Asylum, 307

vaccination, 83, 90, 91–92, 142, 167, 283; after war, 272; failure of, 283–86, 287; and smallpox, 282, 283–88; and USSC, 126, 127, 142, 143
Van Buren, William, 104, 106, 250
Vicksburg, 73–74, 114, 115, 139
Vinovskis, Maris, 53
Virchow, Rudolf, 87
Virginia, 72, 113, 118, 120, 168, 191, 192, 222, 223; Confederate medical director for, 231; food scarcity in, 224–25; organization of medical care in, 192
vital statistics, 11, 78, 185, 272–73, 310–11. *See also* medical research
Von Liebig, Justus, 81

Walker, Mary Edwards, 55–59, 62, 74
ward masters, 194, 230, 231
Washington, DC, 43; general hospitals in, 16, 119, 124, 129, 152; hospitals in, 46, 142, 143, 146, 154, 218; housing in, 129; and smallpox, 282
Washington, George, 91, 184, 282, 311
water, 2, 8, 9, 82, 98, 281, 288; and cholera, 95, 278, 279; and USSC, 126, 296
Waterhouse, Benjamin, 282
Wayne County, Ind., Medical Association, 62
Wayside Hospital / First Louisiana Hospital (Charleston, S.C.), 192–93
West, Nathaniel, 161
Western Reserve Medical School, Cleveland, 54
Western Sanitary Commission (WSC), 52, 74, 138–40
whiskey, 26, 32, 66, 67, 75, 169. *See also* alcohol
Wilder, Burt, 17, 33
Wilderness, battle of the, 42, 166, 214, 233, 240
Wilkerson, James, 220
Williams, Thomas H., 220–21, 230
Wilmington, N.C., 95, 96, 217, 225
Wilson, Henry, 41, 106
Wilson, N.C., 219
Winder, General, 199
Winder Hospital, Richmond, 192, 210, 222, 224, 227
women, 18, 42, 49, 140, 239; affluent, 64, 68, 111; aid societies of, 69–70, 72, 74–75, 109, 111–12, 114, 116, 135, 149, 150, 245; and care for sick, 1–2, 7, 14, 15, 20, 21, 49, 50, 53, 54, 66, 105, 182, 290–91, 310; as delicate, 50, 54, 65; and feminism, 52, 54, 59; health of, 71–74; and hospitals, 21–22, 23–24, 48, 50, 51–52, 64, 65–66, 67–68, 175, 177, 193; and imaginary extended family, 14, 50–51, 52, 182; and medical schools, 11, 54, 62, 75, 311; middle class, 49, 50, 52, 64, 67; as nurses, 7, 10, 12, 14–15, 21, 50, 52, 53, 74; as physicians, 49–50, 51–52, 53–62, 74; and politics, 48, 70, 75; and relief work, 52, 68, 69–71; rights of, 3, 12, 48, 70; southern, 1, 7–8, 15, 66, 71, 75, 193, 229, 230, 242; and unpaid labor for USSC, 136–37; and USSC, 4, 7, 9, 15, 69, 70, 74–75, 102, 103, 105, 108–17, 131, 132–34, 135, 136–37, 149, 150, 245; working-class, 49, 72
Woodward, J. J., 85–86, 93, 99, 210, 211, 296–97
Woodworth, John M., 279–80
Worboys, Michael, 84
Wormeley, Katharine, 108